U0370136

原版影印说明

《材料手册》(8册)是 Springer ***Materials Handbook*** *A Concise Desktop Reference*(2nd Edition)的影印版。为使用方便，由原版1卷分为8册：

1 材料的性能

2 常用的有色金属及其合金

3 不常用的有色金属

4 半导体 超导体 磁性材料 绝缘体 电介质 其他电气材料

5 陶瓷 耐火材料 玻璃 聚合物 弹性体

6 矿物 矿石 宝石 岩石 陨石

7 土壤 肥料 水泥 混凝土 石材 结构材料
林木 燃料 推进剂 炸药 复合材料

8 气体 液体

本手册提供各种材料的物理和化学性质，是一本简洁的手边工具书。第二版与第一版的差别是扩充了新的家用材料，但重点是每一类常见的工业材料。

材料科学与工程图书工作室

联系电话 0451-86412421
0451-86414559

邮　　箱 yh_bj@aliyun.com
xuyaying81823@gmail.com
zhxh6414559@aliyun.com

Springer手册精选原版系列

Materials Handbook

A Concise Desktop Reference

François Cardarelli

2nd Edition

材料手册 5

陶瓷 耐火材料 玻璃
聚合物 弹性体

哈爾濱工業大學出版社
HARBIN INSTITUTE OF TECHNOLOGY PRESS

黑版贸审字08-2014-029号

图书在版编目（CIP）数据

材料手册. 5，陶瓷、耐火材料、玻璃、聚合物、弹性体：英文/（美）卡达雷利主编. —哈尔滨：哈尔滨工业大学出版社，2014. 4
（Springer手册精选原版系列）
ISBN 978-7-5603-4451-5

Ⅰ.①材… Ⅱ.①卡… Ⅲ.①材料科学-技术手册-英文 Ⅳ.①TB3-62

中国版本图书馆CIP数据核字（2013）第304258号

责任编辑 许雅莹 张秀华 杨 桦
出版发行 哈尔滨工业大学出版社
社 址 哈尔滨市南岗区复华四道街10号 邮编 150006
传 真 0451-86414749
网 址 http://hitpress.hit.edu.cn
印 刷 哈尔滨市石桥印务有限公司
开 本 660mm × 980mm 1/16 印张 15
版 次 2014年4月第1版 2014年4月第1次印刷
书 号 ISBN 978-7-5603-4451-5
定 价 76.00元

François Cardarelli

Materials Handbook

A Concise Desktop Reference

2nd Edition

Dedication for the First Edition

The ***Materials Handbook*: *A Concise Desktop Reference*** is dedicated to my father, Antonio, and my mother, Claudine, to my sister, Elsa, and to my spouse Louise Saint-Amour for their love and support. I want also to express my thanks to my two parents and my uncle Consalvo Cardarelli, which in close collaboration have provided valuable financial support when I was a teenager to contribute to my first fully equipped geological and chemical laboratory and to my personal comprehensive scientific library. This was the starting point of my strong and extensive interest in both science and technology, and excessive consumption of scientific and technical literature.

François Cardarelli

Dedication for the Second Edition

The ***Materials Handbook: A Concise Desktop Reference*** is dedicated to my father, Antonio, and my mother, Claudine, to my sister, Elsa, and to my wife Elizabeth I.R. Cardarelli for their love and support. I want also to express my thanks to my two parents and my uncle Consalvo Cardarelli, which in close collaboration have provided valuable financial support when I was a teenager to contribute to my first fully equipped geological and chemical laboratory and to my personal comprehensive scientific library. This was the starting point of my strong and extensive interest in both science and technology, and excessive consumption of scientific and technical literature.

François Cardarelli

Acknowledgements for the First Edition

Mr. Nicholas Pinfield (engineering editor, London), Mr. Jean-Étienne Mittelmann (editor,Paris), Mrs. Alison Jackson (editorial assistant, London), and Mr. Nicolas Wilson (senior production controller, London) are gratefully acknowledged for their valued assistance,patience, and advice.

Acknowledgements for the Second Edition

Mr. Anthony Doyle (senior engineering editor), Mr. Oliver Jackson (associate engineering editor), and Mr. Nicolas Wilson (editorial coordinator) are gratefully acknowledged for their valued assistance, patience, and advice.

Units Policy

In this book the only units of measure used for describing physical quantities and properties of materials are those recommended by the *Système International d'Unités* (SI). For accurate conversion factors between these units and the other non-SI units (e.g., cgs, fps, Imperial, and US customary), please refer to the reference book by the same author:

Cardarelli, F. (2005) *Encyclopaedia of Scientific Units, Weights, and Measures. Their SI Equivalences and Origins*. Springer, London New York. ISBN 978-1-85233-682-1.

Author Biography

Dr. François Cardarelli (Ph.D.)
Born in Paris (France) February 17, 1966
Canadian citizen

Academic Background

- Ph.D., chemical engineering (Université Paul Sabatier, Toulouse, France, 1996)
- Postgraduate degree (DEA) in electrochemistry (Université Pierre et Marie Curie, Paris, 1992)
- M.Sc. (Maîtrise), physical chemistry (Université Pierre et Marie Curie, Paris, 1991)
- B.Sc. (Licence), physical chemistry (Université Pierre et Marie Curie, Paris, 1990)
- DEST credits in nuclear sciences and technologies (Conservatoire National des Arts et Métiers, Paris, 1988)

- Associate degree (DEUG B) in geophysics and geology (Université Pierre et Marie Curie, Paris, 1987)
- Baccalaureate C (mathematics, physics, and chemistry) (CNED, Versailles, France, 1985)

Fields of Professional Activity

The author has worked in the following areas (in chronological order) since 1990.

(1) Research scientist at the Laboratory of Electrochemistry (Université Pierre & Marie Curie, Paris, France) for the development of a nuclear detector device for electrochemical experiments involving radiolabeled compounds;

(2) research scientist at the Institute of Marine Biogeochemistry (CNRS & École Normale Supérieure, Paris, France) for the environmental monitoring of heavy-metal pollution by electroanalytical techniques;

(3) research scientist for the preparation by electrochemistry in molten salts of tantalum protective thin coatings for the chemical-process industries (sponsored by Electricité de France);

(4) research scientist for the preparation and characterization of iridium-based industrial electrodes for oxygen evolution in acidic media at the Laboratory of Electrochemical Engineering (Université Paul Sabatier, Toulouse, France);

(5) registered consultant in chemical and electrochemical engineering (Toulouse, France);

(6) battery product leader in the technology department of ARGOTECH Productions, Boucherville (Québec), Canada, in charge of electric-vehicle, stationary, and oil-drilling applications of lithium polymer batteries;

(7) materials expert and industrial electrochemist in the lithium department of ARGOTECH Productions, involved in both the metallurgy and processing of lithium metal anodes and the recycling of spent lithium polymer batteries;

(8) materials expert and industrial electrochemist in the technology department of AVESTOR, Boucherville (Quebec), Canada, in charge of all strategic raw materials entering into the fabrication of lithium polymer batteries, as well as being in charge of the recycling process of spent lithium batteries;

(9) principal chemist, materials, in the technology department of Rio Tinto Iron and Titanium, Sorel–Tracy (Québec), Canada working on the electrowinning of titanium metal from titania-rich slags and on other novel electrochemical processes;

(10) principal electrochemist at Materials and Electrochemical Research (MER) Corp., Tuscon (Arizona, USA) working on the electrowinning of titanium metal powder from composite anodes and other materials related projects.

Contents

Introduction **15**

10 Ceramics, Refractories, and Glasses **593**
10.1 Introduction and Definitions 593
10.2 Raw Materials for Ceramics, Refractories and Glasses 594
10.2.1 Silica 594
10.2.1.1 Quartz, Quartzite, and Silica Sand 595
10.2.1.2 Diatomite 595
10.2.1.3 Fumed Silica 595
10.2.1.4 Silica Gels and Sol–Gel Silica 595
10.2.1.5 Precipitated Silica 595
10.2.1.6 Microsilica 596
10.2.1.7 Vitreous or Amorphous Silica 596
10.2.2 Aluminosilicates 596
10.2.2.1 Fireclay 597
10.2.2.2 China Clay 598
10.2.2.3 Ball Clay 598
10.2.2.4 Other Refractory Clays 599
10.2.2.5 Andalusite, Kyanite, and Sillimanite 599
10.2.2.6 Mullite 600
10.2.3 Bauxite and Aluminas 600
10.2.3.1 Bauxite 600
10.2.3.2 Alumina Hydrates 603
10.2.3.3 Transition Aluminas (TrA) 606
10.2.3.4 Calcined Alumina 606
10.2.3.5 Tabular Alumina 607
10.2.3.6 White Fused Alumina 608
10.2.3.7 Brown Fused Alumina 608
10.2.3.8 Electrofused Alumina-Zirconia 609
10.2.3.9 High-Purity Alumina 609

10.2.4 Limestone and Lime 610
10.2.5 Dolomite and Doloma 610
10.2.5.1 Dolomite 610
10.2.5.2 Calcined and Dead Burned Dolomite (Doloma) 611
10.2.6 Magnesite and Magnesia 612
10.2.6.1 Magnesite 612
10.2.6.2 Caustic Seawater and Calcined Magnesia 612
10.2.6.3 Dead Burned Magnesia 613
10.2.6.4 Electrofused Magnesia 614
10.2.6.5 Seawater Magnesia Clinker 614
10.2.7 Titania 614
10.2.7.1 Rutile 614
10.2.7.2 Anatase 616
10.2.7.3 Brookite 616
10.2.7.4 Anosovite 616
10.2.7.5 Titanium Sesquioxide 617
10.2.7.6 Titanium Monoxide or Hongquiite 617
10.2.7.7 Titanium Hemioxide 618
10.2.7.8 Andersson–Magnéli Phases 618
10.2.8 Zircon and Zirconia 618
10.2.8.1 Zircon 618
10.2.8.2 Zirconia 618
10.2.9 Carbon and Graphite 623
10.2.9.1 Description and General Properties 623
10.2.9.2 Natural Occurrence and Mining 623
10.2.9.3 Industrial Preparation and Processing 625
10.2.9.4 Industrial Applications and Uses 625
10.2.10 Silicon Carbide 625
10.2.10.1 Description and General Properties 625
10.2.10.2 Industrial Preparation 626
10.2.10.3 Grades of Silicon Carbide 628
10.2.11 Properties of Raw Materials Used in Ceramics, Refractories, and Glasses 628
10.3 Traditional Ceramics 629
10.4 Refractories 630
10.4.1 Classification of Refractories 630
10.4.2 Properties of Refractories 631
10.4.3 Major Refractory Manufacturers 634
10.5 Advanced Ceramics 635
10.5.1 Silicon Nitride 635
10.5.1.1 Description and General Properties 635
10.5.1.2 Industrial Preparation and Grades 635
10.5.2 Silicon Aluminum Oxynitride (SiAlON) 636
10.5.3 Boron Carbide 637
10.5.3.1 Description and General Properties 637
10.5.3.2 Industrial Preparation 637
10.5.3.3 Industrial Applications and Uses 637
10.5.4 Boron Nitride 637
10.5.4.1 Description and General Properties 637
10.5.4.2 Industrial Preparation 638
10.5.4.3 Industrial Applications and Uses 638
10.5.5 Titanium Diboride 638

10.5.5.1 Description and General Properties 638
10.5.5.2 Industrial Preparation and Processing 639
10.5.5.3 Industrial Applications and Uses 639
10.5.6 Tungsten Carbides and Hardmetal 639
10.5.6.1 Description and General Properties 639
10.5.6.2 Industrial Preparation 640
10.5.6.3 Industrial Applications and Uses 640
10.5.7 Practical Data for Ceramists and Refractory Engineers 641
10.5.7.1 Temperature of Color 641
10.5.7.2 Pyrometric Cone Equivalents 641
10.6 Standards for Testing Refractories 643
10.7 Properties of Pure Ceramics (Borides, Carbides, Nitrides, Silicides, and Oxides) 647
10.8 Further Reading 670
10.8.1 Traditional and Advanced Ceramics 670
10.8.2 Refractories 670
10.9 Glasses 671
10.9.1 Definitions 671
10.9.2 Physical Properties of Glasses 671
10.9.3 Glassmaking Processes 671
10.9.4 Further Reading 676
10.10 Proppants 677
10.10.1 Fracturing Techniques in Oil-Well Production 677
10.10.1.1 Hydraulic Fracturing 677
10.10.1.2 Pressure Acidizing 678
10.10.2 Proppant and Frac Fluid Selection Criteria 678
10.10.2.1 Proppant Materials 678
10.10.2.2 Frac Fluids 679
10.10.2.3 Properties and Characterization of Proppants 679
10.10.2.4 Classification of Proppant Materials 679
10.10.2.5 Production of Synthetic Proppants 682
10.10.2.6 Properties of Commercial Proppants 683
10.10.2.7 Proppant Market 687
10.10.2.8 Proppant Producers 687
10.10.3 Further Reading 689

11 Polymers and Elastomers 691
11.1 Fundamentals and Definitions 691
11.1.1 Definitions 691
11.1.2 Additives and Fillers 692
11.1.3 Polymerization and Polycondensation 693
11.2 Properties and Characteristics of Polymers 694
11.2.1 Molar Mass and Relative Molar Mass 694
11.2.2 Average Degree of Polymerization 695
11.2.3 Number-, Mass- and *Z*-Average Molar Masses 695
11.2.4 Glass Transition Temperature 697
11.2.5 Structure of Polymers 697
11.3 Classification of Plastics and Elastomers 697
11.4 Thermoplastics 697
11.4.1 Naturally Occurring Resins 697
11.4.1.1 Rosin 697
11.4.1.2 Shellac 699

11.4.2 Cellulosics....699
11.4.2.1 Cellulose Nitrate....699
11.4.2.2 Cellulose Acetate (CA)....700
11.4.2.3 Cellulose Propionate (CP)....700
11.4.2.4 Cellulose Xanthate....700
11.4.2.5 Alkylcelluloses....701
11.4.3 Casein Plastics....701
11.4.4 Coumarone-Indene Plastics....702
11.4.5 Polyolefins or Ethenic Polymers....702
11.4.5.1 Polyethylene (PE)....702
11.4.5.2 Polypropylene (PP)....703
11.4.5.3 Polybutylene (PB)....704
11.4.6 Polymethylpentene (PMP)....704
11.4.7 Polyvinyl Plastics....704
11.4.7.1 Polyvinyl Chlorides (PVCs)....704
11.4.7.2 Chlorinated Polyvinylchloride (CPVC)....705
11.4.7.3 Polyvinyl Fluoride (PVF)....705
11.4.7.4 Polyvinyl Acetate (PVA)....705
11.4.8 Polyvinylidene Plastics....705
11.4.8.1 Polyvinylidene Chloride (PVDC)....705
11.4.8.2 Polyvinylidene Fluoride (PVDF)....706
11.4.9 Styrenics....706
11.4.9.1 Polystyrene (PS)....706
11.4.9.2 Acrylonitrile Butadiene Styrene (ABS)....706
11.4.10 Fluorinated Polyolefins (Fluorocarbons)....707
11.4.10.1 Polytetrafluoroethylene (PTFE)....707
11.4.10.2 Fluorinated Ethylene Propylene (FEP)....708
11.4.10.3 Perfluorinated Alkoxy (PFA)....708
11.4.10.4 Polychlorotrifluoroethylene (PCTFE)....708
11.4.10.5 Ethylene-Chlorotrifluoroethylene Copolymer (ECTFE)....709
11.4.10.6 Ethylene-Tetrafluoroethylene Copolymer (ETFE)....709
11.4.11 Acrylics and Polymethyl Methacrylate (PMMA)....709
11.4.12 Polyamides (PA)....710
11.4.13 Polyaramides (PAR)....710
11.4.14 Polyimides (PI)....710
11.4.15 Polyacetals (PAc)....711
11.4.16 Polycarbonates (PC)....711
11.4.17 Polysulfone (PSU)....711
11.4.18 Polyphenylene Oxide (PPO)....712
11.4.19 Polyphenylene Sulfide (PPS)....712
11.4.20 Polybutylene Terephthalate (PBT)....712
11.4.21 Polyethylene Terephthalate (PET)....712
11.4.22 Polydiallyl Phthalate (PDP)....713
11.5 Thermosets....713
11.5.1 Aminoplastics....713
11.5.2 Phenolics....714
11.5.3 Acrylonitrile-Butadiene-Styrene (ABS)....714
11.5.4 Polyurethanes (PUR)....715
11.5.5 Furan Plastics....715
11.5.6 Epoxy Resins (EP)....715
11.6 Rubbers and Elastomers....715
11.6.1 Natural Rubber (NR)....716

11.6.2 Trans-Polyisoprene Rubber (PIR) 716
11.6.3 Polybutadiene Rubber (BR) 716
11.6.4 Styrene Butadiene Rubber (SBR) 717
11.6.5 Nitrile Rubber (NR) 717
11.6.6 Butyl Rubber (IIR) 717
11.6.7 Chloroprene Rubber (CPR) 717
11.6.8 Chlorosulfonated Polyethylene (CSM) 718
11.6.9 Polysulfide Rubber (PSR) 718
11.6.10 Ethylene Propylene Rubbers 718
11.6.11 Silicone Rubber 719
11.6.12 Fluoroelastomers 719
11.7 Physical Properties of Polymers 720
11.8 Gas Permeability of Polymers 734
11.9 Chemical Resistance of Polymers 734
11.10 IUPAC Acronyms of Polymers and Elastomers 745
11.11 Economic Data on Polymers and Related Chemical Intermediates 746
11.11.1 Average Prices of Polymers 746
11.11.2 Production Capacities, Prices and Major Producers of Polymers and Chemical Intermediates 747
11.12 Further Reading 750

Index

Introduction

Despite the wide availability of several comprehensive series in materials sciences and metallurgy, it is difficult to find grouped properties either on metals and alloys, traditional and advanced ceramics, refractories, polymers and elastomers, composites, minerals and rocks, soils, woods, cement, and building materials in a single-volume source book.

Actually, the purpose of this practical and concise reference book is to provide key scientific and technical materials properties and data to materials scientists, metallurgists, engineers, chemists, and physicists as well as to professors, technicians, and students working in a broad range of scientific and technical fields.

The classes of materials described in this handbook are as follows:

(i) metals and their alloys;
(ii) semiconductors;
(iii) superconductors;
(iv) magnetic materials;
(v) dielectrics and insulators;
(vi) miscellaneous electrical materials (e.g., resistors, thermocouples, and industrial electrode materials);
(vii) ceramics, refractories, and glasses;
(viii) polymers and elastomers;
(ix) minerals, ores, and gemstones;
(x) rocks and meteorites;
(xi) soils and fertilizers;
(xii) timbers and woods;
(xiii) cement and concrete;
(xiv) building materials;
(xv) fuels, propellants, and explosives;

(xvi) composites;
(xvii) gases;
(xviii) liquids.

Particular emphasis is placed on the properties of the most common industrial materials in each class. The physical and chemical properties usually listed for each material are as follows:

(i) physical (e.g., density, viscosity, surface tension);
(ii) mechanical (e.g., elastic moduli, Poisson's ratio, yield and tensile strength, hardness, fracture toughness);
(iii) thermal (e.g., melting and boiling point, thermal conductivity, specific heat capacity, coefficients of thermal expansion, spectral emissivities);
(iv) electrical (e.g., resistivity, relative permittivity, loss tangent factor);
(v) magnetic (e.g., magnetization, permeability, retentivity, coercivity, Hall constant);
(vi) optical (e.g., refractive indices, reflective index, dispersion, transmittance);
(vii) electrochemical (e.g., Nernst standard electrode potential, Tafel slopes, specific capacity, overpotential);
(viii) miscellaneous (e.g., relative abundances, electron work function, thermal neutron cross section, Richardson constant, activity, corrosion rate, flammability limits).

Finally, detailed appendices provide additional information (e.g., properties of the pure chemical elements, thermochemical data, crystallographic calculations, radioactivity calculations, prices of metals, industrial minerals and commodities), and an extensive bibliography completes this comprehensive guide. The comprehensive index and handy format of the book enable the reader to locate and extract the relevant information quickly and easily. Charts and tables are all referenced, and tabs are used to denote the different sections of the book. It must be emphasized that the information presented here is taken from several scientific and technical sources and has been meticulously checked and every care has been taken to select the most reliable data.

10 Ceramics, Refractories, and Glasses

10.1 Introduction and Definitions

The word "ceramics" is derived from the Greek *keramos*, meaning solid materials obtained from the firing of clays. According to a broader modern definition, ceramics are either crystalline or amorphous solid materials involving only ionic, covalent, or iono-covalent chemical bonds between metallic and nonmetallic elements. Well-known examples are silica and silicates, alumina, magnesia, calcia, titania, and zirconia. Despite the fact that, historically, oxides and silicates have been of prominent importance among ceramic materials, modern ceramics also include borides, carbides, silicides, nitrides, phosphides, and sulfides.

Several processes, namely calcining and firing, are extensively used in the manufacture of raw and ceramic materials, and they must be clearly defined. Calcining consists in the heat treatment of a raw material prior to being used in the final ceramic material. The purpose of calcination is to remove volatile chemically combined constituents and to produce volume changes. Firing or burning is the final heat treatment performed in a kiln to which a green ceramic material is subjected for the purpose of developing a strong chemical bond and producing other required physical and chemical properties.

As a general rule ceramic materials can be grouped into three main groups: traditional ceramics, refractories and castables, and advanced or engineered ceramics.

Before describing each class, a description of the most common raw materials used in the manufacture of traditional and advanced ceramics, refractories, and glasses is presented below.

10.2 Raw Materials for Ceramics, Refractories and Glasses

10.2.1 Silica

Silica, with the chemical formula SiO_2 and relative molar mass of 60.084, exhibits a complex polymorphism characterized by a large number of reversible and irreversible phase transformations (Figure 10.1) usually associated with important relative volume changes ($\Delta V/V$). At low temperature and pressure *beta-quartz* (**β-quartz**) [14808-60-7] predominates, but above 573°C, it transforms reversibly into the high-temperature *alpha-quartz* (**α-quartz**) [14808-60-7] with a small volume change (0.8 to 1.3 vol.%):

β-quartz <—> α-quartz (573°C)

Quartz exhibits a very low coefficient of thermal expansion (0.5 μm/m.K) and an elevated Mohs hardness of seven. Large and pure single crystals of quartz of gem quality called ***lascas*** are used due to their high purity in the preparation of elemental silicon for semiconductors (see Section 5.8.1).

At a temperature of 870°C, α-quartz transforms irreversibly into *alpha-tridymite* (**α-tridymite**, orthorhombic) [15468-32-3] with an important volume change of 14.4 vol.% as follows:

α-quartz —> α-tridymite (870°C)

But in practice, the kinetic of the above reaction is too slow, and tridymite never forms below 1250°C, and hence at 1250°C or 1050°C in the presence of impurities, α-quartz transforms irreversibly into *alpha-cristoballite* (**α-cristoballite**, tetragonal) [14464-46-1] with an important volume change (17.4 vol.%) as follows:

α-quartz —> α-cristoballite (1250°C)

However, if the temperature is raised to 1470°C, α-tridymite transforms also irreversibly into *alpha-cristoballite* (**α-cristoballite**) without any change in volume as follows:

α-tridymite —> α-cristoballite (1470°C)

On cooling α-cristoballite transforms reversibly into *beta-cristoballite* (**β-cristoballite**, cubic) at 260°C with a volume change 0 2.0 to 2.8 vol.%:

α-cristoballite <—> β-cristoballite (260°C)

Finally, α-cristoballite melts at 1713°C while α-tridymite melts at 1670°C. Upon cooling silica melt yields amorphous ***fused silica*** [60676-86-0].

There also exist two high-pressure polymorphs of silica called ***coesite*** and ***stishovite*** (see Section 12.7) that occur in strongly mechanically deformed metamorphic rocks (e.g., impactites), but these two phases are usually not encountered in ceramics, refractories, and glasses.

Industrially, silica products are classified into two main groups: ***natural silica*** products —quartzite, silica sand, and diatomite—and ***specialty silicas*** including fumed silica, silica gel, microsilica, precipitated silica, fused silica, and vitreous silica.

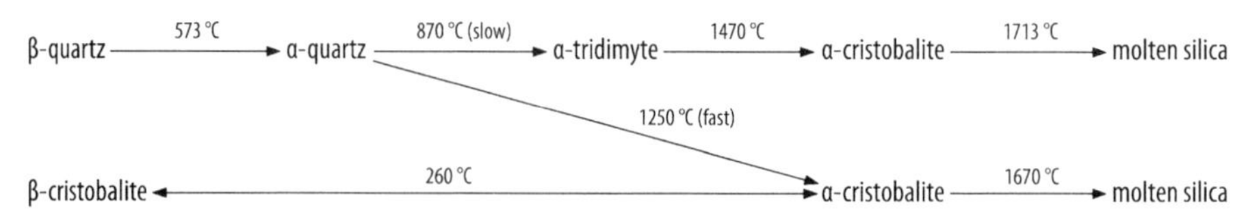

Figure 10.1. Polymorphs of silica (SiO_2)

10.2.1.1 Quartz, Quartzite, and Silica Sand

Quartz is extensively found in nature either as a major mineral in most igneous (e.g., granite), sedimentary (e.g., sand and sandstone), and metamorphic rocks (e.g., quartzite and gneiss). In the case of ceramics, refractories, and glasses, raw quartz is essentially mined as ***round silica sand*** from glacial deposits, beach sands, crushed sandstones, or high-quality quartzite with a silica content of more than 97 wt.% SiO_2. ***Quartzite*** can be either of sedimentary origin with detrital grains of quartz cemented by secondary silica or of metamorphic origin from the contact metamorphism of sandstones or tectonically deformed sandstones. For the most demanding applications, the run-of-mine is even washed with hydrochloric acid to remove traces of iron and aluminum sesquioxides and magnesium and calcium carbonates. Because quartzite consists mainly of beta-quartz, during firing, quartzite is subject to a behavior related to the polymorphism of silica. However, sedimentary quartzite transforms more rapidly than metamorphic equivalent.

Price (2006). Silica sand is priced 15–40 US$/tonne.

10.2.1.2 Diatomite

Diatomaceous earth, or simply ***diatomite***, formerly called ***Kieselguhr***, is a sedimentary rock of biological origin formed by the accumulation at the bottom of the ocean of siliceous skeletons of diatoms, or unicellular algae. Once-calcined diatomite is a white and lightweight material with a mass density ranging from 190 to 275 $kg.m^{-3}$. Diatomite is a highly porous material that exhibits high absorption capabilities and has a good chemical inertness. Major applications are filtering aids, metal polishing, thermal insulation, and Portland cement.

Price (2006). Diatomite is priced 700–800 US$/tonne.

10.2.1.3 Fumed Silica

Fumed silicas are submicrometric particles of amorphous silica produced industrially by burning ***silicon tetrachloride*** or ***tetrachlorosilane*** ($SiCl_4$) using an oxygen-hydrogen burner. The continuous process requires high-purity silicon tetrachloride, which is a byproduct of the carbochlorination of zircon sand for the production of zirconium tetrachloride by companies like Western Zirconium and Wah Chang in the United States or CEZUS in France (see Section 4.3.3, Zirconium and Zirconium Alloys) Fumed silicas usually receives an after-treatment that consists in coating the surface of particles with silanes or silicones in order to enhance hydrophobicity or improve dispersion in aqueous solution. In 2004, the annual production of fumed silica worldwide reached ca. 100,000 tonnes. The German company Degussa-Hüls, with its brand name ***Aerosil®***, is the world leader with half of the world production, followed by Cabot Corp. in the USA.

10.2.1.4 Silica Gels and Sol–Gel Silica

Silica gels are dispersions of colloidal silica obtained by a sol–gel process. The process consists in precipitating colloidal silica from an aqueous solution of sodium silicate by adding hydrochloric or sulfuric acid. The colloidal precipitate or gel consists mainly of hydrated silica ($SiO_2.nH_2O$). After filtration the precipitated silica is washed in order to remove residual salts and stabilized by adding ammonia or sodium hydroxide. The stabilized gel is then dried and later calcined to obtain an activated material, usually in the form of small beads. Major producers are E.I. DuPont de Nemours, Akzo, and Nalco Chemicals Co.

10.2.1.5 Precipitated Silica

Precipitated silica is obtained like silica gel by acidifying an aqueous solution of sodium silicate. Precipitated silica is used as filler in rubber for automobile tires and reinforcement particulate in elastomers, and as a flatting agent in paints and coatings for improving the flatness of coatings. About 850,000 tonnes are produced annually worldwide. Major producers of precipitated silica are PPG Industries and Rhodia.

10.2.1.6 Microsilica

Microsilica, also called ***silica-fume***, is a submicronic amorphous silica with 90 to 98 wt.% SiO_2 and a low bulk density ranging from 200 to 450 $kg.m^{-3}$. It forms most of the dust and other particulates in the off-gases produced during the electrothermal production of ferrosilicon (Fe-Si) or silicon (Si). The dust is collected in baghouses and bagged without further treatment. Due to its high surface area, microsilica reacts readily with hydrated calcium silicates forming strong bonds, and for that reason it is sometimes called reactive silica. Therefore the addition of microsilica to hydraulic cements improves their mechanical strength, reduces their permeability, and enhances their workability, cohesiveness, and flowing properties and hence is extensively used as an additive to cements and monolithic refractories. Annually, ca. 300,000 tonnes of microsilicas are produced worldwide. Major producers are obviously silicon or ferrosilicon producers such as Elkem in Canada and Norway and Fesil in Norway.

10.2.1.7 Vitreous or Amorphous Silica

High-purity amorphous or ***fused silica***, also called ***vitreous silica***, when optically translucent is a high-performance ceramic obtained by electrothermal fusion of high-grade silica sand with a silica content above 99.5 wt.% SiO_2 into an AC electric-arc furnace (EAF) at a temperature of around 1800 to 2100°C. The melt is then rapidly quenched to prevent crystallization.

Fused silica has a mass density of 2200 $kg.m^{-3}$ while ***vitreous silica*** is slightly denser with a density of 2210 $kg.m^{-3}$. Mechanically, fused silica is a relatively strong but brittle material with a tensile strength of 28 MPa, a compressive strength of 1450 MPa, and a Mohs hardness of 5. Both grades exhibit an extremely low coefficient of thermal expansion (e.g., 0.6×10^{-6} K^{-1} from 20 to 1000°C) and a remarkable thermal shock resistance together with a low thermal conductivity. Fused silica, with a dielectric field strength of 16 $MV.m^{-1}$, exhibits also excellent electrical insulation capabilities up to 1000°C. When heated above 1150°C, fused silica converts irreversibly into α-cristoballite as follows:

$$\text{fused silica} \longrightarrow \alpha\text{-cristoballite} \quad (1150°C)$$

Fused silica begins to soften at 1670°C and melt at 1755°C. From a chemical point of view, fused silica possesses an excellent corrosion resistance to most chemicals, especially strong mineral acids, molten metals, and molten glasses.

Common industrial uses for fused silica are steelmaking, coke making, metallurgy, glass production, nonferrous foundries, precision foundries, ceramics, the chemical industry, the nuclear industry, and finally aeronautics.

10.2.2 Aluminosilicates

From a geological point of view, clays are soft, fine-grained, and residual sedimentary rocks resulting from the weathering of feldspars (e.g., orthoclases and plagioclases) and ferromagnesian silicates (e.g., micas, amphiboles) contained in igneous and metamorphic rocks. Hence clays are always made of various hydrated aluminosilicates, mainly kaolinite but also illite and montmorillonite, all exhibiting the typical structure of sheet silicates (i.e., phyllosilicates). When a clay is fired, it loses its absorbed water between 100 and 200°C. Secondly, its major phyllosilicate mineral, ***kaolinite*** [$Al_4(Si_4O_{10})(OH)_8 = 2Al_2O_3{\cdot}4SiO_2{\cdot}4H_2O$], dehydrates between 500 and 600°C, giving off its water to form ***metakaolin*** [$Al_2Si_2O_7 = 2Al_2O_3{\cdot}2SiO_2$]:

$$Al_4[Si_4O_{10}(OH)_8] \longrightarrow 2Al_2Si_2O_7 + 4H_2O \ (500°C < T < 600°C).$$

Above 800°C an important chemical change takes place with the formation of one of the three aluminosilicate polymorphs (Al_2SiO_5), i.e., ***andalusite***, ***kyanite***, or ***sillimanite***, and free silica according to the overall chemical reaction:

$$2Al_2Si_2O_7 \longrightarrow 2Al_2SiO_5 + 2SiO_2 \text{ (at } T > 800°C).$$

If firing is carried out above 1595°C, the highly refractory mineral ***mullite*** then forms (see mullite) with an additional liberation of free silica that melts according to the following chemical reaction:

$$3Al_2SiO_5 \longrightarrow Al_6Si_2O_{13} + SiO_2 \text{ (at } T > 1600°C).$$

Refractory fireclays embrace all types of clays commercially available. Because of the abundant supply of fireclay and its comparative cheapness, refractory bricks made out of it are the most common and extensively used in all places of heat generation. In fact, several technical designations are used in the ceramic industry for classifying refractory clays; these are fire clay, China clay, ball clay, flint clay, and chamotte.

10.2.2.1 Fireclay

Description and general properties. Fireclay denotes a silica-rich natural clay that can withstand a high firing temperature above the ***pyrometric cone equivalent*** (PCE; Table 10.19) of 19 without melting, cracking, deforming, disintegrating, or softening. Typically, a good fireclay should have 24 to 26 vol.% plasticity, and shrinkage after firing should be within 6 to 8 vol.% maximum. Fireclays are mostly made of kaolinite, but some Fe_2O_3 and minor amounts of Na_2O, K_2O, CaO, MgO, and TiO_2 are invariably present depending on the mineralogy and geology of the deposit, making it gray in color. Upon firing, fireclay yields a strong ceramic product with a composition close to the theoretical composition of ***metakaolin*** (i.e., 54.1 wt.% SiO_2 and 45.9 wt.% Al_2O_3), but in practice it contains between 50 and 60 wt.% SiO_2, 24 and 32 wt.% Al_2O_3, no more than 25 wt.% Fe_2O_3 and a loss on ignition of 9 to 12 wt.%. Fireclay is classified under acid refractories, that is, refractories that are not attacked by acid slags. In practice, refractoriness and plasticity are the two main properties required for the manufacture of refractory bricks; hence fireclays are grouped according to the maximum service temperature of the final product before melting in: ***low-duty fireclay*** (max. 870°C, PCE 18 to 28), ***medium-duty fireclay*** (max. 1315°C, PCE 30), ***high-duty fireclay*** (max. 1480°C, PCE 32), and ***super-duty fireclay*** (max. 1480°C–1619°C, PCE 35). In practice, it has been observed that the higher the alumina content in the fireclay, the higher the melting point. All fireclays are not necessarily plastic clays. In such cases, some plastic clay, like ball clay, is added to increase plasticity to a suitable degree. A good fireclay should have 24 to 26% plasticity, and shrinkage after firing should be within 6 to 8% maximum. It should also not contain more than 25% Fe_2O_3.

Industrial preparation. Mined clay is stacked in the factory yard and allowed to weather for about 1 year. For daily production of different types of refractories, this weathered clay is taken and mixed in different percentages with ***grog*** (i.e., ***spent fireclay***). The mixture is sent to the grinding mill from where it is transferred to the pug mill. In the pug mill a suitable proportion of water is added so as to give it proper plasticity. The mold is supplied to different machines for making standard bricks or shapes. Intricate shapes are made by hand. The bricks thus made are then dried in hot floor driers and after drying are loaded in kilns for firing. The firing ranges are, of course, different for different grades of refractories. After firing, the kilns are allowed to cool, then the bricks are unloaded. Upon burning fireclay is converted into a stonelike material that is highly resistant to water and acids, while manufacturing high aluminous fire-bricks bauxite is added along with grog in suitable proportions.

Industrial applications and uses. As a general rule fireclays are used in both ***shaped refractories*** (i.e., bricks) and ***monolithic refractories*** (i.e., castables), while super-duty plastic fireclay is used in the preparation of castable recipes. Therefore, the major applications of fireclays are in power generation, such as in boiler furnaces, in glass-melting furnaces, in chimney linings, in pottery kilns, and finally in blast furnaces where the backup lining is done almost entirely with fireclay bricks. Pouring refractories like sleeves, nozzles, stoppers, and tuyers are also made of fireclay.

10.2.2.2 China Clay

China clay or ***kaolin***, the purest white porcelain discovered and used by the Chinese since ancient times, has always been a much-prized material. Outside of China, a few deposits were found in some parts of Europe and in the United States early in the 18th century. China clay occurs in deposits in the form of china clay rock, a mixture of up to 15 wt.% china clay and up to 10 wt.% mica muscovite, the remainder being free silica as quartz. But the terms china clay and kaolin are not well defined; sometimes they are synonyms for a group of similar clays, and sometimes kaolin refers to those obtained in the United States and china clay to those that are imported. Others use the term china clays for the more plastic of the kaolins. China clays have long been used in the ceramics industry, especially in whitewares and fine porcelains, because they can be easily molded, have a fine texture, and are white when fired. France's clays are made into the famous Sèvres and Limoges potteries. These clays are also used as a filler in making paper. In the United States, deposits are found primarily in Georgia, North Carolina, and Pennsylvania; china clay is also mined in England (Cornwall) and France.

Industrial preparation. The extraction of china clay from its deposits is usually performed in three steps: open-pit mining, mineral processing and beneficiation, and drying. Open-pit operations require the removal of ground overlying the clay (i.e., overburden). The exposed clay is then mined by a hydraulic mining process, that is, a high-pressure water jet from a water cannon called a monitor erodes the faces of the pit. This liberates from the quarry face the china clay, together with sand and mica. The slurry formed flows to the lowest part of the pit or sink, where it is pumped by centrifugal pumps to classifiers, where coarse silica sand is removed. The silica sand is later reused for landscape rehabilitation. The remaining suspension of clay is transported by underground pipeline to the mineral-processing and beneficiation plant, where a series of gravity separation techniques are used to remove particulate materials such as quartz, mica, and feldspars. Sometimes the purified clay slurry undergoes an additional chemical bleaching process that greatly improves its whiteness. The refined clay suspension is then filtrated, and the filtration cake with a moisture content of about 25 wt.% passes through a thermal drier fired by natural gas to yield a final product with 10 wt.% moisture. The end product is normally sold in pelletized form with a pellet size ranging from 6 to 12 mm.

10.2.2.3 Ball Clay

Ball clay, like china clay, is a variety of kaolin. It differs from china clay in having a higher plasticity and less refractoriness. In chemical composition, ball and china clays do not differ greatly except that the former contains a larger proportion of silica. Its name is derived from the practice of removing it in the form of ball-like lumps from clay pits in the UK. The main utility of ball clay is its plasticity, and it is mixed with nonplastic or less plastic clays to make them acquire the requisite plasticity. The high plasticity of ball clay is attributed to the fact that it is fine-grained and contains a small amount of montmorillonite. Over 85% of the particle sizes present in ball clay are of 1 μm or less in diameter. It is light to white in color and on firing may be white buff. The pyrometric cone equivalent to ball clay hardly ever exceeds 33. Usually the following mass fractions of ball clay are commonly used in various types of ceramic compositions: vitreous sanitaryware 10 to 40 wt.%, chinaware 6 to 15 wt.%, floor and wall tiles 12 to 35 wt.%, spark plug porcelain 10 to 35 wt.%, semivitreous whiteware

20 to 45 wt.%, and glass melting-pot bodies 15 to 20 wt.%. The wide use of ball clay is mainly due to its contribution of workability, plasticity, and strength to bodies in drying. Ball clay, on the other hand, also imparts high-drying shrinkage, which is accompanied by a tendency toward warping, cracking, and sometimes even dunting. This undesirable property is compensated by the addition of grog.

Industrial applications. Filler for paper and board, coating clays, ceramics, bone china, hard porcelain, fine earthenware, porous wall tiles, electrical porcelain, semivitreous china, glazes, porcelain, enamels, filler for plastics, rubbers and paints, cosmetics, insecticides, dusting and medicine, textiles, and white cement.

10.2.2.4 Other Refractory Clays

Flint clay or hard clay. This is a hardened and brittle clay material having a conchoidal fracture like flint that resists slacking in water but becomes plastic upon wet grinding.

Chamotte. Chamotte denotes a mixture of calcined clay and spent ground bricks. It is also called fireclay mortar.

Diaspore clay. This is a high-alumina material containing 70 to 80 wt.% Al_2O_3 after firing of a mixture of diasporic bauxite and clay.

10.2.2.5 Andalusite, Kyanite, and Sillimanite

Andalusite, kyanite, and sillimanite are three polymorphs minerals that belong to the nesosilicate minerals. Hence they have the same chemical formula [$Al_2SiO_5 = Al_2O_3{\cdot}SiO_2$] and all contain theoretically 62.92 wt.% Al_2O_3 and 37.08 wt.% SiO_2. They are distinguished from one another by their occurrence and physical and optical properties (see Section 12.7, Minerals and Gemstones Properties Table). Kyanite is easily distinguished from sillimanite or andalusite by its tabular, long-bladed, or acicular habit and by its bluish color and slightly lower hardness than sillimanite and andalusite.

Sillimanite, kyanite, and andalusite are all ***mullite-forming minerals***, that is, on firing they decompose into mullite and vitreous silica (see mullite) according to the chemical reaction:

$$3Al_2SiO_5 \longrightarrow Al_6Si_2O_{13} + SiO_2.$$

However, each polymorph exhibits a different decomposition behavior. Actually, the decomposition of kyanite is unpredictable; it first starts to decompose slowly at 1310°C, and the reaction disrupts at about 1350 to 1380°C with an important volume expansion of 17 vol.%. For that reason, kyanite must always be calcined prior to being incorporated into a refractory in order to avoid blistering and spalling. By contrast, andalusite decomposes gradually from 1380 to1400°C with a low volume increase of 5 to 6 vol.%, while sillimanite does not change into mullite until the temperature reaches 1545°C with a volume expansion of 5 to 6 vol.%.

In nature, these three minerals are originally found in metamorphic rocks, but, due to their high Mohs hardness and relative chemical inertness, they resist weathering processes and are also ubiquitous in mineral sands. For instance, sillimanite is extensively mined as a byproduct of beach mineral sand operations in South Africa and Australia. Sillimanite minerals are predominantly used in refractories and technical porcelains. Sillimanite refractories cut into various shapes and sizes or made out of bonded particles are used in industries like cement, ceramics, glass making, metal smelting, refinery and treatment, tar distillation, coal carbonization, chemical manufacture, and iron foundries. Kyanite in the form of mullite is widely used in the manufacture of glass, burner tips, spark plugs, heating elements, and high-voltage electrical insulations and in the ceramic industry. India is the largest producer of kyanite in the world.

10.2.2.6 Mullite

Mullite [$Al_6Si_2O_{13} = 3Al_2O_3{\cdot}2SiO_2$], with 71.8 wt.% Al_2O_3, is an important silicate mineral that occurs in high-silica alumina refractories. Mullite exhibits a high melting point of 1810°C combined with low thermal expansion coefficients (i.e., 4.5×10^{-6} K^{-1} parallel to the *a*-axis and 5.7×10^{-6} K^{-1} parallel to the *c*-axis), a good mechanical strength with a tensile strength of 62 MPa, and resilience at elevated temperatures that make mullite a highly suitable mineral for highly refractory materials. In nature, mullite is an extremely scarce mineral that occurs only in melted argillaceous inclusions entrapped in lavas from the Cenozoic Era on the Island of Mull, Scotland, but no deposit was found to be economically minable.
Synthetic mullite is formed in high-silica alumina refractories during the firing process at high temperature, the major raw materials being kaolin, alumina, and clays and to a lesser extent kyanite, when available. Actually, when a fire clay is fired, its major phyllosilicate mineral, the kaolinite $Al_4(Si_4O_{10})(OH)_8$, first gives off its water, and above 800°C an important chemical change takes place with the formation of one of the three aluminosilicate polymorphs (Al_2SiO_5), i.e., andalusite, kyanite, or sillimanite, and free silica according to the following chemical reaction:

$$Al_4(Si_4O_{10})(OH)_8 \longrightarrow 2Al_2SiO_5 + 2SiO_2 + 4H_2O \text{ (at } T > 800°C).$$

If firing is carried out above 1595°C, the highly refractory mineral mullite then forms with an additional liberation of free silica that melts according to the following chemical reaction:

$$3Al_2SiO_5 \longrightarrow Al_6Si_2O_{13} + SiO_2 \text{ (at } T > 1600°C).$$

For that reason, high-silica alumina refractories containing less than 71.8 wt.% Al_2O_3 are limited in their use to temperatures below 1595°C. Above 71.8 wt.% Al_2O_3, mullite alone or mullite plus corundum (α-Al_2O_3) coexists with a liquidus at 1840°C. Therefore, the use of ***high-alumina refractories*** is suited for iron- and steelmaking for firebrick and ladles and furnace linings. Two grades of synthetic mullite are available for refractories: ***sintered mullite*** is obtained by calcination of bauxitic kaolin or a blend of bauxite, aluminas, and kaolin or, to a lesser extent, kyanite; ***electrofused mullite*** is made by the electrothermal melting at 2200°C of a mixture of silica sand and bauxite or diasporic clay in an electric-arc furnace. Mullites are formulated to produce dense shapes, some in a glass matrix to yield maximum thermal shock resistance and good mechanical strength. Dense electrofused mullite in a glassy matrix formulated to offer a high-quality economical insulating tubing for thermocouple applications is an extremely versatile and economically viable material. Its workability allows for an extensive range and flexibility in fabrication. It is well suited for the casting of special shapes. Its typical applications are insulators in oxidizing conditions for noble-metal thermocouples used in conditions up to 1450°C, spark plugs, protection tubes, target and sight tubes, furnace muffles, diffusion liners, combustion tubes, radiant furnace tubes, and kiln rollers. Major producers of sintered mullite are C-E Minerals, Andersonville, GA in the USA, followed by several Chinese producers, while Washington Mills Electro Minerals Corp. in Niagara Falls, NY leads the production of electrofused mullite.

10.2.3 Bauxite and Aluminas

10.2.3.1 Bauxite

Bauxite is the major source of ***aluminum sesquioxide*** (***alumina***, Al_2O_3) worldwide. Bauxite is a soft and red clay, rich in alumina, and its name originates from Les Baux de Provence, a small village located in the region of Arles in southeastern France, where it was first discovered in 1821 by P. Berthier. From a geological point of view bauxite is defined as a residual sedimentary rock in the laterite family that results from *in situ* superficial weathering in

Table 10.1. Mineralogy and chemistry of bauxite

Oxide	Chemical composition (wt.%)	Mineralogy
Alumina (Al_2O_3)	35 to 65	Gibbsite, boehmite and diaspore
Silica (SiO_2)	0.5 to 10	Quartz, chalcedony, kaolinite
Ferric oxide (Fe_2O_3)	2 to 30	Goethite, hematite and siderite
Titania (TiO_2)	0.5 to 8	Rutile and anatase
Calcia (CaO)	0 to 5.5	Calcite, magnesite and dolomite

moist tropical climates of clays, clayey limestones, or high-alumina-content silicoaluminous igneous and metamorphic rocks containing feldspars and micas. Around the world there is a restricted number of geographical locations containing bauxite deposits of commercial interest. Their occurrence and origin can be explained by both plate tectonics and climatic conditions. Actually, during weathering water-soluble cations (e.g., Na, K, Ca, and Mg) and part of the silica (SiO_2) are leached by rainwater acidified by the organic decomposition of humus, leaving only insoluble aluminum and iron sesquioxides and a lesser amount of titania (TiO_2). Hence, insoluble cations such as iron (III) and aluminum (III) associated with clays and silica remain in the materials. Bauxite is a sedimentary rock, so it has neither a precise definition nor chemical formula. From a mineralogical point of view, bauxite is mainly composed of hydrated alumina minerals such as ***gibbsite*** [$Al(OH)_3$ or $Al_2O_3.3H_2O$, monoclinic] in recent tropical and equatorial bauxite deposits, while ***boehmite*** [AlO(OH) or $Al_2O_3.H_2O$, orthorhombic] and, to a lesser extent, *diaspore* [AlO(OH) or $Al_2O_3.H_2O$, orthorhombic] are the major minerals in subtropical and temperate bauxite old deposits. The average chemical composition of bauxite is 45 to 60 wt.% Al_2O_3 and 10 to 30 wt.% Fe_2O_3, the remainder consisting of silica, calcia, titanium dioxide, and water. The typical mineralogy and chemical composition of bauxite is presented in Table 10.1. The different types of bauxite are only distinguished according to their mineralogical composition. They are then called gibbsitic, boehmitic, or diasporic bauxite. Gibbsitic bauxite predominates. It is geologically the youngest and situated in tropical or subtropical regions, very close to the ground surface (e.g., laterites). The oldest deposits, which are mainly found in Europe (e.g., Gardanne in France, and Patras in Greece) and in Asia, mainly contain boehmite and diaspore. Most of the time they are underground deposits.

According to the US Geological Survey, the world's bauxite resources are estimated to be 55 to 75 billion tonnes located mainly in South America (33%), Africa (27%), Asia (17%), Oceania (13%), and elsewhere (10%). Today, Australia supplies 35% of the demand worldwide for bauxite, South America 25%, and Africa 15%. The current reserves are estimated at being able to supply worldwide demand for more than two centuries. Note that about 95% of bauxite is of the metallurgical grade and hence used for the production of primary aluminum metal.

Bayer process. Because bauxite exhibits a high alumina content and its worldwide reserves are sufficient to satisfy demand for a few centuries, it is the best feedstock for producing alumina and then aluminum. Actually, today, more than 95% of alumina worldwide is extracted from bauxite using the Bayer process, which was invented in 1887, just one year after the invention of the Hall–Heroult electrolytic process. This Bayer process was implemented for the first time in 1893, in France, at Gardanne. However, the conditions for implementing the process strongly depend on the type of bauxitic ore used. For instance, refractory-type diasporic bauxite must be digested at a higher temperature than gibbsitic bauxite. Therefore, the selection of the type of bauxite to be used is a critical factor affecting the design of the alumina plant. A brief description of the Bayer process is given hereafter.

Table 10.2. Digestion conditions for various bauxitic ore

Bauxitic ore	Digestion reaction	Conditions
Gibbsitic	$2AlO(OH).H_2O + 2NaOH \longrightarrow 2NaAlO_2 + 4H_2O$	Atmospheric pressure 135°C < T < 150°C
Boehmitic	$2AlO(OH) + 2NaOH \longrightarrow 2NaAlO_2 + 2H_2O$	Atmospheric pressure 205°C < T < 245°C)
Diasporic	$2AlO(OH) + 2NaOH \longrightarrow 2NaAlO_2 + 2H_2O$	High pressure (3.5–4 MPa) T > 250°C

Comminution. First, bauxite run-of-mine ore is crushed using a jaw crusher to produce coarse particles below 30 mm in diameter. It is then washed with water in order to remove clay minerals and silica in an operation called desliming. The washed and crushed ore is then mixed with the recycled caustic liquor downstream from the Bayer process, then ground more finely providing a suspension or slurry of bauxite with 90% of particles with a size less than 300 μm (48 mesh Tyler). This grinding step is required to increase the specific surface area of the bauxite in order to improve the digestion efficiency. The recycled liquor comes from the filtration stage of the hydrate after it has been precipitated. This liquor is enriched in caustic soda (i.e., sodium hydroxide, NaOH) and slacked lime [calcium hydroxide, $Ca(OH)_2$] before grinding to meet the digestion conditions and to be more aggressive toward the bauxite. The permanent recycling of the liquor and, more generally, of the water is at the origin of the synonym for the Bayer process, also called the Bayer cycle. The red bauxite-liquor slurry is preheated before being sent to the digesters for several hours.

Digestion of bauxite. The conditions of digestion are strongly related to the mineralogical composition of the bauxitic ore, that is, whether gibbsite, boehmite, or diaspore is the dominant ore. For instance, a gibbsitic bauxite can be digested under atmospheric pressure, whereas high pressures in excess of 1 MPa and temperatures above 250°C are required to digest the alumina contained in refractory diasporic bauxite. The various digestion conditions are summarized in Table 10.2.

Usually, the slurry is heated in an autoclave at 235 to 250°C under a pressure of 3.5 to 4.0 MPa. During the digestion stage, the hydrated alumina is dissolved by a hot and concentrated caustic liquor in the form of ***sodium aluminate*** ($NaAlO_2$). During the dissolution reaction, the sodium hydroxide reacts with both alumina and silica but not with the other impurities such as calcium, iron, and titanium oxides, which remain as insoluble residues. These insoluble residues sink gradually to the bottom of the tank and the resulting red sludge, called ***red mud***, concentrates at the bottom of the digester. The slurry is diluted after digestion to make settling easier. Slowly heating the solution causes the $Na_2Si(OH)_6$ to precipitate out, removing silica. The bottom solution is then pumped out and filtered and washed while the supernatant liquor is filtered to leave only the aluminum-containing $NaAl(OH)_4$. The clear sodium tetrahydroxyaluminate solution is pumped into a huge tank called a precipitator. There are two objectives in washing the red mud that has been extracted from the settler: to recuperate the spent sodium aluminate, which will be reused in the Bayer cycle, and to remove sodium hydroxide from the red mud for safe disposal as an inert mining residue.

Precipitation. The clear sodium aluminate liquor is cooled down, diluted with the water from the red-mud wash, and acidified by bubbling carbon dioxide (CO_2) gas through the solution. Carbon dioxide forms a weak acid solution of carbonic acid, which neutralizes the sodium hydroxide from the first treatment. This neutralization selectively precipitates the aluminum hydroxide [$Al(OH)_3$] but leaves the remaining traces of silica in solution; the precipitation or the crystallization of the hydrate is also called decomposition. The liquor is then sent into huge thickening tanks owing to the extremely slow precipitation kinetics. The alumina hydrate slowly precipitates from tank to tank as the temperature goes down. The floating suspension is recuperated in the last thickening tank. Fine particles of aluminum hydrate are usually added to seed the precipitation process of pure alumina particles as the liquor cools. In fact, 90% of the wet aluminum trihydrate recovered after filtration is recycled and used as a crystallization seed. The liquor is then filtered so as to separate the wet hydrate from the liquor. This liquid is then sent to the bauxite digestion tank, where it will be enriched in soda and in lime. The particles of aluminum hydroxide crystals sink to the bottom of the tank, are removed, and are then vacuum dewatered. The ***alumina trihydrate*** (ATH) or ***gibbsite*** [$Al(OH)_3$] obtained can be commercialized as is or it can be calcined into various grades of alumina (Al_2O_3).

Calcination. To produce ***calcined alumina*** (CA), the alumina trihydrate must be calcined into a rotary kiln or a fluidized-bed calciner operating at 1100 to 1300°C to drive off the chemically combined water. Usually, fluidized-bed calciners are restricted to transition aluminas used in the manufacture of metallurgical-grade alumina, while rotary kilns are used for non-metallurgical-grade alumina. All of the characteristics of calcined alumina are extremely variable and depend on the conditions of calcination. Sodium is the major impurity of the alumina produced in the Bayer process; this can be a hindrance for certain technical applications. Several methods for the removal of sodium exist to produce aluminas with a very low sodium content, such as water leaching or the use of silica to form a soluble sodium silicate phase. These reactions compete with the combination of sodium and alumina to form beta-aluminas. The transformation of gibbsite into alpha-alumina successively gives rise to the following phenomena while the temperature is rising: release of water vapor between 250 and 400°C that fluidizes the alumina and, at around 1000 to 1250°C, the exothermic transformation into alpha-alumina occurs. The appearance of alpha-alumina crystallites modifies the morphology of the grains, which become rough and friable. Completion of the transformation of gibbsite into alpha-alumina requires a residence time of at least 1 h. Some halogenated compounds called mineralizers are used to catalyze the transformation of the alpha-alumina crystallites. The mineralizers also form volatile sodium chloride. Calcined alumina consists of alpha-alumina crystallite clusters with a particle size ranging from 0.5 to 10 μm. The higher the calcinations, the larger the crystallites (Figure 10.2).

10.2.3.2 Alumina Hydrates

Aluminum hydroxides and oxihydroxides, formerly called ***aluminas hydrates***, are all produced during the Bayer process described in the preceding paragraphs. All the aluminum hydroxides exhibit the same molecular unit, which consists of an octahedron made of one hexacoordinated aluminum cation surrounded by six oxygen anions [AlO_6^{9-}]. The great stability of this structure is due to the strong Al-O chemical bonds owing to the high polarization of aluminum cations.

Three crystalline polymorphs of alumina trihydrates (ATH) or aluminum trihydroxide [$Al(OH)_3 = Al_2O_3.3H_2O$] exist: ***gibbsite*** or ***hydrargillite*** [γ-$Al(OH)_3$], ***bayerite*** [α-$Al(OH)_3$], and ***nordstrandite*** [β-$Al(OH)_3$]. The octahedrons form a plane framework of hexagonal crowns of $Al(OH)_3$ forming two planes of oxygen atoms in a compact hexagonal network wrapped around a plane of aluminum atoms two thirds of which is occupied. The three minerals differ by the sequence of these sheets. The sequence is (AB BA AB BA...) for gibbsite, (AB AB AB AB...) for bayerite, which is more compact and hence more stable and dense), and (AB BA BA AB...) for the intermediate case of nordstrandite. The sheets are linked together by hydrogen bonds.

Two crystalline polymorphs of monohydrated alumina or aluminum oxihydroxide [$AlO(OH) = Al_2O_3.H_2O$], where the [AlO_6^{9-}] octahedrons share one edge: ***boehmite*** [γ-$AlO(OH)$];

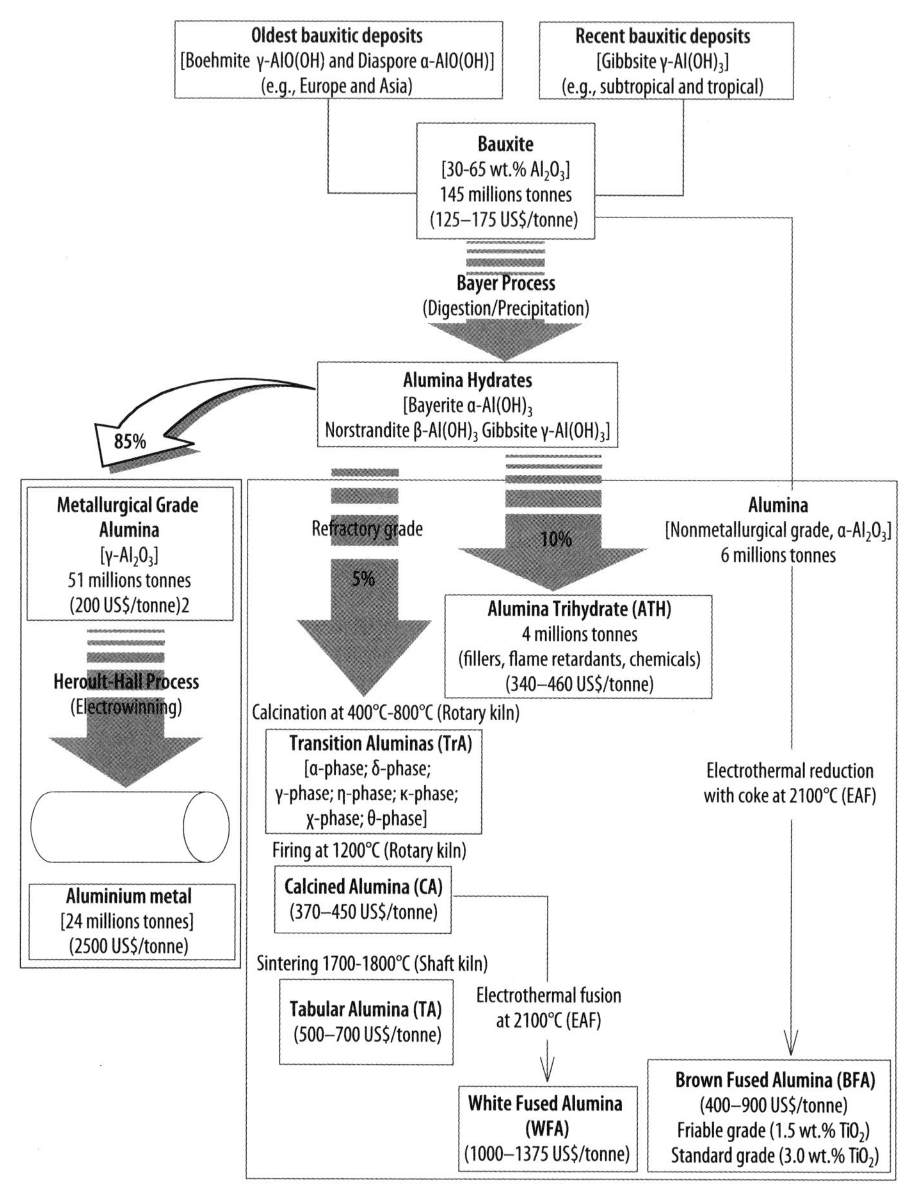

Figure 10.2. Aluminas production flowsheet

and ***diaspore*** [α-AlO(OH)]. The main characteristics of aluminum hydroxides are listed in Table 10.3.

Among all the alumina hydrates, **gibbsite** or gamma-aluminum trihydroxide (ATH) is, after bauxite, by far the most common aluminum commodity. Actually, 85% of the total gibbsite produced by the Bayer process is used to produce ***metallurgical-grade alumina*** for the electrowinning of aluminum metal by the Hall–Heroult process, while 8–10 percent are used for preparing ***non-metallurgical-grade alumina*** required for the manufacture of high

Table 10.3. Alumina hydrates (aluminum trihydroxides and oxihydroxides)

Phase	Chemical formula	Crystal system	Therm. stability range	Density ($kg.m^{-3}$)	Mohs hardness	Tenacity	Average refractive index (n_D)
Gibbsite (hydrargillite)	α-$Al(OH)_3$	Monoclinic	<100	2420	2.5–3.5	Tenacious	1.57–1.59
Bayerite	β-$Al(OH)_3$	Monoclinic	<100	2530	n.d.	Tenacious	1.58–
Nordstrandite	γ-$Al(OH)_3$	Triclinic		2450	3		1.590
Boehmite	γ-AlO(OH)	Orthorhombic	100-350	3010	3.5–4	Highly tenacious	1.65–1.67
Diaspore	α-AlO(OH)	Orthorhombic	100–350	3440	6.5–7	Brittle	1.70–1.75

alumina refractories, abrasives, proppants, and ceramics; the remaining 5 to 7% is used as specialty chemicals such as aluminum chemicals, flocculants, fillers, and flame retardants. High-purity aluminum trihydrate (ATH), with 99.7 wt.% $Al(OH)_3$, is a white solid with a low apparent density of 1200 $kg.m^{-3}$. Due to its low Mohs hardness, ranging between 2.5 and 3.5, it exhibits a low abrasiveness. It is easily flowable and hence can be fluidized. ATH is nonflammable and nonhazardous. It is insoluble in water but becomes soluble in strong mineral acids and bases to form $Al(H_2O)_6^{3+}$ or AlO_2^-, respectively. Its dehydratation reaction is highly endothermic with a specific enthalpy of 1155 $kJ.kg^{-1}$. Gibbsite loses 34.6 wt.% of its mass between 200 and 1200°C. The greatest weight loss occurs between 250 and 400°C, and the fastest dehydration rate occurs at around 350°C. Finally, ATH exhibits good absorption capabilities for aqueous solutions and organic liquids such as oil. ATH absorbs near-ultraviolet radiation with wavelengths below 400 nm. Commercially, ATH is available wet or dry, with wet ATH containing 11 wt.% free moisture. The ten major producers of ATH worldwide are listed in Table 10.4.

The thermal decomposition of alumina hydrates upon firing leads, depending on the intensity of firing, to the formation of four types of aluminas: **transition aluminas, calcined aluminas, tabular aluminas**, and **fused aluminas**. These four main families of alumina are described below.

Table 10.4. Ten major producers of aluminum trihydroxide (2002)[1]

Company	Country	Annual production capacity (tonnes)
Alcoa	United States	1,200,000
Alcan	Canada	390,000
Ajka Alumina Co.	Hungary	331,000
Sherwin Alumina	United States	300,000
Kaiser Aluminum	United States	300,000
Dadco Alumina & Chemicals	Germany	246,000
VAW Aluminium	Germany	164,000
Indian Aluminium Co. (Indal)	India	140,000
Aluminium Pechiney	France	100,000
National Aluminium Co. (Nalco)	India	27,000

[1] Crossley, P. (2002) ATH flexing its strength. *Ind. Min.*, February 2002, pp. 24–43.

Table 10.5. Transition aluminas and precursors

Precursor	Phase-transformation reactions
Gibbsite	Gibbsite —280°C—> χ-phase —800°C—> κ-phase—1000°C—> α-Al_2O_3
Bayerite	Bayerite —280°C—> η-phase —830°C—> θ-phase—1000°C—> α-Al_2O_3
Boehmite	Boehmite—450°C—> γ-phase —800°C—> δ-phase—920°C—>θ-phase—1050°C—> α-Al_2O_3
Diaspore	Diaspore—500°C—> α-Al_2O_3

10.2.3.3 Transition Aluminas (TrA)

During the thermal decomposition of alumina hydrates, the progressive loss of hydratation water leads to the formation of the so-called ***transition aluminas***, denoted by the common acronym TrA. These are metastable aluminas with an intermediate crystallographic structure ranging between that of alumina hydrates and that of alpha-alumina. The family of transition aluminas includes all aluminas that are obtained by the thermal decomposition of aluminum hydroxides or oxihydroxides, with the exception of alpha-alumina. The different transition aluminas generally coexist, and their proportions depend on the type of precursor hydrate and on the decomposition conditions (i.e., temperature, heating rate, relative humidity). As a general rule, each intermediate alumina hydrate exhibits at least two phase transformations when the temperature rises before reaching the final structure of the alpha alumina: a very disordered low-temperature structure produced by the loss of hydratation water and a high-temperature, well-ordered structure (Table 10.5). The most important properties of all transition aluminas are their intrinsic microporosity and their high specific surface area that can reach up to 400 $m^2.g^{-1}$. Because of their high specific surface areas, combined with their adsorptive capabilities, transition aluminas are able to adsorb huge quantities of polar, acidic, or basic compounds, but the adsorption is not selective. Moreover, transition aluminas are also very reactive chemically. Actually, the adsorption of acid in an aqueous medium always leads to the dissolution of part of the alumina and then the adsorption of the salt that has formed. Finally, when they undergo thermal decomposition above a temperature of 1100°C, all the transition aluminas are transformed irreversibly into calcined aluminas.

Four main processes are used for preparing industrial transition aluminas:

(i) The dehydration of gibbsite performed in a rotary kiln at 400°C. The thermal decomposition of gibbsite at 250°C produces a transition alumina having a large specific surface area. If the process is performed under pressure, a hydrothermal transformation occurs, yielding boehmite. Further dehydration of the boehmite produces a gamma transition alumina with a low specific surface area.

(ii) The activation of gibbsite by flash-firing that consists in firing the ATH in a few seconds at around 400°C. The activated alumina thus obtained is amorphous and very reactive.

(iii) The activation of oxihydroxide gels that are first transformed into grains by various processes and are then activated at between 500 and 600°C in a fluidized bed.

(iv) The activation of bayerite, which consists in agglomerating bayerite into a shaped material by means of an appropriate binder.

10.2.3.4 Calcined Alumina

Aluminum sesquioxide, or α-alumina (α-Al_2O_3), also called ***calcined alumina*** (CA) or ***burned alumina*** in the ceramic and refractory industries, is the final product resulting from the thermal decomposition of all aluminum hydroxides. Actually, in the temperature range 1000–1250°C, the exothermic transformation of transition aluminas into α-Al_2O_3 occurs irreversibly. The rate of transformation into α-Al_2O_3 depends on the residence time of the aluminum hydroxide at these temperatures. The transformation is usually complete after a few hours at more than 1250°C. Calcined aluminas are prepared by calcination of aluminum hydroxide performed in rotary kilns, fluidized-bed kilns, or tunnel kilns. The α-Al_2O_3 crystallites give a polycrystalline product that becomes friable. Calcined aluminas are offered in a wide variety of technical grades from those containing 100 wt.% α-Al_2O_3 to grades containing some sodium such as soda-alumina, also called ***beta alumina*** [$NaAl_{11}O_{17} = Na_2O{\cdot}11Al_2O_3$], the remaining component being unreacted transition alumina. Some halogenated compounds (e.g., BF_3, BCl_3) called mineralizers are used to catalyze the nucleation of the α-Al_2O_3 crystallites. The mineralizers also form volatile sodium chloride (NaCl), which removes the sodium. Calcined alumina consists of α-Al_2O_3 crystallite clusters with a particle size ranging from 0.5 to 10 μm, and it is hence a polycrystalline material with grains made of several crystallites. The higher the calcination temperature is, the larger the crystallites are. The morphology of crystallites is strongly influenced by the chemical nature of the mineralizer: fluorine produces tabular crystallites with a hexagonal shape, boron gives rounded crystallites, while boron chloride produces round and dense crystals. By contrast with transition aluminas, crystals in calcined alumina are free from micropores, and hence calcined alumina's specific surface area equals the surface area of its crystallites, and the larger they are, the lower the surface area will be. Technical calcined aluminas are classified according to the particle size of their crystallites, the morphology of the crystallites (i.e., angular, rounded, tabular), their sodium content, and, to a lesser extent, the content of other impurities that result mainly from the Bayer process and bauxite. Commercially, four grades of calcined aluminas are distinguished based on their soda content:

(i) ***Standard calcined aluminas*** with a sodium content of between 3000 and 7000 ppm wt. Na_2O.

(ii) ***Intermediate calcined alumina*** with a sodium content of between 1000 and 3000 ppm wt. Na_2O. The sodium content of this grade has been lowered by modifying the conditions of the precipitation of gibbsite or of the calcination.

(iii) ***Low-sodium calcined aluminas*** with a sodium content of between 300 and 1000 ppm wt. Na_2O. These aluminas are usually obtained by washing the precursor or by the extraction of sodium as a volatile compound with the mineralizer during calcination.

(iv) ***High-purity aluminas*** with an extra-low sodium content below 100 ppm wt. Na_2O. These aluminas obtained from an aluminum hydroxide produced by a process other than the Bayer process. The main applications for calcined aluminas are as feedstocks for refractories, glass and enamel, tiles and porcelain, and advanced ceramics. The diversity of applications for calcined aluminas can be explained by the wide range of properties: refractoriness, sinterability, chemical inertness in both oxidizing and reducing atmosphere and in both acid and alkaline media, hardness, wear and abrasion resistance, dimensional stability, high thermal conductivity, electrical resistivity, low dielectric loss and high permittivity, and high ionic conductivity in the case of beta-alumina.

10.2.3.5 Tabular Alumina

Tabular alumina (TA), also called ***sintered alumina***, is produced by the sintering of calcined alumina, which occurs above 1600°C. Sintering is usually performed industrially in a tall shaft kiln equipped with gas burners in the median zone. First, 20-mm balls are made by pelletizing a mixture of ground calcined alumina, reactive micronized alumina, and an appropriate organic binder to ensure the highest green density. Usually boron trichloride is added for the proper removal of sodium as NaCl upon heating. Prior to being fed into the shaft kiln the balls are always dried. The sintering is performed continuously at a high operating temperature of between 1900°C and 1950°C to obtain the highest mass density of 3550 kg.m^{-3} and a low porosity of 5 vol.% but always below the melting point of α-Al_2O_3 (2050°C). It takes about 15 h for the balls to exit from the bottom of the furnace. After sintering, the balls, which have shrunk by 20 vol.%, are crushed and ground, and iron-rich material is removed by magnetic separation and then sized in several grades. The high purity of tabular alumina (99.8 wt.% Al_2O_3) is due to its low soda content (Na_2O < 1000 ppm wt.) and to the absence of nonvolatile mineral additions in the preparation of green balls. The resulting polycrystalline material exhibits large tabular crystals with a hexagonal shape and with a particle size of between 200 and 300 μm. Moreover, the commercial material contains a finer grain size and additives that lower the melting point in the range of 1700°C to 1850°C. The major properties of tabular alumina are a high density, a low open porosity, refractoriness, hardness, chemical inertness, thermal conductivity and dielectric rigidity at high temperatures, dimensional stability, creep and abrasion resistance, and exceptional resistance to thermal shock. These properties explain its development as a refractory raw material and its use in steelmaking and in electric furnaces, especially in Japan, as well as in ceramics, filters for molten metal, fillers for epoxy resins and polyester, inert catalyst supports, and heat conductors.

10.2.3.6 White Fused Alumina

Above 2050°C, pure alumina (Al_2O_3) melts forming a covalent and nonconducting liquid that upon cooling yields a solidified mass of corundum. Corundum is also called in the ceramics and refractory industries ***white fused alumina***. White fused alumina exhibits a fine-grained microstructure with euhedral crystals. Although the operation can be performed commercially on a small industrial scale by the Verneuil technique to produce kilogram-size single crystals (see Gemstones, Section 12.5), most of the large tonnage production uses a tilting electric-arc furnace with three electrodes operating in an AC mode. Once molten and homogeneous, the alumina melt is poured into molds and allowed to cool slowly until demolding. Beta-alumina represents the major impurity observed in white fused alumina due to the concentration of sodium occurring in certain regions. However, the volatilization of the sodium occurs at 2100°C and creates pores that are beneficial. To improve is mechanical strength, usually 2 wt.% of chromia (Cr_2O_3) is added to the melt. Actually, trivalent chromium substitutes isomorphically for the Al^{3+} increasing the toughness of white fused alumina.

10.2.3.7 Brown Fused Alumina

The electrothermal fusion of bauxite at 2100°C yields an impure and brown electrofused alumina product called ***brown fused alumina***, sometimes simply ***brown corundum***. Brown fused alumina exhibits coarse grains, and the major impurity in brown fused alumina is titania or titanium dioxide (TiO_2) coming from the bauxite ore. Therefore, commercially, two grades of brown fused aluminas can be distinguished according to their titania content: the **friable grade**, with 1.5 wt.% TiO_2, and the **standard grade**, with 3 wt.% TiO_2, which exhibits a greater toughness than the friable grade. The toughness of brown fused alumina is higher than that of white fused alumina, and this is due to the titania content, which reduces the size of the crystallites. Brown electrofused alumina is obtained industrially by the simultaneous electrothermal fusion and reduction by coke in a tilting and triphased electric-arc furnace of a blend of bauxite and spent products of white and brown aluminas. During the process, the reduction of iron oxide, silica, and, to a lesser extent, titania produces a titanium-bearing ferrosilicon alloy (FeSi), which is an important byproduct. The dense droplets of ferrosilicon sink by gravity, settling at the bottom of the crucible, and coalesce to form a pool of liquid FeSi. After several castings of the electrofused brown alumina thus produced, the ferrosilicon that has accumulated at the bottom must be tapped by overturning the crucible; this represents an important byproduct. During electrofusion, the raw materials float over the molten alumina, decreasing the thermal losses by radiation (ca. 900 $kW.m^{-2}$ at 2000°C). Because of the high temperature combined with the corrosiveness of the melt, no container can withstand such melts and skull melting is the only means to contain the molten materials. Actually, a frozen layer forms upon cooling on the inner wall and at the bottom of the crucible that are externally cooled by running water, forming a protective and self-lining skull. In practice, two distinct skulls are formed, on the inner side wall of the furnace the thick skull is made of solidified alumina, while at the bottom a thick skull forms containing titanium carbide (TiC). During the process, the gases resulting from the reduction of various metal oxides with the carbon are mainly CO and SiO. This aspect is of crucial importance both for reasons linked to the process and for safety reasons. Once molten alumina is poured into molds and allowed to cool slowly until demolding, it is crushed and ground while droplets of FeSi are removed by magnetic separation performed with a rotary drum equipped with rare-earth magnets. To ensure that the ferrosilicon is ferromagnetic and hence easily separated from corundum, its silicon content must be less than 21 wt.% Si, but in order to be easily crushed, its silicon content must be at least 13 wt.% Si. Proper operation of the process consists in maintaining a silicon content ranging between these two limits. Moreover, the titanium content is another important parameter to control. If the titanium content is too high, it forms a bed of TiC, reducing the useful depth of the crucible. On the other hand, if the titanium content is too low, the skull at the bottom of the crucible will be too thin. Apart from refractories, brown fused alumina is used in deburring, plunge cut grinding, and sandblasting.

10.2.3.8 Electrofused Alumina-Zirconia

Electrofused alumina-zirconia (EFAZ) is obtained by a process similar to that used for preparing brown fused alumina by electrothermal fusion and reduction by metallurgical coke at 2100°C of a mixture of bauxite ore, zircon sand, and scrap iron in a tilting and triphased electric-arc furnace. After quenching the molten mass, the resulting product obtained is about five times stronger than brown fused alumina.

10.2.3.9 High-Purity Alumina

High-purity alumina contains at least 99.99 wt.% Al_2O_3, with crystallites small in size and morphology. Nearly half the high-purity alumina produced annually is used to manufacture sapphires and, to a lesser extent, as polishing medium for metallographical and optical processes. Four manufacturing processes of ultrapure aluminas are used, using either Bayer gibbsite or aluminum metal.

(i) ***Alum process.*** Gibbsite from the Bayer process is dissolved in an excess of sulfuric acid (H_2SO_4). The resulting liquor is then neutralized by aqueous ammonia and cooled to yield crystals of the double ammonium aluminum sulfate, formerly called ammonium alum [$NH_4Al(SO_4)_2.12H_2O$]. After settling and drying, the dried crystals of alum are calcined above 1000°C, giving a white powder of pure Al_2O_3.

(ii) ***Gel process.*** High-purity aluminum metal is dissolved in an alcoholic solution of KOH into isopropanol. Once dissolved, the aluminum propanolate produced is purified by distillation and hydrolyzed to yield a gel that is later calcined.

(iii) ***Chloride process.*** This process consists in dissolving pure alumina into concentrated hydrochloric acid and precipitating hexahydrated aluminum chloride ($AlCl_3.6H_2O$). After calcination at 1000°C the residue consists of highly pure Al_2O_3.

(iv) ***Alkaline process.*** This process consists in dissolving pure alumina into concentrated sodium hydroxide and precipitating the gibbsite either by Bayer precipitation or by neutralization. Sodium is removed from gibbsite by hydrothermal treatment. All these processes use a tunnel kiln for the final calcination.

10.2.4 Limestone and Lime

Description and general properties. ***Lime*** or ***calcia*** are common names for ***calcium oxide*** [1305-78-8], whose chemical formula is CaO. Lime exhibits a medium density of 3340 $kg.m^{-3}$ and a high melting point of 2899°C. More specifically, ***quicklime*** is calcined calcium oxide (CaO). It reacts vigorously with water according to the following reaction:

$$CaO(s) + H_2O(l) \longrightarrow Ca(OH)_2(s).$$

The hydratation of quicklime is highly exothermic and it releases circa 1.19 MJ per kilogram of lime. If not enough water is added, the heat released can increase the temperature of the water until it reaches its boiling point. Once the reaction is complete, the product obtained is ***calcium hydroxide***, $Ca(OH)_2$ [1305-62-0], also called ***hydrated lime*** or ***slaked lime***. The solution saturated with calcium hydroxide is called ***milk of lime*** and has a pH of 12.25. Hydraulic lime is an impure form of lime that will harden under water. Lime has been used for thousands of years for construction. Archeological discoveries in Turkey indicate lime was used as a mortar as far back as 7000 years ago. Ancient Egyptian civilization used lime to make plaster and mortar.

Industrial preparation. Most lime worldwide is obtained from quarries of carbonated rocks such as limestone, marble, chalk, and dolomite, or even from oyster shells. The suitable raw materials are usually selected because of their low silica and iron contents. After the rock is blasted away, the material is then crushed and sized before being calcined into vertical shaft furnaces (Europe) or rotary kilns (USA) at 1010 to 1345°C. During calcination, the carbon dioxide is driven off and leaves calcium oxide or quicklime according to the following reaction:

$$CaCO_3(s) \longrightarrow CaO(s) + CO_2(g).$$

Theoretically, 100 kg of pure calcium carbonate yields 56 kg of quicklime.

Industrial applications and uses. Today, nearly 90% of lime is used for chemical and industrial purposes. The largest use of lime is in steel manufacturing, where it is used as a flux to remove impurities such as phosphorus and sulfur. Lime is used in power-plant smokestacks to remove sulfur from emissions. Lime is also used in mining to neutralize acid-mine drainage, paper and paper-pulp production, water treatment and purification, and wastewater treatment. It is used in road construction and traditional building construction. ***Limestone*** is a sedimentary carbonated rock essentially made of calcite [$CaCO_3$, rhombohedral] and hence can be used in place of lime for some industrial applications such as agriculture, as a flux in steelmaking, and in sulfur removal. Limestone is much less expensive than lime (60–100 US$/tonne); however, it is not as reactive as lime, so it may not be the best substitute in all cases. Magnesium hydroxide can be used for pH control. Lime resources are plentiful worldwide. Major producers of lime are the United States (Texas, Alabama, Kentucky, Missouri, Ohio, and Pennsylvania), Canada and Mexico, Belgium, Brazil, China, France, Germany, Italy, Japan, Poland, Romania, and the United Kingdom.

10.2.5 Dolomite and Doloma

10.2.5.1 Dolomite

Description and general properties. Dolomite is a massive calcareous sedimentary rock made of the mineral ***dolomite*** [$CaMg(CO_3)_2$, rhombohedral], first identified by the French geologist D. Dolomieu in 1791 and named by H. Saussure after its discoverer. Dolomite occurs as huge geological formations such as in the northeastern Italian Alps called the Dolomiti. Usually dolomite as a rock contains, apart from dolomite, other carbonates (e.g., calcite, magnesite, and siderite), along with some silica and alumina, mostly as clays. For commercial purposes, the mass fraction of combined impurities must be below 7 wt.%, above which it becomes unsuitable for industrial use and is then used only for road ballasts and building stones. When the percentage of calcium carbonate ($CaCO_3$) is above 10 wt.% or more over the theoretical composition, the rock is termed calcitic dolomite, while with a departure from theoretical magnesite content the rock is called dolomitic limestone. With variations in $MgCO_3$ between 5 and 10 wt.%, it is called magnesian limestone, and when it contains up to 5 wt.% $MgCO_3$ or less it is considered limestone for all purposes.

Industrial applications and uses. Pure dolomite, without calcining, is chiefly used as refractory, ramming, and felting material in steel-melting shops and as fluxing material in blast-furnace operations in secondary steel and in the production of ferromanganese. Dolomite for use as flux in iron- and steelmaking should be hard, compact, and fine-grained so that it can withstand the burden of batches in blast furnaces as well as the basic steel converter. Dolomite bricks are kept in backup lining because it exhibits a lower thermal conductivity than magnesite. Chemical impurities must be as low as possible, especially phosphorus and sulfur, while silica and alumina are not deleterious for blast furnaces. Moreover, the magnesia in dolomite acts as desulfurizating agent in molten iron metal. Generally, two grades of dolomite are used, one is called blast furnace grade and the other steel melting shop grade. Dolomite is also used to a lesser extent in the glass industry, especially in sheet-glass manufacture. For that application, dolomite should contain no more than 0.1 wt.% Fe_2O_3. It also finds use in the manufacture of mineral wool. Dolomite is also a useful source for the production of magnesite by reacting calcined dolomite with seawater (Section 10.2.6).

10.2.5.2 Calcined and Dead Burned Dolomite (Doloma)

Description and general properties. Like other carbonates, upon heating above 900°C dolomite decomposes completely into a mixture of calcium and magnesium oxides, and carbon dioxide:

$$CaMg(CO_3)_2(s) \longrightarrow CaO(s) + MgO(s) + 2CO_2(g).$$

The product resulting from this relatively low-temperature calcination is highly porous and reactive and is known as ***calcined dolomite*** or simply ***doloma*** or ***dolime*** (i.e., CaO + MgO). Like lime, most dolime is produced either in vertical shaft kilns (Europe and UK) or rotary kilns (USA). Dolime is used in the extractive metallurgy of magnesium metal by the silicothermic process.

Although pure magnesite decomposes at 700°C and calcite at 900°C, dolime is too porous for most refractory uses. Therefore, prior to use it is calcined at a higher temperature of ca. 1700°C. This harsh treatment allows the material to shrink thoroughly and render it less reactive than calcined dolomite. The product obtained is called ***dead burned dolomite*** and is generally used for the refractory made by firing dolomite, with or without additives, at high temperatures to produce dense, well-shrunk particles.

Industrial preparation. Dead burned refractory dolomite is produced in vertical shaft or rotary kilns. Generally high-purity dolomite, with total impurities of less than 3 wt.%, is selected. As it is difficult to densify high-purity dolomite in a rotary kiln, it is customary to use some mineralizers to facilitate sintering. Iron sesquioxide is a common additive. The manufacturing process varies with the grade of dead burned dolomite needed. Most plants use rotary kilns lined in the hot zone with basic bricks and fired with powdered coal. The temperature reached in the hot zone is ca. 1760°C or above when iron oxide is added. After dead burning, dead burned dolomite is cooled in either rotary or reciprocating recuperative coolers. The air used for cooling gets heated and is again used as secondary air for combustion in the kilns.

There is another product known as ***stabilized refractory dolomite***. It is manufactured by a process similar to that of Portland clinker. Dolomite and serpentine, with small amounts of suitable stabilizing agents, are ground to a slurry in a ball mill. The slurry is fired into a dense mature clinker in a rotary kiln having a temperature on the order of 1760°C.

Industrial applications and uses. Dead burned dolomite exhibits high refractoriness and can withstand temperatures up to 2300°C. It is widely used as a refractory material wherever steel is refined using basic slag. It is used for original hearth installations in open hearth furnaces as well as for hearth maintenance. These hearths are installed using tar-dolomite ramming mixes and rammed dolomite. Dolomite refractories are also used in electrical furnaces and in the cement industry during clinker manufacture.

10.2.6 Magnesite and Magnesia

10.2.6.1 Magnesite

Description and general properties. ***Magnesite*** ($MgCO_3$) is like alumina, that is, it is considered either as an ore for magnesium metal production or as an industrial mineral. When pure, magnesite contains 47.8 wt.% magnesium oxide (MgO) and 52.2 wt.% carbon dioxide. Natural magnesite almost always contains some calcite ($CaCO_3$) and siderite ($FeCO_3$). Magnesium also occurs in dolomite [$FeMg(CO_3)_2$], a sedimentary rock in which $MgCO_3$ constitutes 45.65 wt.% (i.e., 21.7 wt.% MgO) and 54.35 wt.% $CaCO_3$. Magnesite color varies from white, when pure, to yellowish or gray white and brown. Its Mohs hardness ranges from 3.5 to 4.5 and its density varies from 3000 to 3200 $kg.m^{-3}$. A vitreous luster and very slow reaction with cold acids distinguishes magnesite from other carbonates. Magnesite, dolomite, seawater, and lake brines are used as major sources of magnesium metal with the most common source being lake brines and seawater.

Occurrence. Magnesite occurs in two physical forms: as cryptocrystalline or amorphous magnesite and as macrocrystalline magnesite. It occurs in five different ways: as a replacement mineral in carbonate rocks; as an alteration product in ultramafic rocks (e.g., serpentinite, dunite); as a vein-filling material; as a sedimentary rock; and as nodules formed in a lacustrine environment. Replacement-type magnesite deposits involve magnesium-rich hydrothermal fluids entering limestone via openings to produce both magnesite and dolomite. The alteration-type deposits are formed by the action of carbon-dioxide-rich waters on magnesium-rich serpentinite. Sedimentary deposits usually occur as thin layers of variable magnesite quality. Lacustrine magnesite deposits consist of nodules of cryptocrystalline magnesite formed in a lake environment. Both vein filling and sedimentary magnesite occurrences are rarely mined on a large scale.

Mining. All magnesite deposits are mined by open-cut methods. During mining the strip ratio, that is, the quantity of magnesite ore to waste material, may be high. The processing of magnesite ore begins with crushing, screening, and washing. The estimated world economic reserves of magnesite are about 8.60 billion tonnes expressed as $MgCO_3$. China is ranked first, followed by Russia and North Korea.

Industrial applications. Raw magnesite is used for surface coatings, landscaping, ceramics, and as a fire retardant.

10.2.6.2 Caustic Seawater and Calcined Magnesia

Industrial preparation. Raw magnesite coming from the run-of-mine is calcined between 700 and 1000°C in a vertical shaft kiln and decomposes yielding magnesium oxide or magnesia (MgO) and giving off carbon dioxide gas:

$$MgCO_3(s) \longrightarrow MgO(s) + CO_2(g).$$

The product obtained is called ***caustic-calcined magnesia*** (CCM), also called ***natural magnesia.*** The purity of CCM ranges usually between 75 and 96 wt.% MgO, with most of the impurities (e.g., Fe_2O_3, Al_2O_3, SiO_2, etc.) coming from the raw material used.

When a higher-purity magnesia is required, another more energy demanding route consists in preparing directly magnesium oxide from seawater or magnesium-rich brines. Prior to being processed, seawater is pumped and its impurities, mostly carbonates along with dissolved carbon dioxide, are removed. Usually, a hydrotreater removes CO_2 as calcium carbonate by the addition of milk of lime, $Ca(OH)_2$. Afterwards, the addition of hydrochloric acid removes the dissolved CO_2 as gaseous carbon dioxide with an efficiency of 95%. On the other hand, either ***quicklime*** (CaO) obtained from the calcination of pure limestone[2] or, better, ***dolime*** (i.e., CaO and MgO) obtained from the calcination of dolomitic limestone is prepared in a vertical shaft kiln. Once cooled, quicklime or dolime is then slacked with water to yield ***milk lime***, $Ca(OH)_2$. The operation is performed in a rotary slacker from which any traces of calcium carbonate are removed by centrifugation. The milk of lime is added to the decarbonated seawater for precipitating magnesium as brucite [$Mg(OH)_2$]. The milky slurry is filtrated to recover the magnesium hydroxide. The filtration cake is then sintered into a rotary kiln to obtain the so-called ***seawater magnesia clinker***, also called ***synthetic magnesia***, with a purity of at least 97 wt.% MgO.

Both grades of caustic magnesia readily react with water to give ***magnesium hydroxide*** or ***brucite*** [$Mg(OH)_2$], also called slacked or ***spent magnesia***:

$$MgO(s) + H_2O(l) \longrightarrow Mg(OH)_2(s).$$

Due to its alkaline properties and its poor solubility, when an excess of caustic magnesia is mixed with water, it gives a slurry called milk of magnesia with a pH of 10.25, and hence most heavy metals (e.g., Ni) are precipitated as metal hydroxides and then can be either removed by decantation, centrifugation, or filtration or stabilized in situ after drying of the slurry.

Industrial applications and uses. Magnesium hydroxide or brucite is used in sugar refining, as a flame and smoke retardant, in wastewater treatment, and finally in pharmaceuticals. Caustic magnesia is extensively used in acid mine drainage and wastewater treatment to precipitate deleterious metals. On the other hand, caustic-calcined magnesite is used in agriculture as a food supplement in fertilizers, in environmental applications, and in the chemical industry for making magnesium oxychloride and oxysulfate cements that are resilient, fireproof, spark proof, and vermin proof; it is also used as filler in paints, paper, and plastic.

The building industry consumes large quantities of caustic-calcined magnesite for use as a flooring material, in wall boards, and in acoustic tiles. Worldwide annual production is 1,000,000 tonnes of synthetic magnesia.

Prices (2006). Prices are roughly 200 US$/tonne for natural magnesia and up to 400 US$/tonne for synthetic magnesia.

10.2.6.3 Dead Burned Magnesia

When caustic magnesia is further heated at temperatures of between 1530 and 2300°C, the grains of magnesia become sintered and the product obtained is nonhygroscopic and exhibits exceptional stability and strength at high temperatures. This fine-grained product, with a density of 3400 $kg.m^{-3}$, an average periclase grain size above 120 μm that contains at least 97 wt.% MgO, is known as ***dead burned magnesia*** or ***sintered magnesia***. Worldwide circa 7.5 million tonnes of dead burned magnesia are produced annually. Dead burned magnesia and fused magnesia, due to their refractoriness, are used in the manufacture of basic refractories for iron- and steelmaking, nonferrous metallurgy, and finally in cement kilns. Eighty-five percent of the world production is used as ***refractory grade*** dead burned magnesia essentially as a refractory material because of its inertness and high melting point, while the remaining 15% is used in the cement industry, glassmaking, and the metallurgy of nonferrous metals.

[2] In the early days of the Dow Chemical process for producing magnesium metal, tonnes of Oyster shells were used as a source of pure calcium carbonate for the preparation of magnesia from seawater.

10.2.6.4 Electrofused Magnesia

Electrofused magnesia is obtained when dead burned magnesia is melted in an electric-arc furnace at temperatures above the melting point of MgO (2800°C) and pouring the melt into a mold. After cooling and demolding and then crushing, so-called ***electrofused magnesia***, or simply ***fused alumina***, is obtained. It has a higher mechanical strength, high resistance to abrasion, and a higher chemical stability than dead burned magnesia. It is used in the manufacture of premium-grade refractory bricks used in the high wear hot spots of basic oxygen furnaces and electric-arc or similar furnaces where temperatures can approach 950°C.

10.2.6.5 Seawater Magnesia Clinker

Magnesia is also produced from the processing of seawater and magnesium-rich brines. This is a much more complex and energy-demanding process than the processing of natural magnesite.

10.2.7 Titania

Titanium dioxide [13463-67-7], chemical formula TiO_2 and relative molecular molar mass of 79.8788, occurs in nature in three polymorphic crystal forms: ***anatase***, ***rutile***, and ***brookite***. Moreover, under high pressure, the structure of all three polymorphs of titanium dioxide may be converted into that of α-PbO_2. The main properties of the three polymorphs are summarized in Table 10.6.

10.2.7.1 Rutile

Rutile [131-80-2], among other polymorphs of titanium dioxide, is the most thermodynamically stable structure, and thus rutile is the major naturally occurring mineral of pure titanium dioxide and is much more common than either anatase or brookite. It is usually colored red or brownish red by transmitted light owing to trace impurities such as Fe, Nb, Ta, and, to a lesser extent, Sn, Cr, and V. The preparation of single crystals of rutile at the laboratory scale can be performed using Verneuil's flame fusion method,[3] while its large-scale industrial preparation is based on the sulfate and chloride processes (see Titanium). The crystallographic structure of rutile is a flat tetragonal prism where each tetravalent titanium cation is hexacoordinated to six almost equidistant oxygen anions, and each oxygen anion to three titanium anions. The TiO_6^{8-} octahedra are arranged in chains parallel to the *c*-axis. The oxygen atoms are arranged in the form of a somewhat distorted octahedron with each octahedron sharing one edge with adjacent members of the chain. The O-Ti-O bond angles are 90° by symmetry, 80.8°, and 99.2°, respectively. Highly pure rutile is an excellent electrical insulator at room temperature. However, its electrical conductivity, which is highly anisotropic, rises rapidly with temperature owing to the reversible lost of oxygen atoms that leads to a departure from ideal stoichiometry. Hence, upon heating rutile gives an *n*-type semiconductor[4] and its conductivity can increase up to 100 $S.cm^{-1}$ for the composition $TiO_{1.75}$.[5] Expression for the intrinsic conductivity expressed in $S.cm^{-1}$ of single crystals of rutile as a function of temperature have been given by Cronmeyer,[6,7] where the two subscript symbols // and + refer to the *c*-axis.

---

3 Verneuil, A. (1902) *Compt. Rend. Acad. Sci.*, 135, 791.
4 Grant, F.A. (1959) *Rev. Mod. Phys.*, 31, 646.
5 Verwey, E.J.W. (1947) *Philips Tech. Rev.* 9, 46.
6 Cronmeyer, D.C. (1952) *Phys. Rev.*, 87, 876.
7 Cronmeyer, D.C. (1959) *Phys. Rev.*, 113, 1222.

Table 10.6. Properties of titanium-dioxide polymorphic phases

Phase [CAS RN]	Crystal system, space group, and space lattice parameters	Refractive indices (for $\lambda_D = 589.3$ nm)	Miscellaneous properties (density[8], etc.)
Anatase[9] [1317-70-0]	Tetragonal ($I4_1/amd$, Z = 4) $a = 379.3$ pm $c = 951.2$ pm Ti-O: 191 pm (2) – 195 pm (4) Packing fraction = 70%	Uniaxial (–) $n_\varepsilon = 2.4880$ $n_\omega = 2.5612$	Black to red $\rho_{calc.} = 3877$ kg.m^{-3} *trans. temp.* 700°C $\varepsilon_r = 48$ HM = 5.5 – 6.0
Rutile[10] [131-80-2]	Tetragonal ($P4_2/mnm$, Z = 2) $a = 459.37$ pm $c = 296.18$ pm Ti-O: 194.4 pm (4) – 198.8 pm (2) Packing fraction = 77%	Uniaxial (+) $n_\varepsilon = 2.6124$ $n_\omega = 2.8993$	Reddish brown $\rho_{calc.} = 4245$ kg.m^{-3} *m.p.* = 1847°C $\sigma_e = 10^{-14}$ S.cm^{-1} $\chi_m = +74 \times 10^{-9}$ emu[11] $\varepsilon_r = 110 - 117$ HM = 6.0 – 6.5
Brookite[12] [12188-41-9]	Orthorhombic ($Pbca$, Z = 8) $a = 545.6$ pm $b = 918.2$ pm $c = 514.3$ pm Ti-O: 184 pm – 203 pm	Biaxial () $n_\alpha = 2.5831$ $n_\beta = 2.5843$ $n_\gamma = 2.7004$	$\rho_{calc.} = 4130$ kg.m^{-3} *m.p.* = 1900°C $\varepsilon_r = 78$ HM = 5.5 – 6.0
TiO_2 II high-pressure phase (40 kbar, 450°C)[13]	Orthorhombic ($Pbcn$, oP12, Z = 4) $a = 551.5$ pm $b = 549.7$ pm $c = 493.9$ pm Ti-O: 191 pm (4) – 205 pm (2)	n.a.	n.a.
TiO_2 III high-pressure phase	Hexagonal hP48		

$$\ln \sigma_{+} = 7.92 - 17{,}600/T \quad (623.15\ \text{K} - 1123.15\ \text{K})$$
$$= 11.10 - 21{,}200/T \quad (1123.15\ \text{K} - 1673.15\ \text{K})$$

$$\ln \sigma_{//} = 8.43 - 17{,}600/T \quad (773.15\ \text{K} - 1223.15\ \text{K})$$
$$= 11.30 - 21{,}200/T \quad (1223.15\ \text{K} - 1673.15\ \text{K})$$

On the other hand, the electrical conductivity of single crystals of highly pure rutile is strongly affected by the doping of the crystal lattice with traces (i.e., less than 0.1 ppm wt.) of transition-metal cations (e.g., Cr^{3+}, V^{4+}, Nb^{4+}, Nb^{5+}, Fe^{3+}, Co^{2+}, Ni^{2+}, Ni^{3+}, and Cu^{2+}).

From an optical point of view, rutile, which is uniaxial (-), exhibits a high refractive index even higher than that of diamond and is transparent from visible to near-infrared radiation with wavelengths ranging from 408 nm to 5000 nm. However, at the blue end of the visible spectrum the strong absorption band of rutile at 385 nm renders the rutile pigment slightly brighter than anatase, which explains its typical yellow undertone. For that reason it can be used efficiently as sunscreen. When heated in air to ca. 900°C the powdered material

[8] Theoretical density calculated from crystal lattice parameters.

[9] Pascal, P. (1963) Nouveau Traité de Chimie Minérale, Tome IX. Masson & Cie, Paris.

[10] Meagher, E.P.; Lager, G.A. (1979) Polyhedral thermal expansion in the TiO_2 polymorphs: refinement of the crystal structures of rutile and brookite at high temperature. *Can. Mineral.*, **17**, 77–85.

[11] Wide range also reported in the litterature from –300 to +370 × 10^{-9} emu owing to slight departure from stoichiometry and doping with paramgnetic impurities leads to positive susceptibilities.

[12] Weyl, R. (1959) *Z. Krist.* **111**, 401.

[13] Simons, P.Y.; Dachille, F. (1967) *Acta Cryst.*, **23**, 334.

becomes lemon-yellow and exhibits a maximum absorption edge at 476 nm, but coloring disappears on cooling. In addition, doped rutile is phototropic, that is, it exhibits a reversible darkening when exposed to light.[14] On the other hand, rutile exhibits strong photocatalytic properties. As for electrical properties, metallic trace impurities strongly affect the whiteness of rutile. Even minute concentrations on the order of a few parts per million by weight may be sufficient to impart color. Thus in the industrial production of the whitest rutile, it is essential that other chromophoric transition elements not be present in the feedstock or be removed during the processing. Of these, chromium (Cr), vanadium (V), iron (Fe), and, to a lesser extent, niobium (Nb) are particularly deleterious in discoloring rutile. Generally, the colors are too intense to arise from crystal-field effects only but may arise from the excitation of the *d*-electrons of the impurity metal cation into the conduction band of the crystal lattice. From a chemical point of view, titanium dioxide is relatively inert chemically and resists attack from most chemical reagents. This property is further enhanced after titanium dioxide has been calcined at high temperatures.

10.2.7.2 Anatase

The lattice structure of ***anatase*** [1317-70-0] is also tetragonal, but the lower packing fraction of the crystal lattice explains why anatase crystal exhibits both a lower hardness and refractive indices than rutile. Nevertheless, because the crystal lattice energies of the two phases are quite similar, anatase remains metastable over long periods of time despite being less thermodynamically stable. However, above 700°C, the irreversible and rapid monotropic conversion of anatase to rutile occurs. From an optical point of view, anatase exhibits a greater transparency in the near-UV than rutile. The absorption edge being at 385 nm, this explains why anatase absorbs less light at the blue end of the visible spectrum and has a blue tone. Although anatase was the first pigment to be produced commercially and represented a step-change in optical performance over the pigments that preceded it, rutile remains the preferred pigment because of its higher refractive index and lower photocatalytic activity. Actually, rutile ensures greater stability and durability of the paint made from it (less chalking). However, anatase is required in certain applications, especially where low abrasivity may be an issue. Thus anatase pigments were originally the preferred choice for paper filling and coating and also for delustring of synthetic fibers, where the color of the application may degrade by abrasion of metal during frequent rubbing contact with machinery during processing.

10.2.7.3 Brookite

Brookite [12188-41-9], which exhibits an orthorhombic crystal lattice, is more difficult to produce than rutile and anatase, and for that reason it has never been used industrially, especially in the white pigment industry.

Apart from the well-known titanium-dioxide phases mentioned above, other titanium oxides exists such as titanium hemioxide (Ti_2O), titanium monoxide (TiO), titanium sesquioxide (Ti_2O_3), and *anosovite* (Ti_3O_5).

See Table 10.6, page 615.

10.2.7.4 Anosovite

Anosovite [12065-65-5], chemical formula Ti_3O_5, was identified by Ehrlich[15] in the Ti-O binary phase diagram, in the region between 62.3 and 64.3 at.% oxygen. It can be prepared in the following ways:

---

[14] Weyl, W.A.; Forland, T. (1950) *Ind. Eng. Chem.*, 42, 257.

[15] Ehrlich, P.Z. (1939) *Elektrochem.*, 45, 362.

(i) By the hydrogen reduction of solid TiO_2 at temperature around 1300°C[16] according to the following reaction scheme:

$3TiO_2(s) + H_2(g) \longrightarrow T_3O_5(s) + H_2O(g)$ (1300°C)

(ii) By mixing intimately stoichiometric quantities of titanium metal and titanium dioxide in an electric-arc furnace under an argon atmosphere according to the following reaction scheme:

$5TiO_2(s) + Ti(s) \longrightarrow 2T_3O_5(s)$ (1150°C)

followed by annealing in a vacuum of the crushed material for 2 weeks at 1150°C in a sealed silica tube. This oxide is dimorphic with a rapid phase transition from semiconductor to metal occurring at roughly 120°C.

$\alpha\text{-}Ti_3O_5 \longrightarrow \beta\text{-}Ti_3O_5$ (120°C)

The low-temperature form (α-Ti_3O_5), also called ***anosovite type I***, crystallizes with a monoclinic unit cell with the Ti-O bond distances ranging from 178 to 221 pm. The structure can be described in terms of TiO_6^{8-} octahedra joined by sharing the edge and corners to form an infinite three-dimensional network. Anosovite I is obtained by the hydrogen reduction of pure rutile at 1300°C. The high-temperature form (β-Ti_3O_5), also called ***anosovite type II***, is a slightly deformed pseudobrookite structure (AB_2O_5) with the Ti-O bond distances ranging from 191 to 210 pm. The type II is obtained by hydrogen reduction at 1500°C with magnesia as a catalyst. The anasovite type II is similar to that identified in titanium slags.[17] It can be stabilized at room temperature with a small amount of iron.

10.2.7.5 Titanium Sesquioxide

Titanium sesquioxide [1344-54-3], chemical formula Ti_2O_3, exists within a rather narrow range of homogeneity, from 59.4 to 60.8 at.% oxygen ($TiO_{1.49}$–$TiO_{1.51}$). It has a corundum structure and is isomorphous with hematite and ilmenite. It may be prepared by reduction of titania by hydrogen gas at 1000°C or as powder by reacting titanium metal with a stoichiometric amount of TiO_2 at 1600°C as follows:

$Ti(s) + 3TiO_2(s) \longrightarrow 2\,Ti_2O_3(s)$ (1600°C).

10.2.7.6 Titanium Monoxide or Hongquiite

Titanium monoxide or ***hongquiite*** [12137-20-1], chemical formula TiO, exhibits a very wide range of composition, extending approximately from 42 to 54 at. % oxygen ($TiO_{0.64}$ to $TiO_{1.26}$). It may be prepared by direct reduction by mixing stoichiometric amounts of titanium metal and titanium dioxide into a molybdenum crucible at 1600°C or reduction of the titanium dioxide with hydrogen under pressure at 130 atm and 2000°C.

$Ti\,(s) + TiO_2(s) \longrightarrow 2\,TiO(s)$ (1600°C)

On heating the monoxide in air, the compound reverts to other titanium oxides as a function of temperature increase[18]:

$TiO(s) \longrightarrow Ti_2O_3(s)$ (200°C),

$TiO(s) \longrightarrow Ti_3O_5(s)$ (250–350°C),

$TiO(s) \longrightarrow TiO_2(s)$ (350°C).

---

16 Ehrlich, P. (1941) *Z. Anorg. Allgem. Chem.*, 247, 53.

17 Reznichenko, V.A.; Khalimov, F.B. (1959) Reduction of titanium dioxide with hydrogen. *Titan i Ego Splavy*, 2, 11–15.

18 Wyss, R. (1948) *Ann. Chim.* 3, 215.

Its crystal structure has varying proportions of both titanium and oxygen vacancies. Density and y-ray lattice parameter measurements have shown that a third of the oxygen sites are vacant in $TiO_{0.7}$, a quarter of the titanium sites are vacant in $TiO_{1.25}$, and even in stoichiometric TiO about 15% of both sites are vacant. Above 990°C, the vacancies are arranged randomly, giving rise to diffraction patterns typical of the cubic NaCl-type structure.

10.2.7.7 Titanium Hemioxide

Oxygen is soluble in alpha-titanium until the composition $TiO_{0.5}$ (alpha-case) is formed, with the oxygen atoms supposedly being randomly distributed in the octahedral interstices of the hexagonally close-packed titanium lattice.[19]

10.2.7.8 Andersson–Magnéli Phases

In addition, a major series is represented by Andersson–Magnéli's phases[20] that consist of a continuous series of substoichiometric titanium oxides, characterized by the general formula Ti_nO_{2n-1}, where *n* is an integer greater than 4 (i.e., Ti_4O_7, Ti_5O_9, Ti_6O_{11}, Ti_7O_{13}, etc.).
See Table 10.7, page 619.

10.2.8 Zircon and Zirconia

10.2.8.1 Zircon

Zircon [10101-52-7], chemical formula $ZrSiO_4$, is an accessory nesosilicate mineral found in granites and, due to its high Mohs hardness of 7.5 and chemical inertness, it concentrates in the weathering products of mother igneous rocks such as in alluvial placer deposits and beach sands. Because of its chemical inertness and high melting point, zircon is wetted less easily by molten metal, producing smoother surfaces on iron, high alloy steel, aluminum, and bronze casting. The largest use of zircon is as foundry sand, where zircon is used as the basic mold material, as facing material on mold cores, and in ram mixes. Zircon-sand molds have greater thermal shock resistance and better dimensional stability than quartz-sand molds. Zircon grains are usually bonded with sodium silicate. Major producers of zircon sand are Richards Bay Minerals (Rio Tinto plc-BHP Billiton) located on the coastline of the KwaZulu-Natal region of the Republic of South Africa, followed by the Australian mining company Iluka. Both produce zircon sand as coproduct during mineral dressing of weathered ilmenite and rutile from beach mineral sands.

10.2.8.2 Zirconia

Description and general properties. Pure zirconium dioxide (ZrO_2), also called ***zirconia***, is a dense material (5850 $kg.m^{-3}$) that exhibits a high temperature of fusion (2710°C) and a good thermal conductivity (1.8 $Wm^{-1}K^{-1}$). Electrically speaking, zirconia is a dielectric at room temperature but becomes a good ionic conductor at high temperatures. Actually, cubic zirconia is a solid ionic electrolyte that allows oxygen anions to migrate through the crystal structure under an electric field at temperatures above 800°C. Optically, zirconia has a high index of refraction, which allows it to be used for increasing the refractive index of some optical glasses. Chemically, zirconia exhibits an excellent chemical inertness and corrosion resistance to many strong mineral acids, liquid metals, and molten salts up to high temperatures well above the melting point of alumina. Zirconia is not wetted by many metals and is therefore an excellent crucible material when corrosive melts (e.g., molten alumina and titanium slag)

[19] McQuillan, A.D.; McQuillan, M.D. (1956) *Titanium*. Butterworths, London.
[20] Andersson, S.; Collen, B.; Kuylenstierna, U.; Magneli, A. (1957) *Acta Chem. Scand.*, 11, 1641.

Table 10.7. Properties of other titanium oxides

Titanium oxide	Formula	Rel.molar mass (M_r)	wt.% Ti	Color, crystal lattice structure, space group (SG), Pearson symbol, lattice parameters, physical properties
Andersson–Magnéli phases	$Ti_{10}O_{19}$	780.6886	61.1	Triclinic
	Ti_9O_{17}	701.0198	61.2	Triclinic; S.G. P1; aP52
	Ti_8O_{15}	621.3510	61.4	Triclinic, S.G. A1; aC92
	Ti_7O_{13}	541.6822	61.6	Triclinic; S.G. P1; aP40
	Ti_6O_{11}	462.0134	61.9	Triclinic; S.G. A1; aC68
	Ti_5O_9	382.3446	62.3	Triclinic; S.G. P1; aP28
	Ti_4O_7	302.6758	63.0	Triclinic; S.G. P1; aP44
Anosovite [12065-65-5]	Ti_3O_5	223.0070	64.1	Dark blue; dimorphic (120°C) Low T: Anasovite-type I Monoclinic, C2/m, $Z = 4$ $a = 975.2$ pm; $b = 380.2$ pm; $c = 944.2$ pm; $\beta = 91.55°$ High T: Pseudobrookite (orthorhombic) $\rho_{calc.} = 4900$ kg.m^{-3} *m.p.* = 1777°C
Titanium sesquioxide [1344-54-3]	Ti_2O_3	143.3382	66.5	Dark violet to purple violet Corundum type $a = 515.5$ pm; $c = 1361$ pm $\rho_{calc.} = 4486$ kg.m^{-3} *m.p.* = 1839°C $c_p = 679$ J.kg^{-1}.K^{-1} $\chi_m = +63 \times 10^{-6}$ emu
Titanium monoxide or hongquiite [12137-20-1]	TiO	63.6694	74.9	Gold-bronze Halite-type (cubic) $a = 417$ pm $\rho_{calc.} = 4888$ kg.m^{-3} *m.p.* = 1750°C $c_p = 628$ J.kg^{-1}.K^{-1} $\alpha = 9.19$ µ/m.K $\chi_m = +88 \times 10^{-6}$ emu
Titanium hemioxide	Ti_2O	111.3394	85.6	Metallic gray

are absent. It can be used continuously or intermittently at temperatures up to 2200°C in neutral or oxidizing atmospheres. However, above 1600°C, zirconia reacts with alumina, and above 1650°C, in contact with carbon, zirconia forms zirconium carbide (ZrC). It has been used successfully for melting alloy steels and the noble metals. Nevertheless, zirconia in its chemically pure state exhibits poor mechanical and thermal properties that are inappropriate for use in structural and advanced ceramics. Actually, the polymorphism of pure zirconia exhibits deleterious phase transitions between room temperature and its melting point (Figure 10.3). These phase changes, accompanied by important relative volume changes, create a dense population of microcracks for the sake of toughness and thermal shock resistance.

monoclinic-ZrO_2 (baddeleyite) $\xrightarrow{1197\,°C}$ **tetragonal**-ZrO_2 (rutile type) $\xrightarrow{2300\,°C}$ **cubic**-ZrO_2 (fluorite type) $\xrightarrow{2710\,°C}$ **liquid**-ZrO_2

Figure 10.3. Polymorphs of zirconia (ZrO_2)

At room temperature, pure zirconia is essentially made of monoclinic ***baddeleyite*** with a density of 5850 kg.m^{-3} and a coefficient of linear thermal expansion of 6.5×10^{-6} K^{-1}, which is stable up to the transition temperature of 1197°C, at which it transforms into tetragonal zirconia (i.e., rutile type) with a density of 6045 kg.m^{-3}. This inversion in crystalline structure causes an important volume change upon heating (V/V = +7.0 vol.%). Due to the inversion, pure zirconia is highly sensitive to thermal shocks:

$$ZrO_2 \text{ (monoclinic)} \longrightarrow ZrO_2 \text{ (tetragonal)} \qquad T_t = 1197°C.$$

Afterwards, above 2300°C, tetragonal zirconia transforms into high-temperature cubic zirconia (i.e., fluorite type) with a mass density of 5500 kg.m^{-3} and a coefficient of linear thermal expansion of 10.5×10^{-6} K^{-1}:

$$ZrO_2 \text{ (tetragonal)} \longrightarrow ZrO_2 \text{ (cubic)} \qquad T_t = 2300°C.$$

Finally, at 2710°C zirconia melts, giving molten zirconia:

$$ZrO_2\text{(cubic)} \longrightarrow ZrO_2\text{(l)} \qquad T_f = 2710°C.$$

However, to prevent the first disastrous phase transition, it is possible to stabilize high-temperature cubic zirconia introducing foreign bivalent, trivalent, and/or tetravalent cations into its structure. Once stabilized, zirconia is stable from room temperature up to its melting point without any phase changes, and its thermal expansion varies linearly with temperature. The doped material demonstrates superior mechanical, thermal, and electrical properties owing to the modification of its crystal structure. Major lattice stabilizers are, for instance, calcia (CaO), magnesia (MgO), ceria (CeO_2), yttria (Y_2O_3), and lanthania (La_2O_3), which are introduced into the material prior to firing. The stabilized zirconia is then extremely resistant to thermal shock. Actually, white hot parts can be quenched in cold water or liquid nitrogen without break. Usually, calcia is the most widely used addition commercially, not only because it is a cheap raw material but also because the cubic form remains stable at all temperatures, whereas the magnesia-stabilized form may revert to the monoclinic structure at low temperature.

Commercial zirconia grades. Usually ***unstabilized zirconia*** (i.e., fully monoclinic), ***partially stabilized zirconia***, and ***stabilized zirconia*** (i.e., completely cubic) grades exist commercially and are available among advanced ceramic producers worldwide (e.g., Zircoa, Vesuvius, and Degussa), and they are briefly described below.

(i) **Unstabilized zirconia.** As indicated previously, pure zirconia is monoclinic at room temperature and changes to the denser tetragonal form at 1100°C, which involves a large volume change and creates microcracks within its structure. However, pure zirconia is an important constituent of ceramic colors and an important component of lead-zirconia-titanate electronic ceramics. Pure zirconia can be used as an additive to enhance the properties of other oxide refractories. It is particularly advantageous when added to high-fired magnesia and alumina bodies. It promotes sinterability and, with alumina, contributes to abrasive characteristics.

(ii) **Partially stabilized zirconia.** Partially stabilized zirconia (PSZ) is a mixture of various zirconia polymorphs, because insufficient cubic-phase-forming oxide has been added and a cubic plus metastable tetragonal ZrO_2 mixture is obtained. A smaller addition of stabilizer to pure zirconia will bring its structure into a tetragonal phase at a temperature higher than 1100°C, and a mixture of cubic phase and monoclinic or tetragonal phase at a lower temperature. Therefore, partially stabilized zirconia is also called ***tetragonal zirconia polycrystal*** (**TZP**), which is usually a zirconia doped with 2 to 3 mol.% of yttria and which has a fine-grained microstructure (i.e., 0.5 to 0.8 μm) that exhibits most impressive mechanical properties at room temperature. Usually such PSZ consists of more than 8 mol.% MgO (2.77 wt.%), 8 mol% CaO (3.81 wt.%), or 3 to

4 mol.% Y_2O_3 (5.4 to 7.1 wt.%). PSZ is a transformation-toughened material. Microcracks and induced stress may be two explanations for the toughening in partially stabilized zirconia. Microcracks arise due to the difference in the thermal expansion between the cubic-phase particle and monoclinic or tetragonal-phase particles in the PSZ. This difference creates microcracks that dissipate the energy of propagating cracks. The induced-stress explanation depends upon the tetragonal-to-monoclinic transformation, once the application temperature passes the transformation temperature at about 1000° C. The pure zirconia particles in PSZ can metastably retain the high-temperature tetragonal phase. The cubic matrix provides a compressive force that maintains the tetragonal phase. Stress energies from propagating cracks cause the transition from the metastable tetragonal to the stable monoclinic zirconia. The energy used by this transformation is sufficient to slow or stop propagation of the cracks. PSZ has been used where extremely high temperatures are required. PSZ is also used experimentally as heat engine components, such as cylinder liners, piston caps, and valve seats.

(3) **Fully stabilized zirconia (CSZ).** Fully stabilized zirconia, also called cubic stabilized zirconia (CSZ), is essentially a single-phase cubic material with large grain sizes of 10 to 150 μm that result when the stabilizer content and sintering temperature place it entirely in the cubic-phase region. Generally, the addition of more than 16 mol.% CaO (7.9 wt.%), 16 mol.% MgO (5.86 wt.%), or 8 mol.% Y_2O_3 (13.75 wt.%) to a zirconia structure is needed to form a fully stabilized zirconia. Its structure becomes cubic solid solution, which has no phase transformation from room temperature up to 2710°C. Fully yttria-stabilized zirconia (YSZ) is an excellent oxygen anion conductor that has been used extensively either as oxygen sensor or anion exchange membrane in solid oxide fuel cells (SOFCs). Other applications include grinding media and advanced ceramics due to its hardness and high thermal shock resistance.

Preparation of unstabilized zirconia. Zirconia is usually produced from zircon flour. Although the carbochlorination of zircon produces zirconium tetrachloride that can be oxidized to yield zirconia, such a method is only restricted to the production of zirconium metal (see Zirconium and Zirconium alloys). Therefore, to produce zirconia from zircon, the first step is to convert zircon to zirconyl chloride dihydrate. The process starts with the preparation of disodium metazirconate (Na_2ZrO_3) by digesting zircon into molten sodium hydroxide as follows:

$$ZrSiO_4(s) + NaOH(l) \longrightarrow Na_2ZrO_3(s).$$

Then the sodium zirconate is dissovled in concentrated hydrochloric acid:

$$Na_2ZrO_3(s) + 2HCl(l) \longrightarrow ZrOCl_2 \cdot 8H_2O(s).$$

There are two methods used to make zirconia from zirconyl chloride dihydrate ($ZrOCl_2.8H_2O$): thermal decomposition and precipitation.

In the thermal decomposition method, upon heating above 200°C, zirconyl chloride dihydrate loses its hydratation water as follows:

$$ZrOCl_2 \cdot 8H_2O(s) \longrightarrow ZrOCl_2(s) + 8H_2O(g).$$

Afterwards, at a higher temperature anhydrous $ZrOCl_2$ decomposes during calcination into chlorine gas and yields zirconia:

$$ZrOCl_2(s) + 1/2O2(g) \longrightarrow ZrO_2(s) + Cl_2(g).$$

The zirconia lumps obtained from the calcination then undergo a size-reduction process, such as ball milling, into the particle size range needed, usually up to –325 mesh. Thermal decomposition is hence an energy-demanding process from which it is not easy to produce zirconia powders with a high purity and fine particle size.

In the precipitation method, zircornyl chloride dihydrate is dissolved into water, and after addition of aqueous ammonia a precipitate of zirconium hydroxide is obtained:

$$ZrOCl_2 + 4NH_4^+ + 4OH^- \longrightarrow Zr(OH)_4(s) + NH_4^+ + Cl^-.$$

The precipitate of zirconium hydroxide ($Zr(OH)_4$) is washed in order to obtain a chlorine-free product. The solid is recovered by filtration to yield a wet powder that after calcination and quenching into liquid nitrogen yield a high-quality zirconia powder:

$$Zr(OH)_4(s) \longrightarrow ZrO_2(s) + 2H_2O(g).$$

By this method, the grain size, particle shape, agglomerate size, and specific surface area can be modified within a certain degree by controlling the precipitation and calcination conditions. Furthermore, its purity is also more easily controlled. For the applications of zirconiain the slip casting, tape casting, mold injection, particle size, specific surface, etc. are important characteristics. Well-controlled precipitated zirconia powder can be fairly uniform and fine.

Preparation of stabilized zirconia. In order to achieve the requirement of the presence of cubic and tetragonal phases in the microstructure of zirconia, stabilizers, i.e., magnesia, calcia, or yttria must be introduced into pure zirconia powders prior to sintering. Stabilized zirconia can be formed during a process called *in situ* stabilizing. Before the forming processes, such as molding, pressing, or casting, a blend of fine particles of stabilizer and monoclinic zirconia is prepared. Then the mixture is used for forming of green body. The phase conversion is accomplished by sintering the doped zirconia at 1700°C. During the firing (sintering), the phase conversion takes place. On the other hand, high-quality stabilized zirconia powder is made by a coprecipitation process. Stabilizers are then introduced before precipitation of zirconium hydroxide.

Preparation of fused zirconia. Production of electrofused or simply fused zirconia consists in removing silica from zircon by melting zircon sand with coke into an electric arc furnace at temperatures of around 2800 to 3000°C. During the electrothermal process, silica is reduced to volatile silicon monoxide (SiO), which escapes the furnace and leaves molten zirconia. On rapid cooling, a granular material is produced that is screened and crushed. Usually, the monoclinic zirconia produced contains less than 0.2 wt.% silica.

Preparation of zirconia by alkaline leaching. Zircon is roasted with sodium hydroxide and calcia at 600 to 1000°C. During the process silica reacts with calcium and sodium to yield calcium and sodium metasilicates. After acid leaching, the product is dyed and calcined to yield pure zirconia with less than 0.10 wt.% residual silica.

The major producers of zirconia are listed in Table 10.8.

Table 10.8. Major producers of zirconia

Country	Company name	Plant location	Annual nameplate capacity (tonnes)
United States	Washington Mills Electro Minerals Corp.	Niagara Falls, NY	1500
	Ferro Electronic Materials Systems	Niagara Falls, NY	4000
	Universal America	Greeneville, TN	6000
	Saint-Gobain Ceramic Materials	Huntsville, AL	6000
United Kingdom	Unitec Ceramics	Stattford	200
France	SEPZirPro	Le Pontet	8000
Australia	Australia Fused Materials	Rockingham, WA	4000
South Africa	Foskor	Phalaborwa	4000

Table 10.8. *(continued)*

Country	Company name	Plant location	Annual nameplate capacity (tonnes)
Japan	JACO Co.	Osaka	
	Showa Denco Ceramics	Shiojiri	1000
	Fukushima Steel Works	Fukushima	
	JFE Material Co.	Imizu	1000
	IDU Co.	Kochi	1500
China	Shanghai Zirconium Products	Shanghai	
	Yingkou Astron Chemical Co.	Bayuquan	9000
	Zhenzou Fused Zirconia Co.	Zhengzou	5500

10.2.9 Carbon and Graphite

10.2.9.1 Description and General Properties

Graphite [7440-44-0] is one of the two allotropic forms of the chemical element carbon, the other two being diamond (see Section 12.5). Graphite crystallizes in the hexagonal system. It has a black to steel gray color and usually leaves a black streak on the hand when touched because of its extreme softness and greasiness. It is opaque, even in the finest flakes. Graphite exhibits a high thermal conductivity close to that of copper alloys (Table 10.9). An important limitation of this material is its low tensile strength, and all components manufactured from carbon or graphite are highly susceptible to brittle fracture by mechanical shock or vibrations. Graphite is almost completely inert to all but the most severe oxidizing conditions, especially acids. Actually, graphite is recommended for use in 60 wt.% HF, 20 wt.% HNO_3, 96 wt.% H_2SO_4, bromine, fluorine, or iodine. The excellent heat-transfer property of impervious graphite has made it very popular in heat exchangers handling corrosive media, but also for a number of other devices used in the chemical-process industries such as piping, pumps, valves, brick lining for process or storage vessels, anode materials in electrochemical processes, and ring packing for columns. Impervious graphite is also used for liquid metal-handling devices. It has high refractoriness; actually, graphite is highly refractory up to temperatures approaching 3000°C in an inert atmosphere or in a vacuum. However, if oxygen is present, it burns between 620 and 720°C. Graphite has an extraordinarily low coefficient of friction under practically all working conditions. This property is invaluable in lubricants. It diminishes friction and tends to keep the moving surface cool. Dry graphite, as well as graphite mixed with grease and oil, is used as a lubricant for heavy and light bearings. Graphite grease is used as a heavy-duty lubricant where high temperatures may tend to remove the grease. All grades of graphite, especially high-grade amorphous and crystalline graphite, can behave as colloids; for example, in suspension in an oil base, they are used as lubricants. Properties of selected commercial grades of graphite are listed in Table 10.10.

10.2.9.2 Natural Occurrence and Mining

Graphite is usually found in metamorphic rocks as veins, lenses, and pockets and as thin laminae disseminated in gneisses, schists, and phyllites. Depending upon the mode of occurrence and origin, it is graded into three forms: ***flake graphite*** found in metamorphosed rocks as vein deposits, ***crystalline graphite*** found as fissure-filled veins, and ***cryptocrystalline graphite***

Table 10.9. Selected properties of different carbon products[21]

Carbon derivate	Density (ρ/kg.m^{-3})	Young's modulus (E/GPa)	Compressive strength (MPa)	Flexural strength (MPa)	Thermal conductivity (k/W.m^{-1}.K^{-1})	Specific heat capacity (c_P/J.kg^{-1}.K^{-1})	Coeff. linear thermal expansion (α/10^{-6}K^{-1})	Electrical resistivity (ρ/$\mu\Omega$.cm)	Vickers hardness (HV)
Graphite (industrial)	1400–2266	3–12	14–42	5–21	85–350	709	1.3–3.8	1385	HM 1
Pyrolitic carbon (impervious graphite)	1400–2210	16–30	72	32	480–1950	707	4.5	1500	145
Diamond	3514	900	7000	n.a.	900–2300	506	2.16	10^{16}	8000
Vitreous carbon (treated at 1000°C)	1500–1550	28	300	100	4	710	3.2	5500	225
Vitreous carbon (treated at 2500°C)	1500–1550	22	150–200	60–80	8	710	3.2	4500	150–175

Table 10.10. Properties of industrial graphite grades from SGL Carbon

Graphite or carbon grade	Bulk density (ρ/kg.m^{-3})	Open porosity (vol.%)	Average grain size (μm)	Medium pore size (μm)	Dry air permeability at 20°C (cm^2.s^{-1})	Rockwell hardness (HR)	Shore hardness	Flexural strength (MPa)	Compressive strength (MPa)	Young's modulus (E/GPa)	Specific electrical resistivity (ρ/$\mu\Omega$.cm)	Thermal conductivity (k/W.m^{-1}.K^{-1})	Thermal expansion coefficient (20–200°C) (μm/m.K)	Ash content (ppm wt.)
HLM	1700–1780	17–23	1.65	n.a.	n.a.	n.a.	n.a.	15–18	37–44	7.4–10.2	940–1240	105–180	1.4–3.6	1000–3000
MNC	1800	14	200	n.a.	n.a.	n.a.	n.a.	40	55–60	16	1000	130	1.5	1500
MNT	1750	16	400	n.a.	n.a.	n.a.	n.a.	20	28–30	15	1000	130	1	4000
R4340	1720	15	15	2	0.15	80	50	45	90	10.5	1200	90	2.9	200
R4500	1770	13	10	1.5	0.1	70	65	50	120	10.5	1400	80	3.9	200
R4550	1830	10	10	1.5	0.04	95	75	60	125	11.5	1300	100	4	20
R4820	1820	10	20	2.5	0.1	100	65	45	105	11	1150	125	4.2	50
R6300	1730	15	20	2	0.1	80	50	40	90	10	1700	65	3.8	n.a.
R6340	1720	15	15	2	0.15	80	50	45	90	10.5	1200	90	4	
R6500	1770	13	10	1.5	0.1	70	65	50	120	10.5	1400	80	5	
R6510	1830	10	10	1.5	0.04	95	75	60	125	11.5	1300	90	5.1	
R6650	1840	10	7	0.8	0.03	95	75	65	150	12.5	1400	90	5	
R6710	1880	10	3	0.6	0.01	110	80	85	170	13.5	1300	100	5.8	
R6810	1800	11	20	2.5	0.3	95	75	45	100	10	1000	130	5.2	
R6830	1820	9.5	20	2.5	0.1	95	75	50	100	10	1000	130	5	
R8710	1880	10	3	0.6	0.01	110	80	85	170	13.5	1300	100	4.7	200

[21] Technical data from various producers such as Le Carbone Lorraine, Sigradur, SGL, and Tokkai.

formed in metamorphosed coal beds. Natural graphite occurs in many parts of the world in fair abundance and it has been used in various applications. In nature, graphite is found usually in association with feldspars, mica, quartz, pyroxene, rutile, pyrites, and apatite. These impurities are associated with vein graphite. The impurities with amorphous graphite are shale, slate, sandstone, quartz, and limestone. Graphite is found in almost every country, but Ceylon, Madagascar, Mexico, western Germany, and Korea all possess particularly plentiful reserves. Major industrial producers of graphite are South Korea, the largest producer in the world, followed by Austria.

Graphite is usually obtained by underground mining and, to a lesser extent, by hydraulic mining such as in Madagascar. Afterwards, beneficiation of the run-of-mine consists in using the intrinsic floatation ability of natural graphite without having to use a collector. However, the recovery of flake graphite from disseminated deposits is difficult, and several proprietary processes have been developed by many companies. This difficulty arises because fine grinding is not efficient and reduces the size and also lowers the price and value of the graphite.

10.2.9.3 Industrial Preparation and Processing

Impervious graphite is manufactured by processing graphite at temperatures above 2000°C using Acheson furnaces (Section 10.2.10), evacuating the pores, and impregnating with a phenolic resin. The impregnation seals the porosity.

10.2.9.4 Industrial Applications and Uses

Flake graphite containing 80 to 85 wt.% C is used for crucible manufacture; 93 wt.% C and above is preferred for the manufacture of lubricants, and graphite with 40 to 70 wt.% C is used for foundry facings. Natural graphite, refined or otherwise pure, having a carbon content of not less than 95%, is used in the manufacture of carbon rods for dry battery cells. Graphite crucibles are manufactured by pressing a mixture of graphite, clay, and silica sand (formerly called ***plumbago***) and heating the pressed article at a high temperature in an inert atmosphere. Flake graphite is the best material, although crystalline graphite is also used. Crucibles made of graphite are used for melting nonferrous metals, especially brass and aluminum. Coarse-grained flake graphite from Malagasy is regarded as standard for crucible manufacture.

The utility of graphite is dependent largely upon its type, i.e., flake, lumpy, or amorphous. The flake-type graphite is found to possess extremely low resistivity to electrical conductance. The electrical resistivity decreases with an increase in flaky particles. In addition, the bulk density decreases progressively as the particles become more and more flaky. Because of this property in flake graphite, it enjoys widespread use in the manufacture of carbon electrodes, plates, and brushes required in the electrical industry and dry-cell batteries. In the manufacture of plates and brushes, however, flake graphite has been substituted to some extent by synthetic, amorphous, crystalline graphite, and acetylene black. Graphite electrodes serve to give conductivity to the mass of manganese dioxide used in dry batteries.

10.2.10 Silicon Carbide

10.2.10.1 Description and General Properties

Silicon carbide [409-21-2], chemical formula SiC and relative molar mass 40.097, is an important advanced ceramic with a high melting point (2830°C), a high thermal conductivity (135 $Wm^{-1}K^{-1}$), and extremely high Mohs hardness of 9. Silicon carbide is also has a wide band gap for a semiconductor (2.3 eV). The preparation of silicon carbide involves the reaction of silica sand (SiO_2) and carbon (C) at a high temperature (between 1600 and 2500°C).

$$SiO_2(s) + C(s) \longrightarrow SiC(s) + CO(g)$$

The first observation of silicon carbide was made in 1824 by Jöns Jacob Berzelius.[22] It was first prepared industrially in 1893 by the American chemist Edward Goodrich Acheson, who patented both the batch process and the electric furnace for making synthetic silicon-carbide powder.[23] In 1894 he established the Carborundum Company in Monongahela City, PA, to manufacture bulk synthetic silicon carbide commercialized under the trade name ***Carborundum***™. Silicon carbide was initially used to produce grinding wheels, whetstones, knife sharpeners, and powdered abrasives. Despite being extremely rare in nature, when it occurs as a mineral it is called ***moissanite*** after the French chemist Henri Moissan who discovered it in a meteorite[24] in 1905.

Polymorphism and polytypism. Silicon carbide has two polymorphs. At temperatures above 2000°C alpha silicon carbide (α-SiC), with a hexagonal crystal structure, is the more stable polymorph with iridescent and twinned crystals with a metallic luster. At temperatures lower than 2000°C, beta silicon carbide (β-SiC) exhibits a face-centered cubic (fcc) crystal structure.

Moreover, α-SiC exists as different hexagonal polytypes with a carbon atom situated above the center of a triangle of Si atoms and underneath a Si atom belonging to the next layer. The difference between polytypes is the stacking sequence between succeeding double layers of carbon and silicon atoms. If the first double layer is called the A position, the next layer that can be placed according to a close-packed structure will be placed on the B position or the C position. The different polytypes will be constructed by permutations of these three positions. For instance, the 2H-SiC polytype will have a stacking sequence ABAB... The number thus denotes the periodicity and the letter the resulting structure, which in this case is hexagonal. The 3C-SiC polytype is the only cubic polytype and it has a stacking sequence ABCABC... or ACBACB... The cell lattice parameter *a*, between neighboring silicon or carbon atoms, is ca. 308 pm for all polytypes. The carbon atom is positioned at the center of mass of the tetragonal structure outlined by the four neighboring Si atoms so that the distance between the C atom and each of the Si atoms is the same. Geometrical considerations require that the C-Si distance be exactly $a(3/8)^{1/2}$ (189 pm). The distance between two silicon planes is thus $a(2/3)^{1/2}$ (252 pm). The height of a unit cell, *c*, varies between the different polytypes. The ratio (c/a) thus differs from polytype to polytype but is always close to the ideal for a close-packed structure. The actual ratio for the 2H-, 4H- and 6H-SiC polytypes is closed to ideal ratios for these polytypes, that is, $(8/3)^{1/2}$, $2(8/3)^{1/2}$, and $3(8/3)^{1/2}$, respectively.

The different polytypes exhibit different electronic and optical properties. The bandgaps at 4.2 K of the different polytypes range between 2.39 eV for 3C-SiC and 3.33 eV for the 2H-SiC polytype. The important polytypes 6H-SiC and 4H-SiC have bandgaps of 3.02 eV and 3.27 eV, respectively. All polytypes are extremely hard, chemically inert, and have a high thermal conductivity. Properties such as the breakdown voltage, the saturated drift velocity, and the impurity ionization energies are all specific for the different polytypes.

Silicon carbide has long been recognized as an ideal ceramic material for applications where high hardness and stiffness, mechanical strength at elevated temperatures, high thermal conductivity, low coefficient of thermal expansion, and resistance to wear and abrasion are of primary importance. Moreover, because of its low density it offers greater advantages compared to other ceramics.

10.2.10.2 Industrial Preparation

The Acheson process. This process, invented by Edward Goodrich Acheson in 1893, was extensively used for making silicon carbide and was the only industrial process available for

---

[22] Berzelius, J.J. (1824) *Ann. Phys.*, Leipzig, **1**, 169.

[23] Acheson, E.G. (1893) Production of crystalline artificial carbonaceous materials. US Patent 492,767, February 28, 1893.

[24] Moissan, H. (1905) *C.R. Acad. Sci. Paris*, **140**, 405.

making bulk abrasive materials until the mid-1950s. The simplicity of the process makes it useful for production of huge quantities of silicon carbide suitable for grinding and cutting purposes. Some of the material produced by the Acheson process may, however, have adequate quality for electronic device production. A mixture of 50 wt.% silica, 40 wt. coke, 7 wt.% sawdust, and 3 wt.% rocksalt is heated in an electric resisting furnace. The heating is accomplished by a core made of graphite and coke called a resistor placed centrally in the furnace. The mixture of reactants is placed around this core. The mixture is then heated to reach a maximum temperature of ca. 2700°C, after which the temperature is gradually lowered. After the furnace has been fired, the outermost volume, which did not reach such high temperatures, consists of an unreacted mixture. Inside this is a volume where the temperature has not reached 1800°C. In this volume the mixture has reacted to form amorphous SiC. Close to the resistor, where the highest temperatures are obtained, SiC will be produced at first. As the temperature increases in the furnace, this will decompose again into graphite and silicon. The graphite will remain at the core; however, silicon reacts again with carbon to form SiC in colder parts of the furnace. The outer layer of graphite contains SiC in the form of threads of crystallites radiating from the core. The size of the crystallites decreases with increasing distance from the core. The purpose of the sawdust is to make the mixture porous in such a way that the huge amounts of carbon monoxide produced in the reaction may escape. High pressures of gas may locally be built up to form voids and channels to more porous parts of the mixture to eventually find its way out. The common salt serves as a purifier of the mixture. The chlorine reacts with metal impurities to form volatile metal chlorides (e.g., $FeCl_3$, $MgCl_2$), which escape. As a consequence of the furnace geometry, the material formed in the Acheson furnace varies in purity, according to its distance from the graphite resistor that is the heat source. The color changes to blue and black at a greater distance from the resistor, and these darker crystals are less pure and usually doped with aluminum, which increases electrical conductivity. A higher grade of silicon carbide for electronic application can be obtained from a more expensive process described below.

Lely process. A major improvement to the Acheson process is the process developed by J.A. Lely,[25] a scientist at Philips Research Laboratories in Eindhoven, in 1955. The process is similar to that observed in the voids and channels of the Acheson process. Lumps of SiC are packed between two concentric graphite tubes. The inner tube is thereafter withdrawn, leaving a cylinder of SiC lumps inside the outer graphite tube called the crucible. The crucible is closed with a graphite or SiC lid and placed inside a furnace. The crucible is heated to ca. 2500°C in an inert atmosphere of argon at atmospheric pressure. At this temperature SiC sublimes appreciably, leaving a graphite layer at the outermost part of the cylinder, and small platelets start to evolve from the innermost parts of the SiC cylinder. These platelets successively grow to larger sizes during a prolonged heating at this temperature. Each platelet is attached on one edge to an original lump of SiC. On the top and bottom of the cylinder a thick dense layer of SiC is formed. The quality of these crystals can be very high; however, the yield of the process is low, the sizes are irregular, the shape of the crystals is normally hexagonal, and there exists no polytype control. The purity of the crystals is largely governed by the starting material, which may be obtained in a high-purity form. The use of high-quality Lely grown material as substrate for a succeeding epitaxial growth is highly advantageous with regard to the high crystalline quality obtained from these substrates.

The modified Lely process. Despite the high crystalline quality that may be obtained with the Lely method, it has never been considered an important technique for future commercial exploitation on account of the low yield and irregular sizes. In the modified Lely process, which is a seeded sublimation growth process, these problems are overcome, though at the price of a considerably lower crystalline quality. In the modified Lely technique, SiC powder or lumps of SiC are placed inside a cylindrical graphite crucible. The crucible is closed with

[25] Lely, J.A. (1955) Berichte der Deutschen Keramischen Gesellschaft, **32**, 229.

a graphite lid onto which a seed crystal is attached. The crucible is heated to ca. 2200°C normally in an inert argon atmosphere at a reduced pressure. A temperature gradient is applied over the length of the crucible in such a way that the SiC powder at the bottom of the crucible is at a higher temperature than the seed crystal. The temperature gradient is typically kept in a range on the order of 20 to 40°C/cm. The SiC powder sublimes at the high temperature and the volume inside the crucible is filled with a vapor of progressive composition (e.g., Si_2C, SiC_2, Si_2, and Si). Since the temperature gradient is chosen such that the coldest part of the crucible is the position of the seed, the vapor will condense on this and the crystal will grow. The growth rate is largely governed by the temperature, the pressure, and the temperature gradient; however, experiments have shown that also the source-to-seed distance may have some influence. It has also been experimentally confirmed that different growth temperatures and the orientation of the seed crystal give rise to different polytypes.

10.2.10.3 Grades of Silicon Carbide

Several commercial grades of SiC are available on the market:

(i) ***Electrically conductive sintered alpha silicon carbide.*** This is a dense type of SiC and has superior resistance to oxidation, corrosion, wear, and chemical attacks. The single-phase SiC also has high strength and good thermal conductivity.

(ii) ***Black silicon carbide*** (98.5 wt.% SiC) is composed of premium-grade, medium-high-density, high-intensity magnetically treated SiC in which most impurities have been removed from the carbide.

(iii) ***CVD silicon carbide*** (99.9995 wt.% SiC) is a unique type of silicon carbide due to its purity, homogeneity, and chemical and oxidation resistance. It is thermally stable, is very cleanable and polishable, and is dimensionally stable.

(iv) ***Green silicon carbide*** (99.5 wt.% SiC) is an extremely hard synthetic material that possesses very high thermal conductivity. It is also able to maintain its strength at elevated temperatures. General applications of green SiC are in aerospace, blasting, coatings, composites, refractories, compounds, and kiln furniture, and it is used as an abrasive as honing stones, lapping, polishing, sawing silicon and quartz, and in grinding wheels.

Prices (2006). Silicon carbide is priced from 1150 to 1700 US$/tonne.

10.2.11 Properties of Raw Materials Used in Ceramics, Refractories, and Glasses

Table 10.11. Selected properties and prices of raw materials used in ceramics and refractories

Raw material	Apparent density ($kg.m^{-3}$)	Bulk density ($kg.m^{-3}$)	Bond's work index[26] ($kWh/tonne^{-1}$)	Abrasion index[27]	Average price (2006) (US$/tonne)
Alumina (fused)	3480	961	58.18	0.6447	1250–1700
Bauxite (chunk)	2380	1200–1360	9.45	0.1200	125–200
Chrome ore	4060		9.60	0.1200	150–250

[26] The Bond's index is the energy per unit mass of material required to grind it from until 80 wt.% pass 325 mesh, here it is expressed in KWh per short ton (2000 lb.).

[27] The abrasion index is defined as the mass fraction lost by a steel padle beating during 1 hour a charge of 1.6 kg of the material having pellets dimension 3/4 in x 1/2 in with 80 wt.% of the final final product passing 13.25 mm.

Table 10.11. *(continued)*

Raw material	Apparent density ($kg.m^{-3}$)	Bulk density ($kg.m^{-3}$)	Bond's work index ($kWh.tonne^{-1}$)	Abrasion index	Average price (2006) (US$/tonne)
Coke	1510	400–720	20.70	0.3095	
Dolomite (lump)	2820	1440–1600	11.31	0.016	
Feldspar (ground)	2590	1050–1121	11.67	n.a.	60–100
Graphite	1750	450–640	45.03		420–1000
Hematite	5260	3600	12.68		30–50
Ilmenite	4270	2240	13.11		80–100
Kaolin	2600	2600	7.10		
Lime	3340	960–1080			
Limestone	2690	1340–1440	11.61	0.0256	30–40
Magnesite	2980		16.80	0.075	130–180
Silicon carbide	2730		26.17		1200–1700
Quartzite (chunk)	2650		12.77	0.6905	n.a.
Zircon (flour)	4600				700–800

10.3 Traditional Ceramics

Traditional ceramics are those obtained only from the firing of clay-based materials. The common initial composition before firing consists usually of a ***clay mineral*** (i.e., phyllosilicate minerals such as ***kaolinite***, ***montmorillonite***, or ***illite***), ***fluxing agents*** or ***fluxes*** [e.g., feldspars: K-feldspars (orthoclases) and Ca-Na-feldspars (plagioclases)], and ***filler materials*** (e.g., silica, alumina, magnesia). The traditional ceramics can be prepared using two main groups of clays: kaolin or china clays made from the phyllosilicate kaolinite and, to a lesser extent, micas, but free of quartz; and ball clays containing a mixture of kaolinite, montmorillonite, illite, and micas.

Table 10.12. Examples of traditional ceramics

Type	Properties	Applications
Fired bricks	Porosity: 15–30% Firing temperature: 950–1050°C Enameled or not	Bricks, pipes, ducts, walls, ground floors
China	Porosity: 10–15% Firing temperature: 950–1200°C Enamel, opacity	Sanitation, tile
Stoneware	Porosity: 0.5–3% Firing temperature: 1100–1300°C Glassy surface	Crucible, labware, pipe
Porcelain	Porosity: 0–2% Firing temperature: 1100–1400°C Glassy, translucent	Insulators, labware, cookware

The classical procedure for preparing traditional ceramics consists of the following operation sequence: raw material selection and preparation (i.e., grinding, mixing), forming (e.g., molding, extrusion, slip casting, and die pressing), drying, prefiring operations (i.e., glazing), firing, and postfiring operation (e.g., enameling, cleaning, and machining). The common classes of traditional ceramics are ***whitewares*** (e.g., stoneware, china, and porcelain), ***glazes***, ***porcelain enamels***, high-temperature refractories, mortars, cements, and concretes (see Chapter 15).

10.4 Refractories

Refractories perform four basic functions:

(i) they act as a thermal barrier between a hot medium (e.g., flue gases, liquid metal, molten slags, and molten salts) and the wall of the containing vessel;

(ii) they insure a strong physical protection, preventing the erosion of walls by the circulating hot medium;

(iii) they represent a chemical protective barrier against corrosion;

(iv) they act as thermal insulation, insuring heat retention.

As a rule of thumb, an insulating material is considered a refractory material if its melting or solidus temperature is well above the melting point of pure iron (1539°C), i.e., if it exhibits a ***Seger's pyrometric cone*** equivalent of No. 26 or more (Table 10.19). Moreover, the maximum operating temperature of a refractory material is usually 150°C lower than its pyrometric cone equivalent.

10.4.1 Classification of Refractories

The ***classification of refractories*** can be approached in a number of different ways: chemical composition, type of applications, or operating temperature range.

Table 10.13. Classification of refractory by end user

Rank	Refractory industry users	End user
1	Cement and lime production	Building industry
2	Iron and steelmaking	
3	Glass industry	Automotive industry
4	Nonferrous metals production	
5	Oil and gas industries	
6	Waste incineration	Other
7	Basic industries	

Table 10.14. Classification of primary refractories by chemistry

Category (definition)	Description
Basic refractories (i.e., essentially made of calcined magnesite or magnesia, MgO)	Dolomite Dead burned magnesia (min. 95 wt.% MgO) Dead burned magnesia with chromite Fused magnesia Magnesia-carbon bricks
High alumina (i.e., with an alumina content greater than 47.5 wt.% Al_2O_3)	50%, 60%, 70%, 80% Al_2O_3 (±2.5), 85%, 90% Al_2O_3 (±2.0), 99% Al_2O_3 (>97%), Mullite ($3Al_2O_3.2SiO_2$) Chemically bonded bricks (75–85 wt.% Al_2O_3) Alumina-chrome bricks Alumina-carbon bricks
Fireclay (i.e., made of fired aluminum phyllosilicates or clays)	Super-duty (40–44 wt.% Al_2O_3) High-duty Medium-duty Low-duty Semisilica (18–25 wt.% Al_2O_3, 72–80 wt.% SiO_2)
Silica	Silica bricks
Advanced	Graphite and carbon ceramics Silicon carbide Zircon ($ZrSiO_4$) and fused zirconia (ZrO_2) Fused silica Fused alumina (brown and white) Fused and cast refractories

10.4.2 Properties of Refractories

Table 10.15. Selected physical properties of refractories

Refractory materials	Density (ρ/kg.m^{-3})	Melting point (°C)	Thermal conductivity (k/W.m^{-1}.K^{-1})	Specific heat capacity (c_P/J.kg^{-1}.K^{-1})
Alumina brick (64–65 wt.% Al_2O_3)	1842	1650–2030	4.67	
Brick, fireclay	2000		1.0042	753
Brick, hard-fired silica (94–95 wt.% silica)	1800		1.6736	753.10
Brick, high alumina (53 wt.% alumina) (20% porosity)	2330		1.3807	753
Brick, high alumina (83 wt.% alumina) (28% porosity)	2570		1.5062	753
Brick, high alumina (87 wt.% alumina) (22% porosity)	2850		2.9288	753
Brick, kaolin insulating (heavy)	430		0.2510	774
Brick, kaolin insulating (light)	300		0.0837	774
Brick, magnesite (86 wt.% MgO) (17.8% porosity)	2920		3.6819	837
Brick, magnesite (87 wt.% MgO)	2530		3.8493	837

Table 10.15. *(continued)*

Refractory materials	Density (ρ/kg.m^{-3})	Melting point (°C)	Thermal conductivity (k/W.m^{-1}.K^{-1})	Specific heat capacity (c_P/J.kg^{-1}.K^{-1})
Brick, magnesite (89 wt.% MgO)	2670		3.4727	837
Brick, magnesite (90 wt.% MgO) (14.5% porosity)	3080		4.9371	837
Brick, magnesite (93 wt.% MgO) (22.6% porosity)	2760		4.8116	837
Brick, normal fireclay (22% porosity)	1980		1.2970	732
Brick, siliceous (25% porosity)	1930		0.9372	753
Brick, siliceous fireclay (23% porosity)	2000		1.0878	753
Brick, sillimanite (22% porosity)	2310		1.4644	711
Brick, stabilized dolomite (22% porosity)	2700		1.6736	837
Brick, vermiculite	485		0.1674	837
Calcium oxide (pressed)	3030		13.8070	753
Calcium oxide(packed powder)	1700		0.3180	753
Carbon brick (99 wt.% graphite)	1682	3500	3.6	
Carbon brick (fired)	1470		3.5982	707
Chrome brick (100 wt.% Cr_2O_3)	2900–3100	1900	2.3	
Chrome brick (32 wt.% Cr_2O_3)	3200		1.1715	627.60
Chrome-magnesite brick	3000		2.0920	753.10
Chrome-magnesite brick (52 wt.% MgO, 23 wt.% Cr_2O_3)	3100	3045	3.5	
Diabasic glass (artificial)	2400		1.1715	753
Diatomaceous earth brick	440		0.0877	795.00
Diatomaceous earth brick (850°C)	440		0.0921	795.00
Diatomaceous earth brick (fused at 1100°C)	600		0.2218	795.00
Diatomaceous earth brick (high burn)	590		0.2259	795.00
Diatomaceous earth brick (molded)	610		0.2427	795.00
Dolomite (fired) (55 wt.% CaO, 37 wt.% MgO)	2700	2000		
Dolomite brick, stabilized (22 wt.% silica)	2700		1.6736	837
Egyptian fire (64–71 wt.% silica)	950		0.3138	732.20
Egyptian firebrick (64–71 wt.% silica)	950		0.3138	732
Fireclay brick (54 wt.% SiO_2, 40 wt.% Al_2O_3)	2146–2243	1740	0.3–1.6	
Fireclay brick, Missouri	2645		1.0042	960
Fireclay brick, normal (22 wt.% water)	1980		1.2970	732
Fireclay brick, siliceous (23 wt.% water)	2000		1.0878	753
Forsterite brick (58 wt.%MgO, 38 wt.%SiO_2) (20% porosity)	2760		1.0042	795.00
Fused-alumina brick (96 wt.% alumina) (22% porosity)	2900		3.0962	753.10
Fused-alumina brick (96 wt.% alumina) (22% porosity)	2900		3.0962	753.10
High-alumina brick (90–99 wt.% Al_2O_3)	2810–2970	1760–2030	3.12	
Kaolin brick, insulating (dense)	430		0.2511	774

Table 10.15. *(continued)*

Refractory materials	Density (ρ/kg.m^{-3})	Melting point (°C)	Thermal conductivity (k/W.m^{-1}.K^{-1})	Specific heat capacity (c_P/J.kg^{-1}.K^{-1})
Kaolin brick, insulating (light)	300		0.0837	774
Magnesite brick (95.5 wt.% MgO)	2531–2900	2150	3.7–4.4	
Mullite brick (71 wt.% Al_2O_3)	2450	1810	7.1	
Silica brick (95–99 wt.% SiO_2)	1842	1765	1.5	
Silicon carbide brick (80–90 wt.% SiC)	,595	2305	20.5	
Vermiculite brick	485		0.1674	837
Vermiculite insulating powder	270		0.1213	837
Vermiculite, expanded (heavy)	300		0.0690	753
Vermiculite, expanded (light)	220		0.0711	753
Zircon brick (99 wt.% $ZrSiO_4$)	3204	1700	2.6	
Zirconia (stabilized) brick	3925	2650	2.0	

Silica brick. The earliest silica bricks were composed of crushed minerals of 90 wt.% silica, with as much as 3.5 wt.% flux materials (i.e., usually CaO), and fired at about 1010°C. These bricks found use primarily in steel mills and coke-byproduct operations, from the second quarter of this century in chemical service, primarily in strong phosphoric acid exposures where shale and fireclay brick do not long survive. They can serve in higher-temperature service to about 1093°C and are more resistant to thermal shock due to their greater porosity, as high as 16%. When used in chemical service, those of the highest silica content (not below 98 wt.% SiO_2) should be used. The purity of the silica and its percentage of alkali, along with the manufacturing techniques, determine the uniformity or the wideness of ranges of the physical properties. Ranges of chemical composition of silica brick are 98.6 to 99.6 wt.% SiO_2, 0.2 to 0.5 wt.% Al_2O_3, 0.02 to 0.3 wt.% Fe_2O_3, 0.02 to 0.1 wt.% MgO, 0.02 to 0.03 wt.%CaO, and 0.01 to 0.2 wt.% (Na_2O, K_2O, Li_2O). Silica brick serves well and for long periods in acid service, except for hydrofluoric, without noticeable damage, showing greater resistance, especially to strong hot mineral acids, and particularly phosphoric rather than acid brick, and in halogen exposures, except fluorine, solvents, and organic chemical exposures. Silica bricks are not recommended for service in strong alkali environments. They also exhibit better shock resistance than shale or fireclay acid brick, but they have lower strength and abrasion resistance.

Porcelain brick. Porcelain bricks are made from high-fired clays, the temperature of firing depending on the amount of alumina in the clay, 15% to 38% usually at ca. 1200 to 1300°C, 85% at 1500 to 1550°C, and 95 to 98% at 1600 to 1700°C. The bodies of these bricks are extremely dense and nonporous, with zero absorption, and a Mohs hardness ranging from 6 to 9 for 99 wt.% alumina. As alumina content increases, Mohs hardness, the maximum service temperature, and chemical resistance increase. Major uses of porcelain include: the lining of ball mills, where they will outlast almost all other abrasion-resistant linings, and employment (glazed) as pole line hardware for the power industry where, exposed to abrasion, weathering, and cycling temperature changes, they outlast all other materials in similar service. All chemists and laboratory personnel are familiar with glass and porcelain equipment, and so are aware of the fact that they give excellent service in hot chemicals except hydrofluoric acid and acid fluorides and strong sodium or potassium hydroxides, especially in the molten state. Due to the high cost of porcelain brick, they are used sparingly in the process industries, chiefly in dye manufacture, due to their density for the prevention of interbatch contamination and ease of cleaning. The use of porcelain brick is primarily limited by its cost.

10.4.3 Major Refractory Manufacturers

Table 10.16. Major manufacturers of refractories worldwide

Company (Brands)	Location and Address	Refractory applications
Minerals Technologies (MinteQ)	Chrysler Building, 405 Lexington Ave., New York, NY 10174-1901, USA Telephone: +1 (732) 257-1227 E-mail: Minteq.ProductInfo@minteq.com URL: http://www.minteq.com/	Monolithic refractories and castables
Resco Products (Resco and National)	2 Penn Center, West Suite 430, Pittsburgh, PA 15276, USA Telephone: (412) 494-4491 Fax: (412) 494-4571 URL: http://www.rescoproducts.com/	Iron- and steelmaking Nonferrous smelting (Cu, Al) Fireclays and bricks Glassmaking processes Waste incinerators Hydrocarbon processing Power generation
RHI AG (Didier, Radex, and Veitscher)	Wienerbergstrasse 11 A-1100 Vienna, Austria Telephone: +43 (0) 50 213-0 Fax: +43 (0) 50 213-6213 E-mail: rhi@rhi-ag.com URL: http://www.rhi.at/	Iron- and steelmaking Cement and lime kilns Glassmaking processes Nonferrous smelting (Cu, Al, Ni, Sn, Zn) Petrochemical and hydrocarbon processes Oil refineries
SANAC Spa	Viale Certosa, 249, I-20151 Milano, Italy Telephone: (+39) 02307 00335 Fax: (+39) 02380 11158 URL: http://www.sanac.com/	Iron and steelmaking Glassmaking processes Castables Resin-bonded alumina
Shinagawa Refractories	1-7,Kudan-kita 4-chome,Chiyodaku, Tokyo 102-0073 Japan Telephone: +81 3-5215-9700 Fax: +81 3-5215-9720 URL: http://www.shinagawa.co.jp/	Iron- and steelmaking Nonferrous metals (Cu, Zn, Pb,and Al) Cement and lime kilns Refractories for gas, petroleum, and chemical plants Refractories for cement and lime kilns Glassmaking processes and ceramic firing kilns Petrochemical and hydrocarbon processes Oil refineries Waste incinerators
Vesuvius USA Corp.	P.O. Box 4014, Newton Drive 1404 Champaign, IL 61822 USA Telephone: +1 (217) 351-5000 Fax: +1 (217) 351-5031 URL: http://www.vesuvius.com/	Iron and steelmaking Foundry Aluminum smelters Glassmaking processes

10.5 Advanced Ceramics

Advanced ceramics or ***engineered ceramics***, also formerly called ***industrial ceramics***, are various inorganic chemical compounds, not necessarily oxides and silicates, that exhibit improved physical and chemical properties and can be grouped according to their field of application: electrical (e.g., semiconductors, insulators, dielectrics, piezo- and pyroelectrics, and superconductors), optical (e.g., phosphors, lasing crystals, mirrors, and reflectors), magnetics (e.g., permanent magnets), and structural ceramics. The properties of these ceramic materials are extensively described in the section in the book relative to their properties (e.g., insulators in dielectrics materials and superconducting ceramics in the superconductor section). Hereafter, dedicated sections on selected advanced ceramic materials are presented, providing a brief description, the general physical and chemical properties, along with method of preparation, industrial applications, and major producers.

10.5.1 Silicon Nitride

10.5.1.1 Description and General Properties

Silicon nitride [12033-89-5], chemical formula Si_3N_4 and relative molecular mass of 140.284, is a medium-density ceramic material (3290 kg.m^{-3}). The high flexural strength (830 MPa), high fracture toughness (6.1 MPa.m$^{-1/2}$), and creep resistance, even at elevated temperatures, ensure a high temperature strength to silicon nitride. Moreover, its low thermal expansion coefficient (3.3 μm/m.K), combined with a Young's modulus of 310 GPa, confers upon silicon nitride a superior thermal shock resistance compared with most ceramic materials. This set of extreme properties, together with a good oxidation resistance, were the major reasons of its first development in the late 1960s for replacing superalloys in advanced turbine and reciprocating engines to give higher operating temperatures and efficiencies. Although the ultimate goal of ceramic engines has never been achieved, silicon nitride has been used extensively in a number of other industrial applications, such as engine components, bearings, and cutting tools. In general, silicon nitride exhibits higher temperature capabilities than most metals, combining high mechanical strength and creep resistance with oxidation resistance. From a chemical point of view, silicon nitride exhibits an excellent corrosion resistance to numerous molten nonferrous metals such as Al, Pb, Zn, Cd, Bi, Rb, and Sn and molten salts like NaCl-KCl, NaF, and silicate glasses. However, it is corroded by molten Mg, Ti, V, Cr, Fe, and Co, and salts like cryolite, KOH, and Na_2O.

10.5.1.2 Industrial Preparation and Grades

Pure silicon nitride is difficult to produce as a fully dense material because it does not readily sinter and cannot be heated above 1850°C because it decomposes into silicon and nitrogen. Dense silicon nitride can only be made using sintering aids or by the direct nitriding of silicon. Therefore, the final material properties strongly depend on the fabrication method, and hence commercial silicon nitride cannot be considered a single material. Three grades of silicon nitride are available commercially:

(i) ***Reaction bonded silicon nitride*** (RBSN) is a high-purity grade of silicon nitride prepared by direct nitriding of compacted silicon powder. The incomplete nitriding reaction leads to densities of only 70 to 80% of the theoretical density and usually ranges from 2300 to 2700 kg.m^{-3}. It exhibits excellent thermal shock resistance and an outstanding corrosion resistance to molten nonferrous metals, especially aluminum metal. Reaction-bonded silicon nitride represents a cheaper alternative to the fully dense silicon nitride and can be machined to close tolerance (near-net shape) without the need for expensive diamond grinding.

(ii) ***Hot-pressed silicon nitride*** (HPSN) is obtained by applying both heat and pressure through a graphite die using sintering aids. However, most hot-pressed silicon nitride grades can be formulated with a minimum amount of densification aids. These compositions offer the highest mechanical strength of other silicon nitride grades. The major drawbacks are that only simple shaped billets can be produced by this process and the preparation of finished components requires expensive machining by utilizing diamond grinding.

(iii) ***Sintered silicon nitride*** (SSN) consists of a family of fully dense materials having a range of compositions that can be produced in cost-effective, complex net shape. The green compacts, made of powders with a high surface area, are fired under a nitrogen atmosphere, without applying pressure. This grade of silicon nitride has the best combination of properties, making it the leading technical ceramic for a number of structural applications including automotive engine parts, bearings, and ceramic armor.

(iv) ***Hot isostatically pressed silicon nitride*** (HIPSN) is obtained by glass-encapsulated parts that are placed in a high-pressure vessel or autoclave, with heat and pressure applied. The result is a slight decrease in strength but a substantial improvement in reliability. The material is used currently in niche market applications, for example, in reciprocating engine components and turbochargers, bearings, metal cutting and shaping tools, and hot metal handling.

10.5.2 Silicon Aluminum Oxynitride (SiAlON)

Description and general properties. ***Sialon*** is the commercial acronym for ***silicon aluminum oxynitride*** (SiAlON), that is, an alloy of silicon nitride (Si_3N_4) and aluminum oxide (Al_2O_3). Sialon is in fact a fine-grained nonporous advanced ceramic material with less than 1 vol.% open porosities. Sialon is made of a silicon nitride ceramic with a small percentage of aluminum oxide. Its generally adopted chemical formula is $Si_{6-x}Al_xO_xN_{8-x}$. This superior refractory material has the combined properties of silicon nitride, i.e., high-temperature strength, hardness, fracture toughness, and low thermal expansion, and that of aluminum oxide, i.e., corrosion resistance, chemical inertness, high temperature capabilities, and oxidation resistance. Sialon exhibits a medium density of 3400 kg.m^{-3}, a low Young's modulus of 288 MPa, a bulk modulus of 220 GPa, and a shear modulus of 120 GPa with a Poisson ratio of 0.25. Moreover, it has a flexural strength of 760 MPa, an elevated Vickers hardness (1430 to 1850 HV), and a good fracture toughness (6.0 to 7.5 MPa.m$^{-1/2}$). In addition, like pure silicon nitride, the combination of a low Young's modulus of 288 GPa with a low coefficient of linear thermal expansion (3 μm/m.K) ensures an excellent thermal shock resistance together with a low thermal conductivity of 15 to 20 W.m^{-1}K^{-1}. Most refractory products are capable of surviving one or two specific environments that typically involve high temperature, mechanical abuse, corrosion, wear, or electrical resistance. Sialon is perfect for molten-metal applications and high wear or high impact environments up to 1250°C. Moreover, sialon exhibits a good oxidation resistance in air up to 1500°C imparted by its alumina content, and it has an outstanding corrosion resistance to molten nonferrous metals. Moreover, it is neither wetted nor corroded by molten aluminum, brass, bronze, and other common industrial metals.

Industrial applications. Typical uses are as protection sheath for immersion thermocouples used in nonferrous metal melting, immersion heater and burner tubes, degassing and injector tubes in nonferrous metallurgy, metal feed tubes in aluminum die casting, welding, and brazing fixtures and pins.

10.5.3 Boron Carbide

10.5.3.1 Description and General Properties

Boron carbide [12069-32-8], with its chemical formula B_4C and its rhombohedral crystal lattice, is a low-density black solid (2512 kg.m^{-3}) with a metallic luster. Its high refractoriness due to its high melting point (2450°C) allows it to be used for high-temperature applications. On the other hand, it is the hardest manmade solid (HK 3200) after synthetic diamond. Actually under high temperatures above 1300°C, its hardness exceeds that of diamond and cubic boron nitride. A Poisson ratio of 0.21 indicates its high anisotropy. It has a high compressive strength that may vary according to its density and percentage purity. It has a very low thermal conductivity (27 W/m.K). With such a strength-to-density ratio and low thermal conductivity, boron carbide looks promising and ideal for a wide variety of applications. Because of its high hardness, boron carbide succeeded in replacing diamond as a lapping material. Boron carbide is a material with excellent properties. It has a list of important properties such as ultimate strength-to-weight (density) ratio, exceptionally high hardness, and high melting and oxidation temperatures (500°C). In addition, it has a very low thermal expansion coefficient (5.73 μm/mK). However, owing to its high Young's modulus, it possesses less thermal shock resistance. Boron carbide is stable toward dilute and concentrated acids and alkalis and inert to most organic compounds. It is slowly attacked by mixtures of hydrofluoric-sulfuric acids or hydrofluoric-nitric acids. It resists attack by water vapor at 200 to 300°C. However, it is attacked rapidly when put in contact with molten alkali and acidic salts to form borates. In addition, B_4C is characterized by a very high oxidation temperature.

10.5.3.2 Industrial Preparation

Boron carbide is either prepared from boron ores or from pure boron. The process involves the reduction of a boron compound. Usually, boron carbide is obtained by reacting boric acid or boron oxide and carbon at ca. 2500°C in an electric-arc furnace.

$$2B_2O_3(s) + 7C(s) \longrightarrow B_4C(s) + 6CO(g)$$

Its composition is variable over a relatively wide range. The boron/carbon ratio ranges from 3.8 to 10.4. Technical-grade boron carbide values are typically between 3.9 and 4.3. Later, the powder obtained is transformed into dense parts by hot pressing or cold forming and sintering. The cold formed and sintered material is less expensive, but sintering aids or other added bonding agents seriously degrade the material properties.

10.5.3.3 Industrial Applications and Uses

The applications of boron carbide are as wear-resistant components. Armor tiles in military applications such as in light hard bulletproof armor for helicopters and tanks or as thermal shield for the space shuttles. It is used in abrasives as lapping and polishing powders and in raw materials in preparing other boron compounds, notably titanium diboride. It is also us as an insert for spray nozzles and bearing liners and wire drawing guides. Finally, because of its boron content and elevated resistance to high temperatures, boron carbide is used as a shield for neutrons in nuclear reactors.

10.5.4 Boron Nitride

10.5.4.1 Description and General Properties

Boron nitride [10043-11-5], chemical formula BN, exists as three different poly-morphs: ***alpha-boron nitride*** (α-BN), a soft and ductile polymorph (ρ = 2280 kg.m^{-3} and *m.p.* = 2700°C) with a hexagonal crystal lattice similar to that of graphite, also called ***hexagonal boron nitride*** (HBN) or ***white graphite***; ***beta-boron nitride*** (β-BN), the hardest manmade material and densest polymorph (ρ = 3480 kg.m^{-3}, *m.p.* = 3027°C), with a cubic crystal lattice similar to that of diamond, also called cubic boron nitride (CBN) or ***borazon***; and (iii) ***pyrolitic boron nitride*** (PBN). From a chemical point of view, boron nitride oxidizes readily in air at temperatures above 1100°C, forming a thing protective layer of boric acid (H_3BO_3) on its surface that prevents further oxidation as long as it coats the material. Boron nitride is stable in reducing atmospheres up to 1500°C.

10.5.4.2 Industrial Preparation

Cubic BN, or borazon, is produced by subjecting hexagonal BN to extreme pressure and heat in a process similar to that used to produce synthetic diamonds. Melting of either phase is possible only with a high nitrogen overpressure. The alpha-phase decomposes above 2700°C at atmospheric pressure and at ca. 1980°C in a vacuum.

Hexagonal BN is manufactured using hot pressing or pyrolytic deposition techniques. These processes cause orientation of the hexagonal crystals, resulting in varying degrees of anisotropy. There is one pyrolytic technique that forms a random crystal orientation and an isotropic body; however, the density reaches only 50 to 60% of the theoretical density. Both manufacturing processes yield high purity, usually greater than 99 wt.% BN. The major impurity in the hot-pressed materials is boric oxide, which tends to hydrolyze in the presence of water, degrading the dielectric and thermal-shock properties of the material. The addition of calcia reduces the water absorption. Hexagonal hot-pressed BN is available in a variety of sizes and shapes, while the pyrolytic hexagonal material is currently available in thin layers only.

10.5.4.3 Industrial Applications and Uses

The major industrial applications of hexagonal boron nitride rely on its high thermal conductivity, excellent dielectric properties, self-lubrication, chemical inertness, nontoxicity, and ease of machining. These are, for instance, mold wash for releasing molds, high-temperature lubricants, insulating filler material in composite materials, as an additive in silicone oils and synthetic resins, as filler for tubular heaters, and in neutron absorbers. On the other hand, the industrial applications of cubic boron nitride rely on its high hardness and are mainly as abrasives.

10.5.5 Titanium Diboride

10.5.5.1 Description and General Properties

Titanium diboride [12045-63-5], chemical formula TiB_2, is a dense (4520 kg.m^{-3}) and high-melting-point (2980°C) advanced ceramic material. Due to its high elastic modulus of 510 to 575 GPa, titanium diboride exhibits an excellent stiffness-to-density ratio. It is also a hard material with a Vickers hardness (3370) superior to that of tungsten carbide, and its fracture toughness (5 to 7 MPa.m$^{1/2}$) is even greater than that of silicon nitride. The high flexural strength (350 to 575 MPa), combined with a high compressive strength (670 MPa), allows it to be used in military and ballistic applications. As expected from its hexagonal crystal lattice, it is highly anisotropic with a Poisson ratio of 0.18 to 0.20. It retains its mechanical strength up to very high temperatures. By contrast with most ceramics, it is a good electrical conductor, with an electrical resistivity of only 15 μΩ.cm, and has a good thermal conductivity (65 W m^{-1}K^{-1}); its linear coefficient of thermal expansion is 6.4 μm/m K. From a chemical point of view, TiB_2 is not attacked either by concentrated strong mineral acids such as hydrochloric acid or hydroflouric acids. Titanium diboride also has a very good oxidation resistance up to 1400°C. TiB_2 has excellent wettability and stability in liquid metals such as aluminum and zinc and has many applications as a corrosion-resistant material such as crucibles and cutting tools in addition to some military applications.

10.5.5.2 Industrial Preparation and Processing

The most common process for producing large quantities of titanium diboride is by reacting titania (TiO_2) with carbon and boron carbide (B_4C) or boron sesquioxide (B_2O_3) as follows:

$$2TiO_2(s) + C(s) + B_4C(s) \longrightarrow 2TiB_2(s) + 2CO_2(g),$$

$$2TiO_2(s) + 5C(s) + 5B_2O_3(s) \longrightarrow 2TiB_2(s) + 5CO_2(g).$$

The final purity of the powder depends on the purity of the raw materials. Generally, several different grades of TiO_2, carbon, B_4C, and B_2O_3 can be used for the production of a wide panel of TiB_2 products, depending on the required grain size, purity, and price. Vacuum-arc melting is used to produce a fully dense, single-phase titanium diboride. Graphite hearths are commonly used; the molten titanium diboride wets the graphite and exhibits excellent fluidity, and shapes are produced both by gravity and tilt-pour-casting methods. Sintered parts of titanium diboride are usually produced by either hot pressing or pressureless sintering, although hot isostatic pressing HIP has also been used. Quite a number of different sintering methods and sintering aids are used to produce fully dense parts of titanium diboride. Hot pressing of titanium diboride is performed at temperatures above 1800°C in a vacuum or 1900°C in an inert argon atmosphere. However, hot pressing is expensive and the net-shape fabrication is not possible, hence the required shape must still be machined from the hot-pressed billet. Some usual sintering aids used for hot-pressed parts include iron, nickel, cobalt, carbon, tungsten, and tungsten carbide. Pressureless sintering of titanium diboride is a cheaper method for the production of net-shaped parts. Due to the high melting point of titanium diboride, sintering temperatures in excess of 2000°C are often required to promote sintering. Another method, called high-temperature synthesis (HTS), uses a powdered reducing metal such as magnesium or aluminum, and powders of titanium oxide and boron oxide. The materials are mixed and placed in a high-temperature crucible. This mixture is then ignited, and the self-sustaining reaction produces titanium diboride particles dispersed within a matrix of alumina or magnesia. After leaching the reaction mass, it remains as micrometric titanium diboride particles.

The major producers of titanium diboride are Advanced Refractory Technologies, Advanced Ceramics Corp., and Cerac in the United States, H.C. Starck and Electroschmeltzwerk Kempten in Germany, Denka in Japan, and Borides Ceramics and Composites in the UK.

10.5.5.3 Industrial Applications and Uses

Titanium diboride was originally developed to make lightweight armor for US and Soviet army tanks. It also has many commercial applications such as nozzles, seals, cutting tools, dies, wear parts due to its corrosion resistance, and also molten-metal crucibles and electrodes. It is used in crucibles due to its high melting point and chemical inertness.

10.5.6 Tungsten Carbides and Hardmetal

10.5.6.1 Description and General Properties

Tungsten carbide, or ***hardmetal***, was developed in the 1920s for wear-resistant dies to draw incandescent-lamp filament wire. Earlier efforts to manufacture the WC-W_2C eutectic alloy was unsuccessful because of its inherent brittleness; therefore researchers diverted their attention to powder metallurgy techniques. At present, these powder metallurgy techniques are

Table 10.17. Properties of selected hardmetals

Hardmetal (wt.%)	Density (ρ/kg.m^{-3})	Young's modulus (E/GPa)	Transverse rupture strength (MPa)	Compressive strength (MPa)	Vickers hardness (HV/kgf.mm^{-2})	Thermal conductivity (k/Wm^{-1}K^{-1})	Coefficient of linear thermal expansion (10^{-6} K^{-1})	Electrical resistivity (μΩ.cm)
100WC	15,700	707	296–490	2937	1800–2000		5.7–7.2	53
97WC–3Co	15,150	655	979–1175	5778	1600–1700	87.9		
95.5WC–4.5Co	15,050	627	1172–1372	5681	1550–1650	83.7	3.4	
94.5WC–5.5Co	14,800	607	1565–1765	4895	1500–1600	79.5	3.6	20
91WC–9Co	14,600	579	1469–1862	4702	1400–1500	75.3		
89WC–11Co	14,150	565	1565–1958	4502	1300–1400	66.9	3.8	18
87WC–13Co	14,080	545	1662–2,55	4406	1250–1350	58.6		
85WC–15Co	13,800	538	1765–2151	3820	1150–1250		6.0	
80WC–20Co	13,300	490	1958–2544	3330	1050–1150		4.7	
75WC–25Co	13,000	459	1765–2648	3130	900–1000		5.0	
70WC–30Co	12,500				850–950			

being further developed and refined to reduce manufacturing costs and improve performance. Tungsten carbide is in fact a composite material called ***cermet*** or hardmetal made of tungsten carbides in a metal matrix of cobalt. Tungsten carbide is harder than most steels, has greater mechanical strength, transfers heat quickly, and resists wear and abrasion better than other metals. Among the materials that resist severe wear, corrosion, impact, and abrasion, tungsten carbide is superior. Tungsten carbide is a dense (15,630 kg.m^{-3}) and very hard ceramic material (1700–2400 HK). It exhibits outstanding mechanical properties with a Young's modulus of 668 GPa, a tensile strength of 344 MPa, and a compressive strength of 2683–2958 MPa. It has a high melting point of 2777°C and a thermal conductivity at 100°C of 86 W/mK.

10.5.6.2 Industrial Preparation

Most cemented carbides are manufactured by powder metallurgy, which consists in the preparation of the tungsten-carbide powder, powder consolidation, sintering, and postsintering forming. Tungsten-carbide powder is usually obtained by a carburization process and mixed with a relatively ductile matrix material such as cobalt, nickel, or iron and paraffin wax in either an attrition or ball mill to produce a composite powder. Spray drying yields uniform, spheroidized particles that are 100 to 200 mm in diameter. The powder is then consolidated into net and near-net-shape green compacts and billets by pressing and extrusion. Pressed billets can also be machined to shape before sintering. The density of the green compacts is around 45 to 65% of the theoretical. The green parts are then dewaxed at a temperature between 200 and 400°C and are then presintered between 600 and 900°C to impart adequate strength for handling. An alternative technology is a combination sinter-HIP process that combines dewaxing, presintering, vacuum sintering, and low-pressure HIP to speed up the overall cycle time.

10.5.6.3 Industrial Applications and Uses

Tungsten carbide can be used for a wide variety of applications. It has many applications that utilize its corrosion-resistant property such as wear plates, drawing dies, and wear parts for wire wearing machines. There are other applications that make use of its high hardness such as punches, bushings, dies, cylinders, discs, rings, and intricate shapes as well as performs and blanks. There are other minor applications such as in rusticator blades, sander nozzles, air jets, and sander guns. Tungsten carbide is also used primarily and extensively for making drill-tip tunneling, rock crushing, mining, and quarrying purposes, i.e., for most geological activities. Tungsten carbide is also made into tiles for wear and abrasion resistance. It is also very useful in rebuilding worn parts. The application of tungsten carbide on industrial wearing surfaces has been proven to greatly enhance the performance factors for a whole spectrum of industrial applications. The service life of many kinds of machinery can be greatly prolonged by surface coating of wear-prone materials with tungsten carbide.

10.5.7 Practical Data for Ceramists and Refractory Engineers

10.5.7.1 Temperature of Color

In practice, the temperature of an incandescent body can be estimated roughly from the color of radiation emitted according to a practical scale described in Table 10.18.

Table 10.18. Practical color scale for temperature of incandescent body

Color	Temperature range (°C)
Lowest visible red	475
Lowest visible red to dark red	475–650
Dark red to cherry red	650–750
Cherry red to bright cherry red	750–815
Bright cherry red to orange	815–900
Orange to yellow	900–1090
Yellow to light yellow	1090–1315
Light yellow to white	1315–1540
White to dazzling white	1540 and higher

10.5.7.2 Pyrometric Cone Equivalents

The ***pyrometric cone equivalent*** (PCE), a special ceramic material, was introduced by Segers and standardized by Edward Orton, Jr. It is determined by testing the refractory against a series of standardized test pieces, cone shaped and having a ceramic composition with different softening points, one withstanding a slightly higher temperature than the other.

The test pieces are generally made to form triangular pyramids having a height four times the base. The softening point is reached depending upon the temperature and the rate of heat increase. Cones are numbered from 022, 021, 020, 02, 01, 1, and 2 to 42. Where the softening range in cones is too close, for example, in 21, 22, 24, and 25, they are omitted from the series, and where the temperature range is widely spaced, extra cones like 31.5, 32.5, etc. are added. At a temperature increase rate of 20°C per hour, the cones numbering 022 to 01 have softening points between 585 and 1110°C, and those numbered 1 to 35 have softening points between 1125 and 1775°C. Thus, the predetermined pyrometric cone equivalents of standard test pieces are placed along with cones made of the samples being tested in the furnace, and the PCE of the samples are determined by visual comparison. The softening point is noticed when the tip of the cone starts bending with the rise in temperature.

Table 10.19. Temperature equivalents (°C) of pyrometric cones and pyrometric cone equivalents

Cone No.	Heating rate for large cones		Heating rate for small cones	Pyrometric cone equivalent (PCE)
	60°C/h	150°C/h	300°C/h	150°C/h
022	585	600		
021	602	614	643	
020	625	635	666	
019	668	683	723	
018	696	717	752	
017	727	747	784	
016	767	792	825	
015	790	804	843	
014	834	838		
013	869	852		
012	866	884		
011	886	894		
010	887	894	919	
09	915	923	955	
08	945	955	983	
07	973	984	1008	
06	991	999	1023	
05	1031	1046	1062	
04	1050	1060	1098	
03	1086	1101	1131	
02	1101	1120	1148	
01	1117	1137	1178	
1	1136	1152	1179	
2	1142	1162	1179	
3	1152	1168	1196	
4	1168	1186	1209	
5	1177	1196	1221	
6	1201	1222	1255	
7	1215	1240	1264	
8	1236	1260	1300	
9	1260	1280	1317	
10	1285	1305	1330	
11	1294	1315	1366	
12	1306	1326	1355	1337
13	1321	1346		1349
14	1388	1366		1398
15	1424	1431		1430
16	1455	1473		1491

Table 10.19. *(continued)*

Cone No.	Heating rate for large cones		Heating rate for small cones	Pyrometric cone equivalent (PCE)
	60°C/h	150°C/h	300°C/h	150°C/h
17	1477	1485		1512
18	1500	1506		1522
19	1520	1528		1541
20	1542	1549		1564
23	1586	1590		1605
26	1589	1605		1621
27	1614	1627		1640
28	1614	1633		1646
29	1624	1645		1659
30	1636	1654		1665
31	1661	1679		1683
31 1/2	1706			1699
32		1717		1717
32 1/2	1718	1730		1724
33	1732	1741		1743
34	1757	1759		1763
35	1784	1784		1785
36	1798	1796		1804
37				1820
38				1835
39				1865
40				1885
41				1970
42				2015

Reference: Standard Pyrometric Cones. Edward Orton, Jr.Ceramic Foundation, Columbus, OH

10.6 Standards for Testing Refractories

Table 10.20. ASTM standards for testing refractories

ASTM standard	Description
ASTM C-16	Load-testing refractory brick at high temperatures
ASTM C-20	Apparent porosity, water absorption, apparent specific gravity, and bulk density of burned refractory brick and shapes by boiling water
ASTM C-24	Pyrometric cone equivalent of fireclay and high-alumina refractory materials
ASTM C-67	Brick and structural clay-tile testing
ASTM C-92	Sieve analysis and water content of refractory materials
ASTM C-93	Cold crushing strength and modulus of rupture of insulating firebrick

Table 10.20. *(continued)*

ASTM standard	Description
ASTM C-113	Reheat change of refractory brick
ASTM C-133	Cold crushing strength and modulus of rupture of refractories
ASTM C-134	Size and bulk density of refractory brick and insulating firebrick
ASTM C-135	True specific gravity of refractory materials by water immersion
ASTM C-167	Thickness and density of blanket or batt thermal insulations
ASTM C-179	Drying and linear change in refractory plastic and ramming mix specimens
ASTM C-181	Workability index of fireclay and high-alumina plastic refractories
ASTM C-182	Thermal conductivity of insulating firebrick
ASTM C-198	Cold bonding strength of refractory mortar
ASTM C-199	Pier test of refractory mortars
ASTM C-201	Thermal conductivity of refractories
ASTM C-202	Thermal conductivity of refractory brick
ASTM C-210	Reheat change in insulating firebrick
ASTM D-257	DC resistance or conductance of insulating materials
ASTM C-279	Chemical-resistant masonry units
ASTM C-288	Disintegration of refractories in an atmosphere of carbon monoxide
ASTM C-336	Annealing point and strain point of glass by fiber elongation
ASTM C-338	Softening point of glass by fiber elongation
ASTM C-356	Linear shrinkage of preformed high-temperature thermal insulation subjected to soaking heat
ASTM C-357	Bulk density of granular refractory materials
ASTM C-373	Water absorption, bulk density, apparent porosity, and apparent specific gravity of fired whiteware products
ASTM C-417	Thermal conductivity of unfired monolithic refractories
ASTM C-454	Disintegration of carbon refractories by alkali
ASTM C-491	Modulus of rupture of air-setting plastic refractories
ASTM C-559	Bulk density by physical measurements of manufactured carbon and graphite articles
ASTM C-561	Ash in a graphite sample
ASTM C-577	Permeability of refractories
ASTM C-583	Modulus of rupture of refractory materials at elevated temperatures
ASTM C-598	Annealing point and strain point of glass by beam bending
ASTM C-605	Reheat change of fireclay nozzles and sleeves
ASTM C-611	Electrical resistivity of manufactured carbon and graphite articles at room temperature
ASTM C-651	Flexural strength of manufactured carbon and graphite articles using four-point loading at room temperature
ASTM C-695	Compressive strength of carbon and graphite
ASTM C-704	Abrasion resistance of refractory materials at room temperature
ASTM C-747	Modulus of elasticity and fundamental frequencies of carbon and graphite materials by sonic resonance
ASTM C-767	Thermal conductivity of carbon refractories

Table 10.20. *(continued)*

ASTM standard	Description
ASTM C-769	Sonic velocity in manufactured carbon and graphite materials for use in obtaining an approximate Young's modulus
ASTM C-830	Apparent porosity, liquid absorption, apparent specific gravity, and bulk density of refractory shapes by vacuum pressure
ASTM C-831	Residual carbon, apparent residual carbon, and apparent carbon yield in coked-carbon-containing bricks and shapes
ASTM C-832	Measuring the thermal expansion and creep of refractories under load
ASTM C-838	Bulk density of as-manufactured carbon and graphite shapes
ASTM C-860	Determining and measuring consistency of refractory concrete
ASTM C-862	Preparing refractory concrete specimens by casting
ASTM C-863	Evaluating oxidation resistance of silicon carbide refractories at elevated temperatures
ASTM C-865	Firing refractory concrete specimens
ASTM C-885	Young's modulus of refractory shapes by sonic resonance
ASTM C-892	Unfiberized shot content of inorganic fibrous blankets
ASTM C-914	Bulk density and volume of solid refractories by wax immersion
ASTM C-973	Preparing test specimens from basic refractory gunning products by pressing
ASTM C-974	Preparing test specimens from basic refractory castable products by casting
ASTM C-975	Preparing test specimens from basic refractory ramming products by pressing
ASTM C-1025	Modulus of rupture in bending of electrode graphite
ASTM C-1039	Apparent porosity, apparent specific gravity, and bulk density of graphite electrodes
ASTM C-1054	Pressing and drying refractory plastic and ramming mix specimens
ASTM C-1099	Modulus of rupture of carbon-containing refractory materials at elevated temperatures
ASTM C-1100	Ribbon thermal shock testing of refractory materials
ASTM C-1113	Thermal conductivity of refractories by hot wire
ASTM C-1161	Flexural strength of advanced ceramics at ambient temperature
ASTM C-1171	Quantitatively measuring the effect of thermal cycling on refractories
ASTM C-1259	Dynamic Young's modulus, shear modulus, and Poisson ratio for advanced ceramics by impulse excitation of vibration

Table 10.21. ISO standards for testing refractories

ISO standard	Description
ISO 10058: 1992	Magnesites and dolomites – chemical analysis
ISO 10059-1: 1992	Dense, shaped refractory products – determination of cold compressive strength. Part 1: Referee test without packing
ISO 10059-2: 2003	Dense, shaped refractory products – determination of cold compressive strength. Part 2: Test with packing
ISO 10060: 1993	Dense, shaped refractory products – test methods for products containing carbon
ISO 10080: 1990	Refractory products – classification of dense, shaped acid-resisting products
ISO 10081-1: 2003	Classification of dense shaped refractory products. Part 1: Alumina-silica

Table 10.21. *(continued)*

ISO standard	Description
ISO 10081-2: 2003	Classification of dense shaped refractory products. Part 2: Basic products containing less than 7% residual carbon
ISO 10081-3: 2003	Classification of dense shaped refractory products. Part 3: Basic products containing from 7 to 50% residual carbon
ISO 10635: 1999	Refractory products – methods of testing for ceramic fiber products
ISO 1146: 1988	Pyrometric reference cones for laboratory use – specification
ISO 12676: 2000	Refractory products – determination of resistance to carbon monoxide
ISO 12677: 2003	Chemical analysis of refractory products by XRF – fused cast bead method
ISO 12678-1: 1996	Refractory products – measurement of dimensions and external defects of refractory bricks. Part 1: Dimensions and conformity to drawings
ISO 12678-2: 1996	Refractory products – measurement of dimensions and external defects of refractory bricks. Part 2: Corner and edge defects and other surface imperfections
ISO 12680-1: 2005	Methods of testing of refractory products. Part 1: Determination of dynamic Young's modulus (MOE) by impulse excitation of vibration
ISO 13765-1: 2004	Refractory mortars. Part 1: Determination of consistency using the penetrating-cone method
ISO 13765-2: 2004	Refractory mortars. Part 2: Determination of consistency using the reciprocating-flow-table method
ISO 13765-3: 2004	Refractory mortars. Part 3: Determination of joint stability
ISO 13765-4: 2004	Refractory mortars. Part 4: Determination of flexural bonding strength
ISO 13765-5: 2004	Refractory mortars. Part 5: Determination of grain-size distribution (sieve analysis)
ISO 13765-6: 2004	Refractory mortars. Part 6: Determination of moisture content of ready-mixed mortars
ISO 1893: 2005	Refractory products – determination of refractoriness under load – differential method with rising temperature
ISO 1927: 1984	Prepared unshaped refractory materials (dense and insulating) – classification
ISO 20182: 2005	Refractory test-piece preparation – gunning refractory panels by pneumatic-nozzle mixing-type guns
ISO 2245: 1990	Shaped insulating refractory products – classification
ISO 2477: 2005	Shaped insulating refractory products – determination of permanent change in dimensions on heating
ISO 2478: 1987	Dense shaped refractory products – determination of permanent change in dimensions on heating
ISO 3187: 1989	Refractory products – determination of creep in compression
ISO 5013: 1985	Refractory products – determination of modulus of rupture at elevated temperatures
ISO 5014: 1997	Dense and insulating shaped refractory products – determination of modulus of rupture at ambient temperature
ISO 5016: 1997	Shaped insulating refractory products – determination of bulk density and true porosity
ISO 5017: 1998	Dense shaped refractory products – determination of bulk density, apparent porosity, and true porosity
ISO 5018: 1983	Refractory materials – determination of true density
ISO 5019-1: 1984	Refractory bricks – dimensions. Part 1: Rectangular bricks
ISO 5019-2: 1984	Refractory bricks – dimensions. Part 2: Arch bricks

Table 10.21. *(continued)*

ISO standard	Description
ISO 5019-3: 1984	Refractory bricks – dimensions. Part 3: Rectangular checker bricks for regenerative furnaces
ISO 5019-4: 1988	Refractory bricks – dimensions. Part 4: Dome bricks for electric-arc furnace roofs
ISO 5019-5: 1984	Refractory bricks – dimensions. Part 5: Skewbacks
ISO 5019-6: 2005	Refractory bricks – dimensions. Part 6: Basic bricks for oxygen steelmaking converters
ISO 5022: 1979	Shaped refractory products – sampling and acceptance testing
ISO 528: 1983	Refractory products – determination of pyrometric cone equivalent (refractoriness)
ISO 5417: 1986	Refractory bricks for use in rotary kilns – dimensions
ISO 836: 2001	Terminology for refractories
ISO 8656-1: 1988	Refractory products – sampling of raw materials and unshaped products. Part 1: Sampling scheme
ISO 8840: 1987	Refractory materials – determination of bulk density of granular materials (grain density)
ISO 8841: 1991	Dense, shaped refractory products – determination of permeability to gases
ISO 8890: 1988	Dense shaped refractory products – determination of resistance to sulfuric acid
ISO 8894-1: 1987	Refractory materials – determination of thermal conductivity. Part 1: Hot-wire method (cross-array)
ISO 8894-2: 1990	Refractory materials – determination of thermal conductivity. Part 2: Hot-wire method (parallel)
ISO 8895: 2004	Shaped insulating refractory products – determination of cold crushing strength
ISO 9205: 1988	Refractory bricks for use in rotary kilns – hot-face identification marking

10.7 Properties of Pure Ceramics (Borides, Carbides, Nitrides, Silicides, and Oxides)

See Table 10.22, pages 648–669.

Table 10.22. Selected physical properties of advanced ceramics (borides, carbides, nitrides, silicides, and oxides)

IUPAC name (synonyms, common trade names)	Theoretical chemical formula, [CAS RN], relative molecular mass ($^{12}C = 12.000$)	Crystal system, lattice parameters, structure type, Strukturbericht, Pearson, space group, structure type (*Z*)	Density (ρ/kg.m^{-3})	Electrical resistivity (ρ/μΩ.cm)	Dielectric permittivity [1MHz] (ε_r / nil)	Dielectric field strength (E_d/MV.m^{-1})	Dissipation or tangent loss factor ($\tan\delta$)	Melting point (*m.p.*/°C)	Thermal conductivity (k/W.m^{-1}.K^{-1})	Specific heat capacity (c_p/J.kg^{-1}.K^{-1})	Coeff. linear thermal expansion (α/10^{-6}K^{-1})	Young's or elastic modulus (*E*/GPa)	Coulomb's or shear modulus (*G*/GPa)	Bulk or compression modulus (*K*/GPa)	Poisson ratio (ν)	Ultimate tensile strength (σ_{UTS}/MPa)	Flexural strength (τ/MPa)	Compressive strength (σ/MPa)	Fracture toughness (K_{IC}/MPa.m$^{1/2}$)	Vickers or Knoop Hardness (HV or HK/GPa) (/HM)	Other physicochemical Properties, oxidation and corrosion resistance[28], and major uses.
Borides																					
Aluminum diboride	AlB_2 [12041-50-8] 48.604	Hexagonal *a* = 300.50 pm *c* = 325.30 pm C32, *hP3*, P6/mmm, AlB_2 type (*Z* = 1)	3190	n.a.	n.a.	n.a.	n.a.	1654		897.87										9.61 HK	Temp. transition to AlB_{12} at 920°C. Soluble in dilute HCl. Nuclear shielding material
Aluminum dodecaboride	AlB_{12} [12041-54-2] 156.714	Tetragonal *a* = 1016 pm *c* = 1428 pm	2580	n.a.	n.a.	n.a.	n.a.	2421		954.48										23.55–25.50 HK	Soluble in hot HNO_3, insoluble in other acids and alkalis. Neutron shielding material
Barium hexaboride	BaB_6 [] 202.193	Cubic *a* = pm $D2_1$, cP7, Pm3m CaB_6 type (*Z* = 1)	4350	77				2270				385									Black cubic crystals
Beryllium boride	Be_4B [12536-52-6] 46.589		n.a.	n.a.	n.a.	n.a.	n.a.	1160													
Beryllium diboride	BeB_2 [12228-40-9] 30.634	Hexagonal *a* = 979 pm *c* = 955 pm	2420	10,000	n.a.	n.a.	n.a.	1970													Better air oxidation resistance than any other beryllium boride (Be_2B, BeB_6) in temperature range 1000–1200°C
Beryllium hemiboride	Be_2B [12536-51-5]	Cubic *a* = 467.00 pm C1, cF12, Fm3m, CaF_2 type (*Z* = 4)	1890	1000	n.a.	n.a.	n.a.	1520												8.53	
Beryllium hexaboride	BeB_6 [12429-94-6]	Tetragonal *a* = 1016 pm *c* = 1428 pm	2330	10^{13}	n.a.	n.a.	n.a.	2070													
Beryllium boride	BeB [12228-40-9]		n.a.	n.a.	n.a.	n.a.	n.a.	1970													
Boron	β-B [7440-42-8] 10.811	Trigonal (rhombohedral) *a* = 1017 pm α = 65°12' *hR105*, R3m, β-B type	2460	18,000	n.a.	n.a.	n.a.	2190				320								20.15 (HM 11)	Brown or dark powder, unreactive to oxygen, water, acids, and alkalis. $\Delta H_{vap} = 480$ kJ/mol

Calcium hexaboride	CaB_6 [12007-99-7] 104.944	Cubic a = 413.77 pm $D2_1$, cP7, Pm3m CaB_6 type (Z = 1)	2460	222				2235	39.29		6.48	451									
Cerium hexaboride	CeB_6 [12008-02-5] 204.981	Cubic a = 4 pm $D2_1$, cP7, Pm3m CaB_6 type (Z = 1)	4870					2550													Blue hexagonal crystals
Chromium boride	Cr_5B_3 [12007-38-4] 292.414	Orthorhombic a = 302.6 pm b = 1811.5 pm c = 295.4 pm $D8_1$, $tI32$, I4/mcm, Cr_5B_3 type (Z = 4)	6100	n.a.	n.a.	n.a.	n.a.	1900	15.8		13.7										
Chromium diboride	CrB_2 [12007-16-8] 73.618	Hexagonal a = 292.9 pm c = 306.6 pm C32, $hP3$, P6/mmm, AlB_2 type (Z = 1)	5160–5220	21	n.a.	n.a.	n.a.	1850–2200	20–32	712	6.2–7.5	211	n.a.	n.a.	n.a.	n.a.	n.a.	n.a.	n.a.	17.65	Strongly corroded by molten metals such as Mg, Al, Na, Cu, Si, V, Cr, Mn, Fe, and Ni; corrosion resistant to liquid metals Cu, Zn, Sn, Rb, and Bi
Chromium boride	CrB [12006-79-0] 62.807	Tetragonal a = 294.00 pm c = 1572.00 pm B_f, oC8, Cmcm CrB type (Z = 4)	6200	64.0	n.a.	n.a.	n.a.	2100	20.1		12.3						607	1300			
Cobalt hemiboride	Co2B [12045-01-1] 128.677		8100					1280													
Cobalt boride	CoB [12006-77-8] 69.744		7250					1460													
Hafnium diboride	HfB_2 [12007-23-7] 200.112	Hexagonal a = 314.20 pm c = 347.60 pm C32, $hP3$, P6/mmm, AlB_2 type (Z = 1)	11,190	8.8–11	n.a.	n.a.	n.a.	3249	57.1	247.11	6.3–7.56	500			0.12		350			28.44 and 23.54 HK	Gray crystals. HfB_2 exhibits the greatest oxidation resistance of all refractory group IV and V borides above 1090°C. Attacked by hydrogen fluoride gas (HF) but resists fluorine gas up to 590°C and is less resistant than zirconium diboride
Hafnium boride	HfB [] 189.301	Cubic NaCl type	12,800					2899													Gray hexagonal crystals
Iron diboride	FeB_2 [12006-86-9] 122.501	Tetragonal	7300					1389													
Iron boride	FeB [12006-84-7] 66.656	Orthorhombic	7000																		

Table 10.22. *(continued)*

IUPAC name (synonyms, common trade names)	Theoretical chemical formula, [CAS RN], relative molecular mass (^{12}C = 12.000)	Crystal system, lattice parameters, structure type, Strukturbericht, Pearson, space group, structure type (Z)	Density (ρ/kg.m^{-3})	Electrical resistivity (ρ/μΩ.cm)	Dielectric permittivity [1MHz] (ε_r / nil)	Dielectric field strength (E_d/MV.m^{-1})	Dissipation or tangent loss factor ($\tan\delta$)	Melting point ($m.p.$/°C)	Thermal conductivity (k/W.m^{-1}.K^{-1})	Specific heat capacity (c_p/J.kg^{-1}.K^{-1})	Coeff. linear thermal expansion (α/10^{-6}K^{-1})	Young's or elastic modulus (E/GPa)	Coulomb's or shear modulus (G/GPa)	Bulk or compression modulus (K/GPa)	Poisson ratio (ν)	Ultimate tensile strength (σ_{UTS}/MPa)	Flexural strength (τ/MPa)	Compressive strength (σ/MPa)	Fracture toughness (K_{IC}/MPa.m$^{1/2}$)	Vickers or Knoop Hardness (HV or HK/GPa) (/HM)	Other physicochemical Properties, oxidation and corrosion resistance, and major uses.
Lanthanum hexaboride	LaB_6 [12008-21-8] 203.772	Cubic a = 415.7 pm $D2_1$, cP7, Pm3m CaB_6 type (Z = 1)	4760	17.4	n.a.	n.a.	n.a.	2715	47.7		6.4	479					126				Wear-resistant, semiconducting, thermoionic conductor film
Molybdenum boride	Mo_2B_5 [12007-97-5] 245.935	Trigonal a = 301.2 pm c = 2093.7 pm $D8_i$, *hR7*, R3m Mo_2B_5 type (Z = 1)	7480	25–55	n.a.	n.a.	n.a.	1600	50		8.6	672					345				Corroded by molten metals Al, Mg, V, Cr, Mn, Fe, Ni, Cu, Nb, Mo, and Ta; corrosion resistant to molten Cd, Sn, Bi, and Rb
Molybdenum diboride	MoB_2 117.59	Hexagonal a = 305.00 pm c = 311.30 pm C32, *hP3*, P6/mmm, AlB_2 type (Z = 1)	7780	45	n.a.	n.a.	n.a.	2100		527	7.7									12.55	
Molybdenum hemiboride	Mo_2B [12006-99-4] 202.691	Tetragonal a = 554.3 pm c = 473.5 pm C16, *tI12*, I4/mcm, $CuAl_2$ type (Z = 4)	9260	40	n.a.	n.a.	n.a.	2280		377	5.0									(HM 8–9)	Corrosion-resistant film
Molybdenum boride	MoB 106.77	Tetragonal a = 311.0 c = 169.5 B_g, *tI4*, $I4_1$/amd MoB type (Z = 2)	8770	α–MoB 45, β–MoB 25	n.a.	n.a.	n.a.	2180		368										15.40	
Niobium diboride	ε–NbB_2 [12007-29-3] 114.528	Hexagonal a = 308.90 pm c = 330.03 pm C32, *hP3*, P6/mmm, AlB_2 type (Z = 1)	6970	26–65	n.a.	n.a.	n.a.	2900	17–23.5	418	8.0–8.6	637	n.a.	n.a.	n.a.	n.a.	n.a.	n.a.	n.a.	30.70 (HM>8)	Corrosion resistant to molten Ta while corroded by molten rhenium
Niobium boride	δ–NbB [12045-19-1] 103.717	Orthorhombic a = 329.8 pm b = 316.6 pm c = 87.23 pm B_f, oC8, Cmcm, CrB type (Z = 4)	7570	40–64.5	n.a.	n.a.	n.a.	2270–2917	15.6	n.a.	12.9	n.a.	n.a.	n.a.	n.a.	n.a.	n.a.	n.a.			Wear-resistant and semiconductive films, neutron-absorbing layer on nuclear fuel pellets
Silicon hexaboride	SiB_6 [12008-29-6]	Trigonal (rhombohedral)	2430	200,000	n.a.	n.a.	n.a.	1950													

Silicon tetraboride	SiB_4 [12007-81-7] 71.330		2400	n.a.	n.a.	n.a.	n.a.	1870 (dec.)													
Tantalum diboride	TaB_2 [12077-35-1] 202.570	Hexagonal a = 309.80 pm c = 324.10 pm C32, *hP3*, P6/mmm, AlB_2 type (Z = 1)	12,540	33–75	n.a.	n.a.	n.a.	3037–3200	10.9–16.0	237.55	8.2–8.8	257	n.a.	n.a.	n.a.	n.a.	n.a.	n.a.	n.a.	(HM>8)	Gray metallic powder. Severe oxidation in air above 800°C. Corroded by molten metals Nb, Mo, Ta, and Re.
Tantalum boride	TaB [12007-07-7] 191.759	Orthorhombic a = 327.6 pm b = 866.9 pm c = 315.7 pm B_f, oC8, Cmcm CrB type (Z = 4)	14,190	100	n.a.	n.a.	n.a.	2340–3090		246.85	n.a.	n.a.	n.a.	n.a.	n.a.	n.a.	n.a.	n.a.	n.a.	21.57 (HM>8)	Severe oxidation above 1100–1400°C in air
Thorium hexaboride	ThB_6 [12229-63-9] 296.904	Cubic a = 411.2 pm $D2_1$, cP7, Pm3m CaB_6 type (Z = 1)	6800	n.a.	n.a.	n.a.	n.a.	2149	44.8		7.8										
Thorium tetraboride	ThB_4 [12007-83-9] 275.53	Tetragonal a = 725.6 pm c = 411.3 pm $D1_e$, *tP20*, P4/mbm, ThB_4type (Z = 4)	8450	n.a.	n.a.	n.a.	n.a.	2500	25	510	7.9	148					137				
Titanium boride	TiB [] 58.678	Cubic	5260					2060													
Titanium diboride	TiB_2 [12045-63-5] 69.489	Hexagonal a = 302.8 pm c = 322.8 pm C32, *hP3*, P6/mmm, AlB_2 type (Z = 1)	4520	16–28.4	n.a.	n.a.	n.a.	2980–3225	64.4–96	637.22	7.6–8.64	372–551	n.a.	n.a.	0.11	131	240–400	669	6.2–6.7	33.05 (HM>9) 31.87 HK	Gray crystals, superconducting at 1.26 K. High-temperature electrical conductor, used as crucible material for handling molten metals such as Al, Zn, Cd, Bi, Sn, and Rb; strongly corroded by liquid metals such as Ti, Zr, V, Nb, Ta, Cr, Mn, Fe, Co, Ni, and Cu. Begins to be oxidized in air above 1100–1400°C. Corrosion resistance in hot concentrated brines. Maximum operating temperature 1000°C (reducing) and 800°C (oxidizing).
Tungsten hemiboride	W_2B [12007-10-2] 378.491	Tetragonal a = 556.4 pm c = 474.0 pm C16, *tI12*, I4/mcm, $CuAl_2$ type (Z = 4)	16,720	n.a.	n.a.	n.a.	n.a.	2670		168	6.7									23.73 (HM 9)	Black powder
Tungsten boride	WB [12007-09-9] 194.651	Tetragonal a = 311.5 pm c = 1692 pm	15,200–16,000	4.1	n.a.	n.a.	n.a.	2660			6.9									13.63 (HM 9)	Black powder

Table 10.22. *(continued)*

IUPAC name (synonyms, common trade names)	Theoretical chemical formula, [CAS RN], relative molecular mass (^{12}C = 12.000)	Crystal system, lattice parameters, structure type, Strukturbericht, Pearson, space group, structure type (Z)	Density (ρ/kg.m^{-3})	Electrical resistivity (ρ/μΩ.cm)	Dielectric permittivity [1MHz] (ε_r / nil)	Dielectric field strength (E_d/MV.m^{-1})	Dissipation or tangent loss factor ($tan\delta$)	Melting point (*m.p.*/°C)	Thermal conductivity (k/W.m^{-1}.K^{-1})	Specific heat capacity (c_P/J.kg^{-1}.K^{-1})	Coeff. linear thermal expansion (α/10^{-6}K^{-1})	Young's or elastic modulus (E/GPa)	Coulomb's or shear modulus (G/GPa)	Bulk or compression modulus (K/GPa)	Poisson ratio (ν)	Ultimate tensile strength (σ_{UTS}/MPa)	Flexural strength (τ/MPa)	Compressive strength (σ/MPa)	Fracture toughness (K_{IC}/MPa.m$^{1/2}$)	Vickers or Knoop Hardness (HV or HK/GPa) (/HM)	Other physicochemical Properties, oxidation and corrosion resistance, and major uses.
Uranium diboride	UB_2 [12007-36-2] 259.651	Hexagonal a = 313.10 pm c = 398.70 pm C32, *hP3*, P6/mmm, AlB_2 type	12,710	n.a.	n.a.	n.a.	n.a.	2385	51.9		9.0										
Uranium dodecaboride	UB_{12} 367.91	Cubic a = 747.3 pm $D2_f$, *cF52*, Fm3m, UB_{12} type (Z = 4)	5820	n.a.	n.a.	n.a.	n.a.	1500			4.6										
Uranium tetraboride	UB_4 [12007-84-0] 281.273	Tetragonal a = 707.5 pm c = 397.9 pm $D1_e$, *tP20*, P4/mbm, ThB_4type (Z = 4)	5350	n.a.	n.a.	n.a.	n.a.	2495	4.0		7.0	440					413			24.52	
Vanadium diboride	VB_2 [12007-37-3] 72.564	Hexagonal a = 299.8 pm c = 305.7 pm C32, *hP3*, P6/mmm, AlB_2 type (Z = 1)	5070	23	n.a.	n.a.	n.a.	2450–2747	42.3	647.43	7.6–8.3	268								(HM 8–9)	Wear-resistant and semiconductive films
Zirconium diboride	ZrB_2 [12045-64-6] 112.846	Hexagonal a = 316.9 pm c = 353.0 pm C32, *hP3*, P6/mmm, AlB_2 type (Z = 1)	6085	9.2	n.a.	n.a.	n.a.	3060–3245	57.9	392.54	5.5–8.3	343–506	220	n.a.	0.15	n.a.	305	n.a.	n.a.	18.63–33.34 (HM 8)	Gray metallic crystals, excellent thermal shock resistance, greatest oxidation inertness of all refractory hardmetals. Hot-pressed crucible for handling molten metals such as Zn, Mg, Fe, Cu, Zn, Cd, Sn, Pb, Rb, Bi, Cr, brass, carbon steel, cast irons, and molten cryolithe, yttria, zirconia, and alumina. Readily corroded by liquid metals such as Si, Cr, Mn, Co, Ni, Nb, Mo, Ta and attacked by molten salts such as Na_2O, alkali carbonates, and NaOH. Severe oxidation in air occurs above 1100–1400°C. Stable above 2000°C in inert or reducing atmosphere.

Zirconium dodecaboride	ZrB_{12} 283.217	Cubic a = 740.8 pm $D2_f$, *cF52*, Fm3m, UB_{12} type (Z = 4)	3630	60–80	n.a.	n.a.	n.a.	2680		523										–	
Carbides																					
Aluminum carbide	Al_4C_3 [1299-86-1] 143.959	Trigonal (rhombohedral) a = 333 pm c = 2494 pm D71	2360	n.a.	n.a.	n.a.	n.a.	2798	n.a.	n.a.	n.a.	n.a.	n.a.	n.a.	n.a.	n.a.	n.a.	n.a.	n.a.	n.a.	Decomposes in water with evolution of CH_4
Beryllium hemicarbide	Be_2C [506-66-1] 30.035	Cubic a = 433 pm C1, *cF12*, Fm3m, CaF_2 type (Z = 4)	1900	n.a.	n.a.	n.a.	n.a.	2100	21.0	1397	10.5	314.4	n.a.	96.5	0.100	155	n.a.	723	n.a.	23.63 HK	Brick-red or yellowish-red octahedra. Nuclear reactor cores.
Boron carbide (Norbide®)	B_4C [12069-32-8] 55.255	Hexagonal a = 560 pm c = 1212 pm $D1_g$, *hR15*, R3m, B_4C type	2512	4500	n.a.	n.a.	n.a.	2350–2427	27	1854	2.63–5.6	448–470	184	490	0.207	310–350	n.a.	1400–2900	3.2–4.2	31.38–34.32 HK (HM 9.32)	Hard black shiny crystals, fourth hardest material known after diamond, cubic boron nitride, and boron oxide. Does not burn in an O_2 flame if temperature is maintained below 983°C. Maximum operating temperature 2000°C (inert, reducing) or 600°C (oxidizing). Not attacked by hot HF or chromic acid. Used as abrasive, crucible container for molten salts except molten alkali hydroxides. In form of molded shape, used for pressure-blast nozzles, wire-drawing dies, and bearing surfaces for gauges. For grinding and lapping application available mesh sizes cover range 240 to 800.
Chromium carbide	Cr_7C_3 400.005	Hexagonal a = 1398.02 pm c = 453.20 pm	6992	109.0	n.a.	n.a.	n.a.	1665	n.a.	n.a.	11.7	n.a.	n.a.	n.a.	n.a.	n.a.	n.a.	n.a.	n.a.	13.10	Resists oxidation in range 800–1000°C. Corroded by molten metals Ni and Zn.
Chromium carbide	Cr_3C_2 [12012-35-0] 180.010	Orthorombic a = 282 pm b = 553 pm c = 1147 pm $D5_{10}$, *oP20*, Pbnm, Cr_3C_2 type (Z = 4)	6680	75.0	n.a.	n.a.	n.a.	1895	19.2	n.a.	10.3	386	n.a.	0.78	0.280	n.a.	n.a.	1041–1350	n.a.	25.99	Corroded by molten metals Ni, Zn, Cu, Cd, Al, Mn, and Fe. Corrosion resistant in molten Sn and Bi.

Table 10.22. *(continued)*

IUPAC name (synonyms, common trade names)	Theoretical chemical formula, [CAS RN], relative molecular mass (^{12}C = 12.000)	Crystal system, lattice parameters, structure type, Strukturbericht, Pearson, space group, structure type (Z)	Density (ρ/kg.m^{-3})	Electrical resistivity (ρ/μΩ.cm)	Dielectric permittivity [1MHz] (ε_r / nil)	Dielectric field strength (E_d/MV.m^{-1})	Dissipation or tangent loss factor ($tan\delta$)	Melting point ($m.p.$/°C)	Thermal conductivity (k/W.m^{-1}.K^{-1})	Specific heat capacity (c_P/J.kg^{-1}.K^{-1})	Coeff. linear thermal expansion (α/10^{-6}K^{-1})	Young's or elastic modulus (E/GPa)	Coulomb's or shear modulus (G/GPa)	Bulk or compression modulus (K/GPa)	Poisson ratio (ν)	Ultimate tensile strength (σ_{UTS}/MPa)	Flexural strength (τ/MPa)	Compressive strength (σ/MPa)	Fracture toughness (K_{IC}/MPa.m$^{1/2}$)	Vickers or Knoop Hardness (HV or HK/GPa) (/HM)	Other physicochemical Properties, oxidation and corrosion resistance, and major uses.
Diamond	C [7782-40-3] 12.011	Cubic a = 356.683 pm A4, *cF8*, Fd3m, diamond type (Z =8)	3515.24	>10^{16} (I, IIa) >10^{3} (IIb)	n.a.	n.a.	n.a.	3550	900 (I) 2400 (IIa)	n.a.	2.16	930	n.a.	n.a.	n.a.	n.a.	n.a.	7000	5.3–6.7	78.45 HK (HM 10)	Exists in two major varieties: those bearing nitrogen as an impurity (Type I) and those without nitrogen (Type II). These two subgroups are further subdivided into Types Ia, Ib, IIa, and IIb. Type Ia diamonds are the most common type of naturally occurring diamond; they exhibit 0.1 to 0.2 wt.% nitrogen present in small aggregates, including platelets. By contrast, nitrogen in Type Ib diamonds is dispersed substitutionally. Of the two Type II diamond types, Type IIb is a semiconductor due to minute amounts of boron impurities and exhibits a blue color, whereas Type IIa diamonds are comparatively pure. Electric insulator (E_g = 7 eV.). Burns in oxygen.
Graphite	C [7782-42-5] 12.011	Hexagonal a = 246 pm b = 428 pm c = 671 pm A9, *hP4*,P6$_3$/mmc, graphite type (Z =4)	2250	1385	n.a.	n.a.	n.a.	3650	n.a.	n.a.	0.6–4.3	6.9	n.a.	n.a.	n.a.	28	n.a.	n.a.	n.a.	(HM 2)	High-temperature lubricant, crucible container for handling molten metals such as Mg, Al, Zn, Ga, Sb, and Bi
Hafnium carbide	HfC [12069-85-1] 190.501	Cubic a = 446.0 pm B1, *cF8*, Fm3m, rock salt type (Z = 4)	12,670	45.0	n.a.	n.a.	n.a.	3890–3950	22.15	n.a.	6.3	424	179	n.a.	0.17	n.a.	n.a.	n.a.	n.a.	18.34–28.44	Dark gray brittle solid, most refractory binary material known. Controls rods in nuclear reactors, crucible container for melting HfO_2 and other oxides. Corrosion resistant to liquid metals such as Nb, Ta, Mo, and W. Severe oxidation in air above 1100–1400°C and stable up to 2000°C in helium.

Lanthanum dicarbide	LaC_2 [12071-15-7] 162.928	Tetragonal $a = 394.00$ pm $c = 657.20$ pm C11a, *tI6*, I4/mmm, CaC_2 type ($Z = 2$)	5290	68.0	n.a.	n.a.	n.a.	2360–2438	n.a.	n.a.	12.1	n.a.	n.a.	n.a.	n.a.	n.a.	n.a.	n.a.	n.a.	n.a.	Decomposed by H_2O
Molybdenum hemicarbide	β–Mo_2C [12069-89-5] 203.891	Hexagonal $a = 300.20$ pm $c = 427.40$ pm L'3, *hP3*, $P6_3$/mmc, Fe_2N type ($Z = 1$)	9180	71.0	n.a.	n.a.	n.a.	2687	n.a.	29.4	7.8	221	n.a.	n.a.	n.a.	n.a.	n.a.	n.a.	n.a.	14.70 (HM>7)	Gray powder. Wear-resistant film. Oxidized in air at 700–800°C. Corroded in molten metals Al, Mg, V, Cr, Mn, Fe, Ni, Cu, Zn, and Nb. Corrosion resistant in molten Cd, Sn, and Ta
Molybdenum carbide	MoC [12011-97-1] 107.951	Hexagonal $a = 290$ pm $c = 281$ pm B_k, $P6_3$/mmc, BN type ($Z = 4$)	9150	50.0	n.a.	n.a.	n.a.	2577	n.a.	n.a.	5.76	197	n.a.	n.a.	0.204	n.a.	n.a.	n.a.	n.a.	17.65 (HM>9)	Oxidized in air at 700–800°C
Niobium hemicarbide	Nb_2C [12011-99-3] 197.824	Hexagonal $a = 312.70$ pm $c = 497.20$ pm L'3, *hP3*, $P6_3$/mmc, Fe_2N type ($Z = 1$)	7800	n.a.	n.a.	n.a.	n.a.	3090	n.a.	n.a.	n.a.	n.a.	n.a.	n.a.	n.a.	n.a.	n.a.	n.a.	n.a.	20.82	
Niobium carbide	NbC [12069-94-2] 104.917	Cubic $a = 447.71$ pm B1, *cF8*, Fm3m, rock salt type ($Z = 4$)	7820	51.1–74.0	n.a.	n.a.	n.a.	3760	14.2	n.a.	6.84	340	n.a.	n.a.	n.a.	n.a.	n.a.	n.a.	n.a.	24.22 (HM>9)	Lavender gray powder, soluble in HF-HNO_3 mixture. Wear-resistant film, coating graphite in nuclear reactors. Oxidation in air becomes severe only above 1000°C.
Silicon carbide (Carbolon®, Crystolon®, Carborundum®)	β–SiC [409-21-2] 40.097	Cubic $a = 435.90$ pm B3, *cF8*, F43m, ZnS type ($Z = 4$)	3160	107–200	n.a.	n.a.	n.a.	2093–2400	135	1205	4.5	262–468	168	414	0.192	550	n.a.	1000	3.1	26.48–32.85 (HM 9.5)	Green to bluish black, iridescent crystals. Soluble in fused alkali hydroxides. Abrasives best suited for grinding low-tensile-strength materials such as cast iron, brass, bronze, marble, concrete, stone and glass, optical structural, and wear-resistant components. Corroded by molten metals such as Na, Mg, Al, Zn, Fe, Sn, Rb, and Bi. Resistant to oxidation in air up to 1650°C. Maximum operating temperature of 2000°C in reducing or inert atmosphere.
Silicon carbide (moissanite, Carbolon®, Crystolon®, Carborundum®)	α–SiC [409-21-2] 40.097	Hexagonal $a = 308.10$ pm $c = 503.94$ pm B4, *hP4*, $P6_3$/mmc, wurtzite type ($Z = 2$)	3160	410,000	10.2	n.a.	n.a.	2093 trans.	42.5	690–715	4.3–4.6	386–414	179	n.a.	0.16	450–520	359	500	3.1	23.54–24.52 (HM 9.2)	Semiconductor (Eg=3.03 eV) soluble in fused alkali hydroxides

Table 10.22. *(continued)*

IUPAC name (synonyms, common trade names)	Theoretical chemical formula, [CAS RN], relative molecular mass (^{12}C = 12.000)	Crystal system, lattice parameters, structure type, Strukturbericht, Pearson, space group, structure type (*Z*)	Density (ρ/kg.m^{-3})	Electrical resistivity (ρ/μΩ.cm)	Dielectric permittivity [1MHz] (ε_r / nil)	Dielectric field strength (E_d/MV.m^{-1})	Dissipation or tangent loss factor ($tan\delta$)	Melting point (*m.p.*/°C)	Thermal conductivity (k/W.m^{-1}.K^{-1})	Specific heat capacity (c_P/J.kg^{-1}.K^{-1})	Coeff. linear thermal expansion (α/10^{-6}K^{-1})	Young's or elastic modulus (*E*/GPa)	Coulomb's or shear modulus (*G*/GPa)	Bulk or compression modulus (*K*/GPa)	Poisson ratio (ν)	Ultimate tensile strength (σ_{UTS}/MPa)	Flexural strength (τ/MPa)	Compressive strength (σ/MPa)	Fracture toughness (K_{1C}/MPa.m$^{1/2}$)	Vickers or Knoop Hardness (HV or HK/GPa) (/HM)	Other physicochemical Properties, oxidation and corrosion resistance, and major uses.
Tantalum hemicarbide	Ta_2C [12070-07-4] 373.907	Hexagonal a = 310.60 pm c = 493.00 pm L'3, *hP3*, $P6_3/mmc$, Fe_2N type (Z = 1)	15,100	80.0	n.a.	n.a.	n.a.	3327	n.a.	n.a.	n.a.	n.a.	n.a.	n.a.	n.a.	n.a.	n.a.	n.a.	n.a.	16.80–19.61	
Tantalum carbide	TaC [12070-06-3] 194.955	Cubic a = 445.55 pm B1, *cF8*, Fm3m, rock salt type (Z = 4)	14,800	30–42.1	n.a.	n.a.	n.a.	3880–3920	22.2	190	6.64–8.4	364	n.a.	n.a.	0.172	n.a.	n.a.	n.a.	n.a.	15.68–17.65 (HM 9–10)	Golden brown crystals, soluble in HF-HNO_3 mixture. Crucible container for melting ZrO and similar oxides with high melting point. Corrosion resistant to molten metals such as Ta and Re. Readily corroded by liquid metals such as Nb, Mo, and Sn. Burning occurs in pure oxygen above 800°C. Severe oxidation in air above 1100–1400°C. Maximum operating temperature of 3760°C in helium
Thorium dicarbide	α–ThC_2 [12071-31-7] 256.060	Tetragonal a = 585 pm c = 528 pm C11a, *tI6*, I4/mmm, CaC_2 type (Z = 2)	8960–9600	30.0	n.a.	n.a.	n.a.	2655	23.9	n.a.	8.46	n.a.	n.a.	n.a.	n.a.	n.a.	n.a.	n.a.	n.a.	5.88	α–β transition at 1427°C and β–γ at 1497°C. Decomposed by H_2O with evolution of C_2H_6
Thorium carbide	ThC [12012-16-7] 244.049	Cubic a = 534.60 pm B1, *cF8*, Fm3m, rock salt type (Z = 4)	10,670	25.0	n.a.	n.a.	n.a.	2621	28.9	n.a.	6.48	n.a.	n.a.	n.a.	n.a.	n.a.	n.a.	n.a.	n.a.	9.807	Readily hydrolyzes in water evolving C_2H_6.
Titanium carbide	TiC [12070-08-5] 59.878	Cubic a = 432.8 pm B1, *cF8*, Fm3m, rock salt type (Z = 4)	4938	52.5	n.a.	n.a.	n.a.	2940–3160	17–21	841	7.5–7.7	310–462	172	0.854	0.182	275–450	n.a.	1310	1.7–3.8	25.69–31.38 (HM 9–10)	Gray crystals. Superconducting at 1.1 K. Soluble in HNO_3 and aqua regia. Resistant to oxidation in air up to 450°C. Maximum operating temperature 3000°C in helium. Crucible container for handling molten metals such as Na, Bi, Zn, Pb, Sn, Bi, Rb, and Cd. Corroded by liquid metals Mg, Al, Si, Ti, Zr, V, Nb, Ta, Cr, Mo, Mn, Fe, Co, and Ni. Attacked by molten NaOH

Tungsten carbide (Widia®)	WC [12070-12-1] 195.851	Hexagonal $a = 290.63$ pm $c = 283.86$ pm L'3, *hP3*, $P6_3/mmc$, Fe_2N type ($Z = 1$)	15,630	19.2	n.a.	n.a.	n.a.	2870	121	n.a.	6.9	710	n.a.	0.58	0.26	n.a.	n.a.	530	7.5–8.9	26.48 (HM>9)	Gray powder, dissolved by HF-HNO_3 mixture. Cutting tools, wear-resistant semiconductor film. Corroded by molten metals Mg, Al, V, Cr, Mn, Ni, Cu, Zn, Nb, and Mo. Corrosion resistant to molten Sn and Ta
Tungsten hemicarbide	W_2C [12070-13-2] 379.691	Hexagonal $a = 299.82$ pm $c = 472.20$ pm L'3, *hP3*, $P6_3/mmc$, Fe_2N type ($Z = 2$)	17,340	81.0	n.a.	n.a.	n.a.	2730	n.a.	n.a.	3.84	421	n.a.	n.a.	n.a.	n.a.	n.a.	n.a.	n.a.	29.42	Black. Resistant to oxidation in air up to 700°C. Corrosion resistant to Mo
Uranium carbide	U_2C_3 [12076-62-9]	Cubic $a = 808.89$ pm D5c, *cI40*, I43d, Pu_2C_3 type ($Z = 8$)	12,880	n.a.	n.a.	n.a.	n.a.	1777	n.a.	n.a.	11.4	179–221	n.a.	n.a.	n.a.	n.a.	n.a.	434	n.a.	n.a.	
Uranium dicarbide	UC_2 [12071-33-9] 262.051	Tetragonal $a = 352.24$ pm $c = 599.62$ pm C11a, *tI6*, I4/mmm, CaC_2 type ($Z = 2$)	11,280	n.a.	n.a.	n.a.	n.a.	2350–2398	32.7	147	14.6	n.a.	n.a.	n.a.	n.a.	n.a.	n.a.	n.a.	n.a.	5.88	Transition tetragonal to cubic at 1765°C. Decomposes in H_2O, slightly soluble in alcohol. Used in microsphere pellets to fuel nuclear reactors
Uranium carbide	UC [12070-09-6] 250.040	Cubic $a = 496.05$ pm B1, *cF8*, Fm3m, rock salt type ($Z = 4$)	13,630	50.0	n.a.	n.a.	n.a.	2370–2790	23.0	n.a.	11.4	172.4	66.9	n.a.	0.29	n.a.	n.a.	351.6	n.a.	7.35–9.17 (HM>7)	Gray crystals with metallic appearance, reacts with oxygen. Corroded by molten metals Be, Si, Ni, and Zr
Vanadium hemicarbide	V_2C [12012-17-8] 113.89	Hexagonal $a = 286$ pm $c = 454$ pm L'3, *hP3*, $P6_3/mmc$, Fe_2N type ($Z = 2$)	5750	n.a.	n.a.	n.a.	n.a.	2166	n.a.	n.a.	n.a.	n.a.	n.a.	n.a.	n.a.	n.a.	n.a.	n.a.	n.a.	29.42	Corroded by molten Nb, Mo, and Ta
Vanadium carbide	VC [12070-10-9] 62.953	Cubic $a = 413.55$ pm B1, *cF8*, Fm3m, rock salt type ($Z = 4$)	5770	65.0–98.0	n.a.	n.a.	n.a.	2810	24.8	n.a.	4.9	614	435	n.a.	n.a.	790–825	n.a.	613	n.a.	20.50	Black crystals soluble in HNO_3 with decomposition. Wear-resistant film, cutting tools. Resistant to oxidation in air up to 300°C
Zirconium carbide	ZrC [12020-14-3] 103.235	Cubic $a = 469.83$ pm B1, *cF8*, Fm3m, rock salt type ($Z = 4$)	6730	68	n.a.	n.a.	n.a.	3540–3560	20.61	205	6.82	345	123	338	0.257	110	n.a.	1641	n.a.	17.95–28.73 (HM >8)	Dark gray brittle solid, soluble in HF solutions containing nitrate or peroxide ions. UC-nuclear power reactor, crucible container for handling molten metals such as Bi, Cd, Pb, Sn, Rb, and molten zirconia ZrO_2. Corroded by liquid metals Mg, Al, Si, V, Nb, Ta, Cr, Mo, Mn, Fe, Co, Ni, and Zn. In air oxidizes rapidly above 500°C. Maximum operating temperature of 2350°C in helium

Table 10.22. *(continued)*

IUPAC name (synonyms, common trade names)	Theoretical chemical formula, [CAS RN], relative molecular mass ($^{12}C = 12.000$)	Crystal system, lattice parameters, structure type, Strukturbericht, Pearson, space group, structure type (*Z*)	Density (ρ/kg.m^{-3})	Electrical resistivity (ρ/μΩ.cm)	Dielectric permittivity [1MHz] (ε_r / nil)	Dielectric field strength (E_d/MV.m^{-1})	Dissipation or tangent loss factor ($tan\delta$)	Melting point (*m.p.*/°C)	Thermal conductivity (*k*/W.m^{-1}.K^{-1})	Specific heat capacity (c_p/J.kg^{-1}.K^{-1})	Coeff. linear thermal expansion ($\alpha/10^{-6}K^{-1}$)	Young's or elastic modulus (*E*/GPa)	Coulomb's or shear modulus (*G*/GPa)	Bulk or compression modulus (*K*/GPa)	Poisson ratio (ν)	Ultimate tensile strength (σ_{UTS}/MPa)	Flexural strength (τ/MPa)	Compressive strength (σ/MPa)	Fracture toughness (K_{1C}/MPa.m$^{1/2}$)	Vickers or Knoop Hardness (HV or HK/GPa) (/HM)	Other physicochemical Properties, oxidation and corrosion resistance, and major uses.
Nitrides																					
Aluminum nitride	AlN [24304-00-5] 40.989	Hexagonal $a = 311.0$ pm c = 497.5 pm B4, *hP8*, $P6_3mc$, wurtzite type ($Z = 2$)	3255	10^{17}	n.a.	n.a.	n.a.	2230	29.96	820	5.3	346	n.a.	0.28	n.a.	270	n.a.	2068	2.79	11.77 (HM 9–10)	Insulator (E_g=4.26 eV). Decomposes with water, acids, and alkalis to $Al(OH)_3$ and NH_3. Crucible container for GaAs crystal growth
Beryllium nitride	α-Be_3N_2 [1304-54-7] 55.050	Cubic $a = 814$ pm $D5_3$, *cI80*, Ia3, Mn_2O_3 type ($Z = 16$)	2710	n.a.	n.a.	n.a.	n.a.	2200	n.a.	1221	n.a.	n.a.	n.a.	n.a.	n.a.	n.a.	n.a.	n.a.	n.a.	n.a.	Hard white or grayish crystal. Oxidizes in air above 600°C. Slowly decomposes in water, quickly in acids and alkalis with evolution of NH_3
Boron nitride	BN [10043-11-5] 24.818	Hexagonal $a = 250.4$ pm $c = 666.1$ pm B_k, *hP8*, *$P6_3/mmc$*, BN type ($Z = 2$)	2250	10^{19}	n.a.	n.a.	0.0002	2730 (dec.)	15.41	711	7.54	85.5	n.a.	0.11	n.a.	41–62	n.a.	310	5.0	2.26 (HM 2.0)	Insulator (E_g=7.5 eV). Crucibles for melting molten metals such as Na, B, Fe, Ni, Al, Si, Cu, Mg, Zn, In, Bi, Rb, Cd, Ge, and Sn. Corroded by molten metals U, Pt, V, Ce, Be, Mo, Mn, Cr, V, and Al. Attacked by molten salts PbO_2, Sb_2O_3, AsO_3, Bi_2O_3, KOH, and K_2CO_3.Used in furnace insulation-diffusion masks and passivation layers
Boron nitride (Borazon®, CBN)	BN 24.818	Cubic $a = 361.5$ pm	3430	1900 (2000°C)	2.54	n.a.	0.0002	1540	n.a.	n.a.	n.a.	n.a.	n.a.	n.a.	n.a.	n.a.	n.a.	7000	n.a.	46.09–49.00 (HM 10)	Tiny reddish to black grains. Used as abrasive for grinding tool and die steels and high-alloy steels when chemical reactivity of diamonds is a problem
Chromium heminitride	Cr_2N [12053-27-9] 117.999	Hexagonal $a = 274$ pm $c = 445$ pm L'3, *hP3*, $P6_3/mmc$, Fe_2N type ($Z = 1$)	6800	76	n.a.	n.a.	n.a.	1661	22.5	630	9.36	n.a.	n.a.	n.a.	n.a.	n.a.	n.a.	n.a.	n.a.	11.77–15.40	
Chromium nitride	CrN [24094-93-7] 66.003	Cubic $a = 415.0$ pm B1, *cF8*, Fm3m, rock salt type ($Z = 4$)	6140	640	n.a.	n.a.	n.a.	1499 (dec.)	12.1	795	2.34	n.a.	n.a.	n.a.	n.a.	n.a.	n.a.	n.a.	n.a.	10.69	

Hafnium nitride	HfN [25817-87-2] 192.497	Cubic $a = 451.8$ pm B1, *cF8*, Fm3m, rock salt type ($Z = 4$)	13,840	33	n.a.	n.a.	n.a.	3310	21.6	210	6.5	n.a.	n.a.	n.a.	n.a.	n.a.	n.a.	n.a.	n.a.	16.08 (HM 8–9)	Most refractory of all nitrides
Molybdenum heminitride	Mo_2N [12033-31-7] 205.887	Cubic $a = 416$ pm L'1, *cP5*, Pm3m, Fe_4N type ($Z = 2$)	9460	19.8	n.a.	n.a.	n.a.	760–899	17.9	293	6.12	n.a.	n.a.	n.a.	n.a.	n.a.	n.a.	n.a.	n.a.	16.68	Temperature transition at 5.0 K
Molybdenum nitride	MoN [12033-19-1] 109.947	Hexagonal $a = 572.5$ pm $c = 560.8$ pm B_h, *hP2*, P6/mmm, WC type ($Z = 1$)	9180	n.a.	n.a.	n.a.	n.a.	1749	n.a.	n.a.	n.a.	n.a.	n.a.	n.a.	n.a.	n.a.	n.a.	n.a.	n.a.	650	
Niobium nitride	NbN [24621-21-4] 106.913	Cubic $a = 438.8$ pm B1, *cF8*, Fm3m, rock salt type ($Z = 4$)	8470	78	n.a.	n.a.	n.a.	2575	3.63	n.a.	10.1	n.a.	n.a.	n.a.	n.a.	n.a.	n.a.	n.a.	n.a.	13.73 (HM>8)	Dark gray crystals. Transition temperature 15.2 K. Insoluble in HCL, HNO_3, and H_2SO_4 but attacked by hot caustic, lime, or strong alkalis evolving NH_3
Silicon nitride	β-Si_3N_4 [12033-89-5] 140.284	Hexagonal $a = 760.8$ pm $c = 291.1$ pm P6/3m	3170	10^6	n.a.	n.a.	n.a.	1850	28	713	2.25	55	n.a.	n.a.	0.25	n.a.	850	n.a.	6.1	17 (HM>9)	Gray amorphous powder or crystals. Resistant to thermal shock. Good oxidation resistance up to 1500°C but decomposes into nitrogen and silicon above 1850°C. Excellent corrosion resistance to molten nonferrous metals such as Al, Pb, Zn, Cd, Bi, Rb, and Sn, and molten salts like NaCl-KCl, NaF, and silicate glasses. However, corroded by molten Mg, Ti, V, Cr, Fe, and Co, cryolite, KOH, and Na_2O
Silicon nitride (Nitrasil®)	α-Si_3N_4 [12033-89-5] 140.284	Hexagonal $a = 775.88$ pm $c = 561.30$ pm P31c	3184	10^{19}	9.4	n.a.	n.a.	1900 (sub.)	17	700	2.5–3.3	304	n.a.	n.a.	0.26	400–580	750	n.a.	7.0–8.3	26–35 (HM>9)	
Tantalum heminitride	Ta_2N 375.901	Hexagonal $a = 306$ pm $c = 496$ pm L'3, *hP3*, $P6_3$/mmc, Fe_2N type (Z=1)	15,600	263	n.a.	n.a.	n.a.	2980	10.04	126	5.2	n.a.	n.a.	n.a.	n.a.	n.a.	n.a.	n.a.	n.a.	31.38	Decomposed by KOH with evolution of NH_3
Tantalum nitride (ε)	TaN [12033-62-4] 194.955	Hexagonal $a = 519.1$ pm $c = 290.6$ pm	13,800	128–135	n.a.	n.a.	n.a.	3093	8.31	210	3.2	n.a.	n.a.	n.a.	n.a.	n.a.	n.a.	n.a.	n.a.	10.89 (HM>8)	Bronze or black crystals. Transition temperature 1.8 K. Insoluble in water, slowly attacked by aqua regia, HF, and HNO_3
Thorium nitride	ThN [12033-65-7] 246.045	Cubic a = 515.9 B1, *cF8*, Fm3m, rock salt type ($Z = 4$)	11,560	20	n.a.	n.a.	n.a.	2820	7.38	n.a.	n.a.	n.a.	n.a.	n.a.	n.a.	n.a.	n.a.	n.a.	n.a.	5.88	Gray solid; slowly hydrolyzed by water

Table 10.22. *(continued)*

IUPAC name (synonyms, common trade names)	Theoretical chemical formula, [CAS RN], relative molecular mass (^{12}C = 12.000)	Crystal system, lattice parameters, structure type, Strukturbericht, Pearson, space group, structure type (*Z*)	Density (ρ/kg.m^{-3})	Electrical resistivity (ρ/μΩ.cm)	Dielectric permittivity [1MHz] (ε_r / nil)	Dielectric field strength (E_d/MV.m^{-1})	Dissipation or tangent loss factor ($tan\delta$)	Melting point (*m.p.*/°C)	Thermal conductivity (k/W.m^{-1}.K^{-1})	Specific heat capacity (c_P/J.kg^{-1}.K^{-1})	Coeff. linear thermal expansion (α/10^{-6}K^{-1})	Young's or elastic modulus (*E*/GPa)	Coulomb's or shear modulus (*G*/GPa)	Bulk or compression modulus (*K*/GPa)	Poisson ratio (ν)	Ultimate tensile strength (σ_{UTS}/MPa)	Flexural strength (τ/MPa)	Compressive strength (σ/MPa)	Fracture toughness (K_{IC}/MPa.m$^{1/2}$)	Vickers or Knoop Hardness (HV or HK/GPa) (/HM)	Other physicochemical Properties, oxidation and corrosion resistance, and major uses.
Thorium nitride	Th_3N_4 [12033-90-8]	Hexagonal a = 388 pm c = 618 pm $D5_2$, *hP5*, 3m1, La_2O_3 type (*Z* = 1)	10,400	n.a.	n.a.	n.a.	n.a.	1750	n.a.	n.a.	n.a.	n.a.	n.a.	n.a.	n.a.	n.a.	n.a.	n.a.	n.a.	n.a.	
Titanium nitride	TiN [25583-20-4] 61.874	Cubic a = 424.6 pm B1, *cF8*, Fm3m, rock salt type (*Z* = 4)	5430	21.7	n.a.	n.a.	n.a.	2930 (dec.)	29.1	586	9.35	248	n.a.	n.a.	n.a.	n.a.	n.a.	972	5.0	18.63 (HM 8–9)	Bronze powder. Transition temperature 4.2 K. Corrosion resistant to molten metals such as Al, Pb, Mg, Zn, Cd, and Bi. Corroded by molten Na, Rb, Ti, V, Cr, Mn, Sn, Ni, Cu, Fe, and Co. Dissolved by boiling aqua regia, decomposed by boiling alkalis evolving NH_3
Tungsten dinitride	WN_2 [60922-26-1] 211.853	Hexagonal a = 289.3 pm c = 282.6 pm	7700	n.a.	n.a.	n.a.	n.a.	600 (dec.)	n.a.	n.a.	n.a.	n.a.	n.a.	n.a.	n.a.	n.a.	n.a.	n.a.	n.a.	n.a.	Brown crystals
Tungsten heminitride	W_2N [12033-72-6] 381.687	Cubic a = 412 pm L'1, *cP5*, Pm3m, Fe_4N type (*Z* = 2)	17,700	n.a.	n.a.	n.a.	n.a.	982	n.a.	n.a.	n.a.	n.a.	n.a.	n.a.	n.a.	n.a.	n.a.	n.a.	n.a.	n.a.	Gray crystals
Tungsten nitride	WN [12058-38-7]	Hexagonal	15,940	n.a.	n.a.	n.a.	n.a.	593	n.a.	n.a.	n.a.	n.a.	n.a.	n.a.	n.a.	n.a.	n.a.	n.a.	n.a.	n.a.	
Uranium nitride	UN [25658-43-9] 252.096	Cubic a = 489.0 pm B1, *cF8*, Fm3m, rock salt type (*Z* = 4)	14,320	208	n.a.	n.a.	n.a.	2900	12.5	188	9.72	149	60	n.a.	0.24	n.a.	n.a.	n.a.	n.a.	n.a.	
Uranium nitride	U_2N_3 [12033-83-9] 518.259	Cubic a = 1070 pm $D5_3$, *cI80*, Ia3, Mn_2O_3 type (*Z* = 16)	11,240	n.a.	n.a.	n.a.	n.a.	n.a.	n.a.	n.a.	n.a.	n.a.	n.a.	n.a.	n.a.	n.a.	n.a.	n.a.	n.a.	4.46	
Vanadium nitride	VN [24646-85-3] 64.949	Cubic a = 414.0 pm B1, *cF8*, Fm3m, rock salt type (*Z* = 4)	6102	86	n.a.	n.a.	n.a.	2360	11.25	586	8.1	n.a.	n.a.	n.a.	n.a.	n.a.	n.a.	n.a.	n.a.	14.91 (HM 9–10)	Black powder. Transition temperature 7.5 K. Soluble in aqua regia

Zirconium nitride	ZrN [25658-42-8] 105.231	Cubic $a = 457.7$ pm B1, *cF8*, Fm3m, rock salt type ($Z = 4$)	7349	13.6	n.a.	n.a.	n.a.	2980	20.9	377	7.24	n.a.	n.a.	n.a.	n.a.	n.a.	n.a.	979	n.a.	14.51 (HM>8)	Yellow solid. Transition temperature 9 K. Corrosion resistant to steel, basic slag, and cryolithe and molten metals such as Al, Pb, Mg, Zn, Cd, and Bi. Corroded by molten Be, Na, Rb, Ti, V, Cr, Mn, Sn, Ni, Cu, Fe, and Co. Soluble in concentrated HF, slowly soluble in hot H_2SO_4
Silicides																					
Chromium disilicide	$CrSi_2$ [12018-09-6] 108.167	Hexagonal $a = 442$ pm $c = 635$ pm C40, *hP9*, $P6_222$, $CrSi_2$ type ($Z = 3$)	4910	1400	n.a.	n.a.	n.a.	1490–1550	106	n.a.	13.0	n.a.	n.a.	n.a.	n.a.	n.a.	n.a.	n.a.	n.a.	9.77–11.08	
Chromium silicide	Cr_3Si [12018-36-9] 184.074	Cubic $a = 456$ pm A15, *cP8*, Pm3n, Cr_3Si type ($Z = 2$)	6430	45.5	n.a.	n.a.	n.a.	1770	n.a.	n.a.	10.5	n.a.	n.a.	n.a.	n.a.	n.a.	n.a.	n.a.	n.a.	9.86	
Hafnium disilicide	$HfSi_2$ [12401-56-8] 234.66	Orthorhombic $a = 369$ pm $b = 1446$ pm $c = 364$ pm C49, *oC12*, Cmcm, $ZrSi_2$ type ($Z = 4$)	8030	n.a.	n.a.	n.a.	n.a.	1699	n.a.	n.a.	n.a.	n.a.	n.a.	n.a.	n.a.	n.a.	n.a.	n.a.	n.a.	8.48–9.12	
Molybdenum disilicide	$MoSi_2$ [12136-78-6] 152.11	Tetragonal $a = 319$ pm $c = 783$ pm C11b, *tI6*, I4/mmm, $MoSi_2$ type ($Z = 2$)	6260	21.5	n.a.	n.a.	n.a.	1870	58.9	n.a.	8.12	407	163	0.344	0.165	276	n.a.	2068–2415	n.a.	12.36	The compound is thermally stable in air up to 1000°C. Corrosion resistant to molten metals such as Zn, Pd, Ag, Bi, and Rb. Corroded by liquid metals Mg, Al, Si, V, Cr, Mn, Fe, Ni, Cu, Mo, and Ce
Niobium disilicide	$NbSi_2$ [12034-80-9] 149.77	Hexagonal $a = 479$ pm $c = 658$ pm C40, *hP9*, $P6_222$, $CrSi_2$ type ($Z = 3$)	5290	50.4	n.a.	n.a.	n.a.	2160	n.a.	n.a.	n.a.	n.a.	n.a.	n.a.	n.a.	n.a.	n.a.	n.a.	n.a.	10.30	
Tantalum disilicide	$TaSi_2$ [12039-79-1] 237.119	Hexagonal $a = 477$ pm $c = 655$ pm C40, *hP9*, $P6_222$, $CrSi_2$ type ($Z = 3$)	9140	8.5	n.a.	n.a.	n.a.	2299	n.a.	n.a.	8.8–9.54	n.a.	n.a.	n.a.	n.a.	n.a.	n.a.	n.a.	n.a.	11.77–15.69	Corroded by molten Ni
Tantalum silicide	Ta_5Si_3 [12067-56-0] 988.992	Hexagonal $a = 747.4$ pm $c = 522.5$ pm $P6_3$/mcm	13,060		n.a.	n.a.	n.a.	2499	n.a.	n.a.	n.a.	n.a.	n.a.	n.a.	n.a.	n.a.	n.a.	n.a.	n.a.	11.77–14.71	The compound is thermally stable in air up to 400°C

Table 10.22. *(continued)*

IUPAC name (synonyms, common trade names)	Theoretical chemical formula, [CAS RN], relative molecular mass ($^{12}C = 12.000$)	Crystal system, lattice parameters, structure type, Strukturbericht, Pearson, space group, structure type (Z)	Density (ρ/kg.m^{-3})	Electrical resistivity (ρ/μΩ.cm)	Dielectric permittivity [1MHz] (ε_r / nil)	Dielectric field strength (E_d/MV.m^{-1})	Dissipation or tangent loss factor ($\tan\delta$)	Melting point ($m.p.$/°C)	Thermal conductivity (k/W.m^{-1}.K^{-1})	Specific heat capacity (c_p/J.kg^{-1}.K^{-1})	Coeff. linear thermal expansion (α/10^{-6}K^{-1})	Young's or elastic modulus (E/GPa)	Coulomb's or shear modulus (G/GPa)	Bulk or compression modulus (K/GPa)	Poisson ratio (ν)	Ultimate tensile strength (σ_{UTS}/MPa)	Flexural strength (τ/MPa)	Compressive strength (σ/MPa)	Fracture toughness (K_{IC}/MPa.m$^{1/2}$)	Vickers or Knoop Hardness (HV or HK/GPa) (/HM)	Other physicochemical Properties, oxidation and corrosion resistance, and major uses.
Thorium disilicide	$ThSi_2$ [12067-54-8] 288.209	Tetragonal a =413 pm c =1435 pm Cc, tI12, I4/amd, $ThSi_2$ type (Z = 4)	7790	n.a.	n.a.	n.a.	n.a.	1850	n.a.	n.a.	n.a.	n.a.	n.a.	n.a.	n.a.	n.a.	n.a.	n.a.	n.a.	10.98	Corrosion resistant to molten Cu; corroded by molten Ni
Titanium disilicide	$TiSi_2$ [12039-83-7] 104.051	Orthorhombic a = 360pm b = 1376 pm c = 360 pm C49, *oC12*, Cmcm, $ZrSi_2$ type (Z = 4)	4150	123	n.a.	n.a.	n.a.	1499	n.a.	n.a.	10.4	n.a.	n.a.	n.a.	n.a.	n.a.	n.a.	n.a.	n.a.	10.19	
Titanium trisilicide	Ti_5Si_3 [12067-57-1] 323.657	Hexagonal a = 747 pm c = 516 pm $D8_8$, *hP16*, $P6_3$/mcm, Mn_5Si_3 type (Z = 2)	4320	55	n.a.	n.a.	n.a.	2120	n.a.	n.a.	11.0	n.a.	n.a.	n.a.	n.a.	n.a.	n.a.	n.a.	n.a.	9.67	
Tungsten disilicide	WSi_2 [12039-88-2] 240.01	Tetragonal a = 320 pm c = 781 pm C11b, *tI6*, I4/mmm, $MoSi_2$ type (Z = 2)	9870	33.4	n.a.	n.a.	n.a.	2165	n.a.	n.a.	8.28	n.a.	n.a.	n.a.	n.a.	n.a.	n.a.	n.a.	n.a.	10.69	
Tungsten silicide	W_5Si_3 [12039-95-1] 1003.46	Hexagonal a = 719 pm c = 485 pm $P6_3$/mcm	12,210	n.a.	n.a.	n.a.	n.a.	2320	n.a.	n.a.	n.a.	n.a.	n.a.	n.a.	n.a.	n.a.	n.a.	n.a.	n.a.	7.55	Corroded by molten Ni
Uranium disilicide	USi_2 294.200	Tetragonal a = 397 pm c = 1371 pm Cc, *tI12*, I4/amd, $ThSi_2$ type (Z = 4)	9250	n.a.	n.a.	n.a.	n.a.	1700	n.a.	n.a.	n.a.	n.a.	n.a.	n.a.	n.a.	n.a.	n.a.	n.a.	n.a.	6.86	
Uranium silicide	β–U_3Si_2 770.258	Tetragonal a = 733 pm c = 390 pm D5a, *tP10*, P4/mbm, U_3Si_2 type (Z = 2)	12,200	150	n.a.	n.a.	n.a.	1666	14.7	n.a.	14.8	77.9	33.1	n.a.	0.170	n.a.	n.a.	n.a.	n.a.	7.81	
Vanadium disilicide	VSi_2 [12039-87-1] 107.112	Hexagonal a = 456 pm c = 636 pm C40, *hP9*, $P6_222$, $CrSi_2$ type (Z = 3)	5100	9.5	n.a.	n.a.	n.a.	1699	n.a.	n.a.	11.2	n.a.	n.a.	n.a.	n.a.	n.a.	n.a.	n.a.	n.a.	13.73	

Vanadium silicide	V_3Si [12039-76-8] 180.9085	Cubic $a = 471$ pm A15, *cP8*, Pm3n, Cr_3Si type ($Z = 2$)	5740	203	n.a.	n.a.	n.a.	1732	n.a.	n.a.	8.0.	n.a.	n.a.	n.a.	n.a.	n.a.	n.a.	n.a.	n.a.	14.71	
Zirconium disilicide	$ZrSi_2$ [12039-90-6] 147.395	Orthorhombic $a = 372$ pm $b = 1469$ pm $c = 366$ pm C49, *oC12*, Cmcm, $ZrSi_2$ type ($Z = 4$)	4880	161	n.a.	n.a.	n.a.	1604	n.a.	n.a.	8.6	n.a.	n.a.	n.a.	n.a.	n.a.	n.a.	n.a.	n.a.	10.10	
Oxides																					
Aluminum sesquioxide (alumina, corundum, saphir)	α–Al_2O_3 [1344-28-1] [1302-74-5] 101.961	Trigonal (rhombohedral) $a = 475.91$ pm $c = 1298.94$ pm $D5_1$, *hR10*, R-3c, corundum type (Z=2)	3987	$2 \infty 10^{23}$	9.1–9.8	28–47	0.0005	2054	35.6–39	795.5–880	7.1–8.3	365–416	162–184	234–496	0.231–0.254	206–255	282–1084	2549–3103	3.5–5.0	20.59–29.42 (HM 9)	White and translucent hard material used as abrasive for grinding. Excellent electric insulator and also wear resistant. Insoluble in water and in strong mineral acids, readily soluble in strong alkali hydroxides, attacked by HF or NH_4HF_2. Owing to its corrosion resistance, in inert atmosphere, in molten metals such as Mg, Ca, Sr, Ba, Mn, Sn, Pb, Ga, Bi, As, Sb, Hg, Mo, W, Co, Ni, Pd, Pt, and U it is used as crucible container for these liquid metals. Alumina is readily attacked in an inert atmosphere by molten metals such as Li, Na, Be, Al, Si, Ti, Zr, Nb, Ta, and Cu. Maximum service temperature 1950°C
Beryllium monoxide (beryllia)	BeO [1304-56-9] 25.011	Trigonal (hexagonal) $a = 270$ pm $c = 439$ pm B4, *hP4*, $P6_3mc$, wurtzite type ($Z = 2$)	3008–3030	$1.0 \infty 10^{22}$	6.8–7.66	11.8	0.0004	2550–2565	245–250	996.5	7.5–9.7	296.5–345	n.a.	n.a.	0.340	103.4	241–250	1551	3.68	14.71 (HM 9)	It is the only material with diamond that combines both excellent thermal shock resistance, high electrical resistivity, and high thermal conductivity, and hence is used for heat sinks in electronics. Beryllia is very soluble in water, but slowly in concentrated acids and alkalis. Highly toxic. Exhibits outstanding corrosion resistance to liquid metals Li, Na, Al, Ga, Pb, Ni, and Ir. Readily attacked by molten metals such as Be, Si, Ti, Zr, Nb, Ta, Mo, and W. Maximum service temperature 2400°C.

Table 10.22. *(continued)*

IUPAC name (synonyms, common trade names)	Theoretical chemical formula, [CAS RN], relative molecular mass (^{12}C = 12.000)	Crystal system, lattice parameters, structure type, Strukturbericht, Pearson, space group, structure type (*Z*)	Density (ρ/kg.m^{-3})	Electrical resistivity (ρ/μΩ.cm)	Dielectric permittivity [1MHz] (ε_r / nil)	Dielectric field strength (E_d/MV.m^{-1})	Dissipation or tangent loss factor ($\tan\delta$)	Melting point (*m.p.*/°C)	Thermal conductivity (k/W.m^{-1}.K^{-1})	Specific heat capacity (c_P/J.kg^{-1}.K^{-1})	Coeff. linear thermal expansion (α/10^{-6}K^{-1})	Young's or elastic modulus (*E*/GPa)	Coulomb's or shear modulus (*G*/GPa)	Bulk or compression modulus (*K*/GPa)	Poisson ratio (ν)	Ultimate tensile strength (σ_{UTS}/MPa)	Flexural strength (τ/MPa)	Compressive strength (σ/MPa)	Fracture toughness (K_{IC}/MPa.m$^{1/2}$)	Vickers or Knoop Hardness (HV or HK/GPa) (/HM)	Other physicochemical Properties, oxidation and corrosion resistance, and major uses.
Calcium oxide (calcia, lime)	CaO [1305-78-8] 56.077	Cubic a = 481.08 pm B2, *cP2*, Pm3m, CsCl type (*Z*=1)	3320	1.0 ∞ 10^{14}	11.1	n.a.	n.a.	2927	8–16	753.1	3.88	n.a.	n.a.	n.a.	n.a.	n.a.	n.a.	n.a.	n.a.	5.49 (HM 4.5)	White or grayish ceramics. Readily absorbs CO_2 and water from air to form spent lime and calcium carbonate. Reacts readily with water to give $Ca(OH)_2$. Volumic expansion coefficient 0.225 × 10^{-9}.K^{-1}. Exhibits outstanding corrosion resistance to liquid metals Li and Na
Cerium dioxide (ceria, cerianite)	CeO_2 [1306-38-3] 172.114	Cubic a = 541.1 pm C1, *cF12*, Fm3m, fluorite type (*Z* = 4)	7650	10^{10}	n.a.	n.a.	n.a.	2340	n.a.	389	10.6	181	70.3	n.a.	0.311	n.a.	n.a.	589	n.a.	(HM 6)	Pale yellow cubic crystals. Abrasive for polishing glass, interference filters, antireflection coating. Insoluble in water, soluble in H_2SO_4 and HNO_3 but insoluble in diluted acid
Chromium oxide (eskolaite)	Cr_2O_3 [1308-38-9] 151.990	Trigonal (rhombohedral) a = 538 pm, 54°50' $D5_1$, *hR10*, R3c, corundum type (*Z* = 2)	5220	1.3 ∞ 10^{9} (346°C)	n.a.	n.a.	n.a.	2330	n.a.	921.1	10.90	103	n.a.	n.a.	n.a.	n.a.	268	n.a.	3.9	29 (HM >8)	
Dysprosium oxide (dysprosia)	Dy_2O_3 [1308-87-8] 373.00	Cubic $D5_3$, *cI80*, Ia3, Mn_2O_3 type (*Z* = 16)	8300	n.a.	n.a.	n.a.	n.a.	2408	n.a.	n.a.	7.74.	n.a.	n.a.	n.a.	n.a.	n.a.	n.a.	n.a.	n.a.	n.a.	
Europium oxide (europia)	Eu_2O_3 [1308-96-9] 351.928	Cubic $D5_3$, *cI80*, Ia3, Mn_2O_3 type (*Z* = 6)	7422	n.a.	n.a.	n.a.	n.a.	2350	n.a.	n.a.	7.02.	n.a.	n.a.	n.a.	n.a.	n.a.	n.a.	n.a.	n.a.	n.a.	
Gadolinium oxide (gadolinia)	Gd_2O_3 [12064-62-9] 362.50	Cubic $D5_3$, *cI80*, Ia3, Mn_2O_3 type (*Z* =1 6)	7630	n.a.	n.a.	n.a.	n.a.	2420	n.a.	276	10.44	124	n.a.	n.a.	n.a.	n.a.	n.a.	n.a.	n.a.	4.71	
Hafnium dioxide (hafnia)	HfO_2 [12055-23-1] 210.489	Monoclinic [1790°C] a = 511.56 pm b = 517.22 pm c = 529.48 pm C43, *mP12*, $P2_1/c$, baddeleyite type (Z = 4)	9680	5 ∞10^{15}	n.a.	n.a.	n.a.	2900	1.14	121	5.85	57	n.a.	n.a.	n.a.	n.a.	n.a.	n.a.	n.a.	7.65–10.30	Monoclinic (baddeleyite) below 1790°C, tetragonal above 1790°C

Lanthanum dioxide (lanthania)	La_2O_3 [1312-81-8] 325.809	Trigonal (hexagonal) $D5_2$, *hP5*, P3m1), lanthania type (*Z* = 1)	6510	1 ∞ 10^{14} (550°C)	n.a.	n.a.	n.a.	2315	n.a.	288.89	11.9	n.a.	n.a.	n.a.	n.a.	n.a.	n.a.	n.a.	n.a.	n.a.	Insoluble in water, soluble in diluted strong mineral acids
Magnesium monoxide (magnesia, periclase)	MgO [1309-48-4] 40.304	Cubic *a* = 420 pm B1, *cF8*, Fm3m, rock salt type (*Z* = 4)	3581	1.3 ∞ 10^{15}	9.65–9.8	n.a.	n.a.	2852	50–75	962.3	11.52	303.4	117–130	n.a.	0.33–0.36	200–300	441	1300–1379	1.8	7.35 (HM 5.5–6)	White ceramics, with a high reflective index in the visible and near-UV regions. Used as linings in steel furnaces. Crucible container for fluoride melts. Very slowly soluble in pure water but soluble in diluted strong mineral acids. Exhibits outstanding corrosion resistance in liquid metals Mg, Li, and Na. Readily attacked by molten metals Be, Si, Ti, Zr, Nb, and Ta. MgO reacts with water, CO_2, and diluted acids. Maximum service temperature 2400°C. Transmittance of 80% and *n*=1.75 in IR region 7 to 300 μm
Niobium pentaoxide (columbite, niobia)	Nb_2O_5 [1313-96-8] 265.810	Trigonal (rhombohedral) *a* = 211.6 pm *b* = 382.2 pm *c* = 193.5 pm columbite type	4470	5.5 ∞ 10^{12}	n.a.	n.a.	n.a.	1520	n.a.	502.41	n.a.	n.a.	n.a.	n.a.	n.a.	n.a.	n.a.	n.a.	n.a.	14.71	Dielectric used in film supercapacitors. Insoluble in water, soluble in HF and in hot concentrated H_2SO_4
Samarium oxide (samaria)	Sm_2O_3 [12060-58-1] 348.72	Cubic $D5_3$, *cI80*, Ia3, Mn_2O_3 type (*Z* = 16)	7620	n.a.	n.a.	n.a.	n.a.	2350	2.07	331	10.3	183	n.a.	n.a.	n.a.	n.a.	n.a.	n.a.	n.a.	4.30	
Silicium dioxide (silica, α–quartz)	α–SiO_2 [7631-86-9] [14808-60-7] 60.085	Trigonal (rhombohedral) *a* = 491.27 pm *c* = 540.46 pm C8, *hP9*, R-3c, α–quartz type (Z=3)	2202–2650	1∞10^{20}	3.79	50	0.0002	1710	1.38	787	0.55	72.95	29.9	n.a.	0.170	69–276	310	690–1380	0.9–1.2	5.39–12.36 (HM 7)	Colorless amorphous (i.e., fused silica) or crystalline (i.e., quartz) material having a low thermal expansion coefficient and excellent optical transmittance in far UV. Silica is insoluble in strong mineral acids and alkalis except HF, concentrated H_3PO_4, NH_4HF_2, concentrated alkali metal hydroxides. Owing to its good corrosion resistance to liquid metals such as Si, Ge, Sn, Pb, Ga, In, Tl, Rb, Bi, and Cd, it is used as crucible container for melting these metals, while silica is readily attacked in an inert atmosphere by molten metals such as Li, Na, K Mg, and Al. Quartz crystals are piezoelectric and pyroelectric. Maximum service temperature 1090°C

Table 10.22. *(continued)*

IUPAC name (synonyms, common trade names)	Theoretical chemical formula, [CAS RN], relative molecular mass (^{12}C = 12.000)	Crystal system, lattice parameters, structure type, Strukturbericht, Pearson, space group, structure type (*Z*)	Density (ρ/kg.m^{-3})	Electrical resistivity (ρ/μΩ.cm)	Dielectric permittivity [1MHz] (ε_r / nil)	Dielectric field strength (E_d/MV.m^{-1})	Dissipation or tangent loss factor ($\tan\delta$)	Melting point (*m.p.*/°C)	Thermal conductivity (k/W.m^{-1}.K^{-1})	Specific heat capacity (c_p/J.kg^{-1}.K^{-1})	Coeff. linear thermal expansion (α/10^{-6}K^{-1})	Young's or elastic modulus (*E*/GPa)	Coulomb's or shear modulus (*G*/GPa)	Bulk or compression modulus (*K*/GPa)	Poisson ratio (ν)	Ultimate tensile strength (σ_{UTS}/MPa)	Flexural strength (τ/MPa)	Compressive strength (σ/MPa)	Fracture toughness (K_{1C}/MPa.m$^{1/2}$)	Vickers or Knoop Hardness (HV or HK/GPa) (/HM)	Other physicochemical Properties, oxidation and corrosion resistance, and major uses.
Tantalum pentaoxide (tantalite, tantala)	Ta_2O_5 [1314-61-0] 441.893	Trigonal (rhombohedral) columbite type	8200	1.0 ∞ 10^{12}	n.a.	n.a.	n.a.	1882	n.a.	301.5	n.a.	n.a.	n.a.	n.a.	n.a.	n.a.	n.a.	n.a.	0.9	n.a.	Dielectric used in film supercapacitors. Tantalum oxide is a high-refractive-index, low-absorption material used in making optical coatings in the near-UV (350 nm) to IR (8 μm). Insoluble in most chemicals except HF, HF-HNO_3 mixtures, oleum, fused alkali hydroxides (e.g., NaOH, KOH), and molten pyrosulfates
Thorium dioxide (thoria, thorianite)	ThO_2 [1314-20-1] 264.037	Cubic a = 559.52 pm C1, *cF12*, Fm3m, fluorite type (*Z*= 4)	9860	4 ∞ 10^{19}	n.a.	n.a.	n.a.	3390	14.19	272.14	9.54	144.8	94.2	n.a.	0.280	96.5	n.a.	1475	1.07	9.27 (HM 6.5)	Corrosion-resistant container material for molten metals Na, Hf, Ir, Ni, Mo, Mn, Th, and U. Corroded by liquid metals Be, Si, Ti, Zr, Nb, and Bi. Radioactive
Titanium dioxide (Anatase)	TiO_2 [13463-67-7] [1317-70-0] 79.866	Tetragonal a = 379.3 pm c = 951.2 pm C5, *tI12*, $I4_1$/amd, Anatase type (Z = 4) Ti-O: 191pm (2) 195 pm (4) Packing fraction: 70%	3900 [3890]	n.a.	n.a.	n.a.	n.a.	700°C (rutile)	n.a.	n.a.	n.a.	n.a.	n.a.	n.a.	n.a.	n.a.	n.a.	n.a.	n.a.	(HM 5.5–6)	Metastable over long periods of time despite being less thermodynamically stable than rutile. However, above 700°C, the irreversible and rapid monotropic conversion of anatase to rutile occurs. It exhibits a greater transparency in the near-UV than rutile. With an absorption edge at 385 nm, anatase absorbs less light at the blue end of the visible spectrum and has a blue tone
Titanium dioxide (brookite)	TiO_2 [13463-67-7] [12188-41-9] 79.866	Orthorhombic a = 545.6 pm b = 918.2 pm c = 514.3 pm C21, *oP24*, Pbca brookite type, Z = 8 Ti-O: 184 pm – 203 pm	4130					1750												(HM 5.5–6.0)	

Titanium dioxide (rutile, titania)	TiO_2 [13463-67-7] [1317-80-2] 79.866	Tetragonal $a = 459.37$ pm $c = 296.18$ pm C4, *tP6*, P4/mnm rutile type ($Z = 2$) Ti-O: 194.4 pm (4) 198.8 pm (2), packing fraction 77%	4240 [4250]	10^{19}	110–117	769	n.a.	1847	10.4 (// c) 7.4 (⊥ c)	711	7.14	248–282	111	206–282	0.278	69–103	340	800–940	2.8	10.89 (HM 7–7.5)	White solid that exhibits a high refractive index, even higher than that of diamond. Transparent from visible to near-infrared radiation (i.e., 408 nm to 5000 nm). On the blue end of the visible spectrum the strong absorption band at 385 nm renders rutile powder slightly brighter than anatase, explaining its typical yellow undertone. When heated in air to 900°C the powdered material becomes lemon-yellow and exhibits a maximum absorption edge at 476 nm but coloring disappears on cooling. Doped rutile is phototropic, i.e., it exhibits a reversible darkening when exposed to light. Readily soluble in HF and in concentrated H_2SO_4. Reacts rapidly in molten alkali hydroxides and fused alkali carbonates. Corrosion resistant to liquid Ni and Mo. Readily attacked in an inert atmosphere by molten Be, Si, Ti, Zr, Nb, and Ta
Titanium monoxide (hongquiite)	TiO [12137-20-1] 63.6694	Cubic $a = 417$ pm B1, *cF8*, Fm3m rock salt type ($Z = 4$)	4888					1750		628	9.19										Gold-bronze solid. Prepared by mixing stoichiometric amounts of Ti and TiO_2 heated in a Mo-crucible at 1600°C or by the reduction of TiO_2 with H_2 under pressure at 130 atm and 2000°C. Slightly paramagnetic solid with $\chi_m = +88 \times 10^{-6}$ emu
Titanium sesquioxide	Ti_2O_3 [1344-54-3] 143.3382	Trigonal (rhombohedral) $a = 515.5$ pm $c = 1361$ pm $D5_1$, *hR10*, R-3c corundum type ($Z=2$)	4486					1839		679											Dark-violet to purple-violet solid. It can be prepared by mixing stoichiometric amounts of Ti and TiO_2 heated in a Mo-crucible at 1600°C. Slightly paramagnetic solid with $\chi_m = +63 \times 10^{-6}$ emu

Table 10.22. *(continued)*

IUPAC name (synonyms, common trade names)	Theoretical chemical formula, [CAS RN], relative molecular mass (^{12}C = 12.000)	Crystal system, lattice parameters, structure type, Strukturbericht, Pearson, space group, structure type (Z)	Density (ρ/kg.m^{-3})	Electrical resistivity (ρ/μΩ.cm)	Dielectric permittivity [1MHz] (ε_r / nil)	Dielectric field strength (E_d/MV.m^{-1})	Dissipation or tangent loss factor ($\tan\delta$)	Melting point ($m.p.$/°C)	Thermal conductivity (k/W.m^{-1}.K^{-1})	Specific heat capacity (c_P/J.kg^{-1}.K^{-1})	Coeff. linear thermal expansion (α/10^{-6}K^{-1})	Young's or elastic modulus (E/GPa)	Coulomb's or shear modulus (G/GPa)	Bulk or compression modulus (K/GPa)	Poisson ratio (ν)	Ultimate tensile strength (σ_{UTS}/MPa)	Flexural strength (τ/MPa)	Compressive strength (σ/MPa)	Fracture toughness (K_{1C}/MPa.m$^{1/2}$)	Vickers or Knoop Hardness (HV or HK/GPa) (/HM)	Other physicochemical Properties, oxidation and corrosion resistance, and major uses.
Trititanium pentoxide (anasovite)	Ti_3O_5 [12065-65-5] 223.0070	Dimorphic (120°C) Low temperature: anasovite type monoclinic C2/m (Z = 4) mC32 a = 975.2 pm b = 380.2 pm c = 944.2 pm β = 91.55° High temperature: pseudobrookite orthorhombic Cc mC32	4900					1777													Dark blue crystals. Anasovite Type II is similar to that identified in titania slags.[29] Can be stabilized at room temperature with a small amount of iron
Uranium dioxide (uraninite)	UO_2 [1344-57-6] 270.028	Cubic a = 546.82 pm C1, $cF12$, Fm3m, fluorite type (Z= 4)	10,960	3.8×10^{10}	n.a.	n.a.	n.a.	2880	10.04	234.31	11.2	145	74.2	n.a.	0.302	n.a.	n.a.	n.a.	n.a.	5.88 (HM 6–7)	Used in nuclear power reactors as nuclear-fuel-sintered element containing either natural or enriched uranium
Yttrium oxide (yttria)	Y_2O_3 [1314-36-9] 225.81	Trigonal (Hexagonal) $D5_2$, $hP5$, P3m1, lanthania type (Z = 1)	5030	n.a.	n.a.	n.a.	n.a.	2439	n.a.	439.62	8.10	114.5	48.3	n.a.	0.186	n.a.	n.a.	393	0.71	6.86	Yttria is a medium-refractive-index, low-absorption material used for optical coating in the near-UV (300 nm) to IR (12 μm) regions and hence used to protect Al and Ag mirrors. Used for crucibles containing molten lithium
Zirconium dioxide (baddeleyite, monoclinic zirconia)	ZrO_2 [1314-23-4] [12036-23-6] 123.223	Monoclinic a = 514.54 pm b = 520.75 pm c = 531.07 pm 99.23° C43, $mP12$, $P2_1/c$, baddeleyite type (Z = 4)	5850	2.3×10^{10}	n.a.	n.a.	n.a.	2710	n.a.	711	7.56	241	97	n.a.	0.337	n.a.	2068	n.a.	9.2	11.77 (HM 6.5)	Monoclinic zirconia (baddeleyite structure) stable below 1197°C, tetragonal zirconia (rutile structure) stable between 1197 and 2300°C, cubic zirconia (fluorine structure) stable above 2300°C or at lower temperature if stabilized by addition of magnesia, calcia or yttria. Maximum service temperature 2400°C. Zirconia starts to act as an oxygen anion conductor at 1200°C. Highly
Zirconium dioxide [tetragonal zirconia phase (TZP), >1170°C]	ZrO_2 [1314-23-4] 123.223	Tetragonal C4, $tP6$ $P4_2/mnm$, rutile type (Z=2)	5680–6050	7.7×10^{7}	n.a.	n.a.	n.a.	2710	n.a.	n.a.	10–11	200–210	n.a.	n.a.	0.310	n.a.	800–1200	>2900	7–12	12.5	

Zirconium dioxide [yttria-stabilized zirconia (YSZ) with 8–10 mol.% Y_2O_3]	ZrO_2 [1314-23-4] [64417-98-7] 123.223	Cubic C1, *cF12*, Fm3m, fluorite type (*Z*= 4)	6045	n.a.	n.a.	n.a.	n.a.	n.a.	n.a.	n.a.	n.a.	n.a.	n.a.	n.a.	n.a.	n.a.	n.a.	n.a.	n.a.	n.a.	corrosion resistant to molten metals such as Bi, Hf, Ir, Pt, Fe, Ni, Mo, Pu, and V. Strongly attacked by liquid metals Be, Li, Na, K, Si, Ti, Zr, and Nb. Insoluble in water, but slowly soluble in HCl and HNO_3; soluble in boiling concentrated H_2SO_4 and alkali hydroxides but readily attacked by HF
Zirconium dioxide [partially stabilized zirconia (PSZ) with MgO]	ZrO_2 [1314-23-4] [64417-98-7] 123.223	Cubic C1, *cF12*, Fm3m, fluorite type (*Z* = 4)	5800–6045	n.a.	24.7	400–480	n.a.	2710	1.8	400	10.1	200	n.a.	n.a.	0.230	700	690	1850	9.0–9.5	15.69	
Zirconium dioxide TTZ (stabilized Y_2O_3)	ZrO_2 [1314-23-4] [64417-98-7] 123.223	Cubic C1, *cF12*, Fm3m, fluorite type (*Z* = 4)	6045	n.a.	n.a.	n.a.	n.a.	2710	n.a.	n.a.	n.a.	n.a.	n.a.	n.a.	n.a.	n.a.	n.a.	n.a.	9.2–10	n.a.	

[28] Corrosion data in molten salts from: Geirnaert, G. (1970) Céramiques et métaux liquides: compatibilités et angles de mouillages. *Bull. Soc. Fr. Ceram.* **106**, 7–50.

[29] Reznichenko, V.A.; Khalimov, F.B. (1959) Reduction of titanium dioxide with hydrogen. *Titan i Ego Splavy*, **2**, 11–15.

10.8 Further Reading

10.8.1 Traditional and Advanced Ceramics

ALPER, A.M. (ed.) (1970–1971) *High Temperature Oxides, 4 volumes.* Academic, New York.
ARONSSON, B.; LUNDSTROM, T.; RUNDQUIST, S. (1965) *Borides, Silicides, and Phosphides.* Methuen, London.
BILLUPS, W.E.; CIUFOLINI, M.A. (1993) *Buckminsterfullerenes.* VCH, Weinheim.
BLESA, M.A.; MORANDO, P.J.; REGAZZONI, A.E. (1994) *Chemical Dissolution of Metal Oxides.* CRC Press, Boca Raton, FL.
BRADSHAW, W.G.; MATTHEWS, C.O. (1958) *Properties of Refractory Materials: Collected Data and References.* Lockheed Aircraft, Sunnyvale, CA, U.S. Government Report AD 205 452.
BRIXNER, L.H. (1967) *High Temperature Materials and Technology.* Wiley, New York.
FREER, R. (1989) *The Physics and Chemistry of Carbides, Nitrides and Borides.* Kluwer, Boston.
GOODENOUGH, J.B.; LONGO, J.M. (1970) *Crystallographic and Magnetic Properties of Perovskite and Perovskite related Compounds.* Springer, Berlin Heidelberg New York.
KOSOLAPOVA, T.A. (1971) *Carbides, Properties, Productions, and Applications.* Plenum, New York.
MATKOVICH, V.I. (ed.) (1977) *Boron and Refractory Borides.* Springer, Berlin Heidelberg New York.
MATKOVICH, V.I.; SAMSONOV, G.V., HAGENMULLER, P.; LUNDSTROM, T. (1977) *Boron and Refractory Borides.* Springer, Berlin Heidelberg New York.
PIERSON, H.O. (1996) *Handbook of Refractory Carbides and Nitrides: Properties, Characteristics, Processing and Applications.* Noyes, Westwood, NJ.
SAMSONOV, G.V. (1974) *The Oxides Handbook.* Plenum, New York.
SINGER, F.; SINGER, S.S. (1963) *Industrial Ceramics.* Chemical Publishing Company, New York.
STORMS, E.K. (1967) *The Refractory Carbides.* Academic, New York.
TOTH, L.E. (1971) *Transition Metals Carbides and Nitrides.* Academic, New York.
TOROPOV, N.A. (ed.) *Phase Diagrams of Silicates Systems Handbook.* Document NTIS AD 787517.

10.8.2 Refractories

ANDREW, W. (1992) *Handbook of Industrial Refractories: Technology, Principles, Types and Properties.* Noyes, Westwood, NJ.
BANERJEE, S. (1998) *Monolithic Refractories: A Comprehensive Handbook.* World Scientific, Singapore.
CAMPBELL, I.E.; SHERWOOD, E.M. (ed.) (1967) *High-temperature Materials and Technology.* Wiley, New York.
CARNIGLIA, S.L.; BARNA, G.L. (1992) *Handbook of Industrial Refractories Technology: Principles, Types, Properties, and Applications.* Noyes, Park Ridge, NJ.
CHESTERS, J.H. (1974) *Refractories for Iron and Steelmaking.* Metals Society, London.
CHESTERS, J.H. (1973) *Refractories: Production and Properties.* Iron and Steel Institute (ISI), London.
Collective (1984) *Technology of Monolithic Refractories.* Plibrico Japan Company, Tokyo.
JOURDAIN, A. (1966) *La technologie des produits céramiques réfractaires.* Gauthier-Villars, Paris.
KUMASHIRO, Y. (2000) *Electric Refractory Materials.* Marcel Dekker, New York.
LETORT, Y.; HALM, L. (1953) *Produits réfractaires et isolants: nature, fabrication, et utilisation.* Centre d'études supérieures de la sidérurgie (CESS), Metz.
NORTON, F.H. (1968) *Refractories,* 4th ed. McGraw-Hill, New York.
OATES, J.A.H. (1998) *Lime and Limestone: Chemistry and Technology, Production and Uses.* Wiley-VCH, Weinheim.
PINCUS, A.G. (1980) *Refractories in the Glass Industry.* Books for Industry, Glass Industry Magazine, New York.
SCHACHT, C. (2004) *Refractories Handbook.* CRC Press, Boca Raton, FL.
SCHWARZKOPF, P.; KIEFFER, R. (eds.) (1953) *Refractory Hard Metals: Borides, Carbides, Nitrides, and Silicides.* Macmillan, New York.
STORMS, E.K. (1967) *The Refractory Carbides.* Academic, New York.
TAKAMIYA,Y.; ENDO,Y.; HOSOKAWA, S. (1998) *Refractories Handbook.* American Ceramic Society (ACerS) Westerville, OH.

10.9 Glasses

10.9.1 Definitions

Glass is, from a thermodynamic point of view, a ***supercooled liquid***, i.e., a molten liquid cooled at a rate sufficiently rapid to fix the random microscopic organization of a liquid and avoid the crystallization process to operate. Therefore, by contrast with crystallized solids, glasses do not exhibit a clear melting temperature and the structural change is only reported by an inflection in the temperature-time curve. This change is called the ***glass transition temperature.*** As a general rule, glasses are amorphous inorganic solids usually made of silicates, but other inorganic or organic compounds can exhibit a vitreous structure (e.g., sulfides, polymers). As a general rule, commercial glasses are hard but both brittle and thermal-shock-sensitive materials, excellent electrical insulators, optically transparent media, and exhibit for certain particular chemical compositions (e.g., ***Vycor®*** and borosilicated glasses, such as ***Pyrex®***) an excellent corrosion resistance to a wide range of chemicals except hydrofluoric acid, ammonium fluoride, and strong alkali-metal hydroxides and other strong alkalis. Owing to their good transmission in the visible range, glasses are extensively used for optical lenses, sight lenses, and windows. while corrosion-resistant glasses are widely used for cookware and laboratory glassware. The basic components of silicate glasses are silica, SiO_2 (e.g., from siliceous sand), lime, CaO (i.e., from fired limestone, $CaCO_3$), and soda, Na_2O (i.e., from soda ash, Na_2CO_3). Other oxides are used for special purposes such as boric acid (B_2O_3), potash (K_2O), baria (BaO), and lithia (Li_2O), while colored glasses require minute additions of transition-metal oxides (e.g., FeO, Co_2O_3).

Silicate glasses can be grouped into the following categories: **A-glass** (i.e., high alkali or soda-lime), **C-glass** (i.e., chemical resistant), **E-glass** (i.e., calcium alumino borosilicate or borosilicated glasses), and **S-glass** (i.e., high strength magnesium alumino silicate).

10.9.2 Physical Properties of Glasses

See Table 10.23, pages 672–675.

10.9.3 Glassmaking Processes

The majority of industrial glass is produced by continuous melting processes, while batch-type processes are restricted to customized formulations for special purposes.

Large-scale production of industrial glasses utilizes huge melting crucibles with a rectangular shape called glass tanks that are heated from the bottom and sidewalls by natural gas or oil burners; sometimes auxiliary electric heaters immersed in the melt (i.e., booster electrodes) are used to provide additional heat. The temperature of the melt can be as high as 1660°C to ensure the complete melting of alumina-rich raw materials (Ca-feldspars); the specific energy consumption is about 2.8 kWh/kg of glass. Commercial glass tanks can hold up to 1200 tonnes of molten glass. The thick bottom and sidewalls are built with refractory materials, usually mullite bricks, while electrofused alumina-silica-zirconia bricks are used for the inner layer, which is in direct contact with the melt. The vault or cupola is usually made of silica bricks. The glass tank is divided into two distinct sections called the ***melting end*** where the feed (i.e., cullet and raw materials) is introduced, while in the second section, called the ***working or refining end***, the molten glass reaches its working viscosity. The division between the two sections can be either permanent with a refractory barrier or using mobile baffles.

Table 10.23. Physical properties of selected commercial glasses

Glass trade name	Chemical composition (wt.%)	Density (kg.m^{-3})	Young's modulus (E/GPa)	Poisson ratio (ν/nil)	Knoop hardness (HK)[30]	Thermal conductivity (k/W.m^{-1}.K^{-1})	Specific heat capacity (c_P/J.kg^{-1}.K^{-1})	Coefficient linear thermal expansion (0–300°C) (10^{-6}K^{-1})	Strain point (/°C)[31]	Annealing point (°C)[32]	Softening point (°C)[33]	Working point (°C)[34]	Continuous operating temperature (°C)	Refractive index at 589.3nm (n_D/nil)	Relative permittivity at 1MHz (ε_r/nil)	Dielectric field strength (E_d/MV.m^{-1})	Loss factor (tanδ/nil)	Electrical volume resistivity (Ω.m)
Corning® 0080 (light bulb)	$73SiO_2$-$17Na_2O$-5CaO-4MgO-$1Al_2O_3$	2470	71	0.22	465	n.a.		9.35	473	514	696	1005	110	1.512	7.2	n.a.	0.009	$10^{12.8}$
Corning® 0120 (potash soda lead glass)	$56SiO_2$-29PbO-$9K_2O$-$4Na_2O$-$2Al_2O_3$	3050	59	0.22	382			8.95	395	435	630	985		1.560	6.7		0.008	10^{17}
Corning® 0137 (potash soda lead glass)	$52.5SiO_2$-28PbO-$13K_2O$-5SrO-$1Al_2O_3$-$0.5Na_2O$	3180	n.a.	0.22				9.70	436	478	661	977		1.570				
Corning® 0138 (potash soda lead glass)	$54SiO_2$-23PbO-$8K_2O$-$6Na_2O$-5SrO-$2Al_2O_3$-3CaO-2MgO	3020	n.a.	0.22				9.85	450	490	670			1.563				
Corning® 0160 (crystal glass)	$56SiO_2$-31PbO-$8Na_2O$-5SrO-$4K_2O$-$1Al_2O_3$-$1Li_2O$-$1Sb_2O_3$-$1As_2O_3$	3090	n.a.	0.22				9.30	367	405	583			1.569				
Corning® 0281 (glassware)	$74SiO_2$-$14Na_2O$-9CaO-$1Al_2O_3$-$1B_2O_3$-$0.3Sb_2O_3$-$0.1As_2O_3$	2570		0.22				8.7	500	540	719			1.515				
Corning® 0317 (aircraft window)	$61SiO_2$-$17Al_2O_3$-$13Na_2O$-$3K_2O$-3MgO-$1TiO_2$-$1As_2O_3$-0.4CaO	2480	73	0.22				8.7	574	624	871			1.506				
Corning® 0320 (tape reel)	$63SiO_2$-$12Al_2O_3$-$13B_2O_3$-$6Na_2O$-$5Li_2O$	2450		0.22				7.07	463	493	638							
Corning® 0331 (centrifuge tubes)	$66SiO_2$-$21Al_2O_3$-$9Na_2O$-$4Li_2O$-1MgO-$0.2K_2O$	2380		0.22				7.55	510	548								
Corning® 6720 (tableware)	$60SiO_2$-$10Al_2O_3$-10ZnO-$8Na_2O$-5CaO-$2K_2O$-$1B_2O_3$	2410																
Corning® 7570 (high leaded glass)	74PbO-$12B_2O_3$-$11Al_2O_3$-$3SiO_2$	5420	56	0.28	n.a.	n.a.		8.40	342	363	440	558	100	1.860	15		0.0022	10^{17}

Corning® 8078 (ophthalmic glass)	$44SiO_2$-24PbO-21BaO-$6Na_2O$-4ZnO-1CaO-5ZrO2-$3TiO_2$-$2La_2O_3$-$0.1Sb_2O_3$-$0.1As_2O_3$																	
E-Glass (electrical glass)	$54SiO_2$-$14Al_2O_3$-$10B_2O_3$-17.5CaO-4.5MgO	2600	72					5.0					600		6.1			
C-Glass (chemically resistant glass)	$65SiO_2$-$4Al_2O_3$-$5.5B_2O_3$-14CaO-3MgO-$8Na_2O$-$0.5K_2O$																	
S-Glass (high-strength glass)	$65SiO_2$-$25Al_2O_3$-10MgO	2490	87					5.6					760		5.2			
Corning® 9025 (cathodic ray tube panel)	$68SiO_2$-$12B_2O_3$-$4Al_2O_3$-$6K_2O$-$5Na_2O$-$3Li_2O$-$1TiO_2$-$1CeO_2$																	
Corning® 9068 (color TV panel)	$65SiO_2$-10SrO-$9K_2O$-$7Na_2O$-$2Al_2O_3$-2CaO-2BaO-2PbO-1MgO-$1TiO_2$-$1CeO_2$																	
Float glass (soda lime glass)	$74SiO_2$-$15Na_2O$-5CaO-4MgO-$1Al_2O_3$	2530	72	0.23	n.a.	0.937		8.90	514	546	726	n.a.	230	1.523	n.a.	n.a.	n.a.	n.a.
Kimble®EG11	Leaded glass	2850	n.a	n.a.	n.a.			10.80	394	434	626	980	n.a.	1.540	n.a.	n.a.	n.a.	10^9
Macor® (machinable glass)	$46SiO_2$-17MgO-$16Al_2O_3$-$10K_2O$-$7B_2O_3$-4F (55 wt.% fluorophlogopite and 45 wt.% borosilicate glass)	2520	66.9	0.29		1.46	790	9.30					800 (1000)		6.03	9.40	0.0047	10^{16}
Pyrex® 0211 (microsheet glass)	$65SiO_2$-$2Al_2O_3$-5.9CaO-7ZnO-$9B_2O_3$-$7Na_2O$-$7K_2O$-$3TiO_2$	2570	76	0.22	593	0.98		7.38	508	550	720	1008		1.523	6.7		0.005	
Pyrex® 7059 (substrate glass)	$49SiO_2$-25BaO-$15B_2O_3$-$10Al_2O_3$-$1As_2O_3$	2760						4.60	593	639	844			1.533				
Pyrex® 7070	$72SiO_2$-$25B_2O_3$-$1Al_2O_3$-$1K_2O$-$0.5Li_2O$-$0.5Na_2O$	2130	52	0.22	n.a.			3.20	456	496	n.a.	n.a.	230	1.469	4.1	n.a.	0.006	10^{17}
Pyrex®7740 (Labware)	$80.6SiO_2$-$13B_2O_3$-$4Na_2O$-$2.3Al_2O_3$-$0.1K_2O$	2230	76	0.20	418	1.13		3.25	510	560	821	1252	230	1.474	4.6	n.a.	0.005	10^{17}
Pyrex®7789	$81SiO_2$-$13B_2O_3$-$3Na_2O$-$2Al_2O_3$-$1K_2O$	2220	64.3	n.a.	n.a.	n.a.		3.25	510	560	815	n.a.	n.a.	1.474				
Pyrex®7799	$70SiO_2$-$10B_2O_3$-$9Na_2O$-$6Al_2O_3$-2BaO-$1K_2O$-1CaO-0.5MgO-0.5ZnO	2470						6.20	525	560	740							
Pyrex®7800 (pharmaceutical glass)	$72SiO_2$-$11B_2O_3$-$7Na_2O$-$6Al_2O_3$-2CaO-$1K_2O$-1BaO	2340	n.a.	n.a.														

Table 10.23. *(continued)*

Glass trade name	Chemical composition (wt.%)	Density (kg.m^{-3})	Young's modulus (E/GPa)	Poisson ratio (ν/nil)	Knoop hardness (HK)[30]	Thermal conductivity (k/W.m^{-1}.K^{-1})	Specific heat capacity (c_p/J.kg^{-1}.K^{-1})	Coefficient linear thermal expansion (0–300°C) (10^{-6}K^{-1})	Strain point (/°C)[31]	Annealing point (°C)[32]	Softening point (°C)[33]	Working point (°C)[34]	Continuous operating temperature (°C)	Refractive index at 589.3nm (n_D/nil)	Relative permittivity at 1MHz (ε_r/nil)	Dielectric field strength (E_d/MV.m^{-1})	Loss factor ($\tan\delta$/nil)	Electrical volume resistivity (Ω.m)
Pyrex®7913 (Vycor® HT)	96.5SiO_2-3B_2O_3-0.5Al_2O_3	2180	89	0.19	487	0.19		0.552–0.75	890	1020	1530	n.a.	900	1.458	3.8	n.a.	0.0015	10^{17}
Pyrex®plus		2302			438			5.70	467	502	676			1.492				
Robax® (fire-resistant glass)	Pyroceramics	2580	93	0.25	n.a.	1.6	800	0.5			n.a.		680					
Sapphire glass	fused Al_2O_3	3980	379	0.29	1500	16–23		n.a.							9–11		48	10^{18}
Schott®BaK1 (barium crown)		3190	73	0.252	530	0.795	687	8.60						1.5725				
Schott®BK1 (borosilicate crown)		2460	74	0.210	560	1.069	825	8.80						1.5101				
Schott®BK7 (borosilicate crown)		2510	82	0.206	610	1.114	858	8.30						1.5168				
Schott®FK3 (fluoro crown)		2270	46	0.243	380	0.90	840	9.40						1.4650				
Schott®FK5 (fluoro crown)		2450	62	0.232	520	0.925	808	10.00						1.4875				
Schott®FK51 (fluoro crown)		3730	81	0.293	430	0.911	636	15.30						1.4866				
Schott®FK52 (fluoro crown)		3640	78	0.291	400	0.861	716	16.00						1.4861				
Schott®FK54 (fluoro crown)		3180	76	0.286	390	n.a.	n.a.	16.50						1.4370				
Schott®K5 (crown)		2590	71	0.224	530	0.950	783	9.60						1.5225				

Schott®KF9 (crown flint)		2710	67	0.202	490	1.160	490	7.90						1.5474				
Schott®LaK9 (lanthanum flint)		3510	110	0.285	700	0.908	649	7.50						1.6910				
Schott®LF5 (lanthanum flint)		3220	59	0.223	450	0.866	657	10.60						1.5814				
Schott®PK3 (phosphate crown)		2590	84	0.209	640	1.193	779	8.30						1.5254				
Schott®PK50 (phosphate crown)		2590	66	0.235	430	0.772	812	10.30						1.5205				
Schott®PSK3 (heavy phosphate crown)		2910	84	0.226	630	0.990	682	7.30						1.5523				
Schott®SF11 (heavy flint)		4740	66	0.235	450	0.744	450	6.80						1.7845				
Schott®SF63 (heavy flint)		4620	58	0.235	390	0.744	431	9.00						1.7484				
Schott®SK2 (heavy crown)		3550	78	0.263	550	0.776	595	7.10						1.6074				
Schott®ZKN7		2490	70	0.214	530	1.042	770	5.20						1.5085				

[30] Microhardness 100-g-force load
[31] For a dynamic viscosity of $10^{15.5}$Pa.s
[32] For a dynamic viscosity of 10^{13}Pa.s
[33] For a dynamic viscosity of $10^{8.6}$Pa.s
[34] For a dynamic viscosity of 10^{4}Pa.s

Float glass (annealed glass). Historically, two techniques were used to produce sheets of glass. **Flat glass** was obtained by extruding and rolling a softened mass of glass, while **cylinder glass** was obtained by blowing molten glass into a cylindrical iron mold. The ends were cut and removed while a cut was made on the overall length of the cylinder. The cut cylinder was then placed in an oven, where the cylinder bent flat into a glass sheet. In both processes, from an optical point of view, the surfaces were rarely parallel, leading to optical distortions. By contrast, today 90% of the flat glass produced worldwide is obtained by the ***float glass process*** invented in the 1950s by Sir Alastair Pilkington of Pilkington Glass Co. In the float glass process, molten glass exiting a melting furnace is poured onto a bath of molten tin metal. The glass floats on the specular surface of the molten tin and levels out as it spreads along the bath, providing a smooth finish on both sides. The glass cools and slowly solidifies as it travels over the molten tin and leaves the tin bath in a continuous ribbon. The glass is then fire-polished. The finished product has near-perfect parallel surfaces.The only drawback of annealed glass is that upon mechanical stress it breaks into large and sharp pieces that can cause serious injury. For that reason, building codes worldwide prohibit the use of annealed glass where there is a high risk of breakage and injury.

Tempered glass (toughened glass, safety glass). Tempered glass is obtained after applying a thermal tempering process to annealed glass. The glass is cut to the required size and any required processing such as polishing or drilling is carried out before the tempering process begins. The hot glass at 600°C coming from an annealing furnace is placed onto a roller table. The glass is then quenched with forced cold air convection. This rapidly cools the glass surface below its annealing point, causing it to harden and contract, while the inner core of the glass remains free to flow for a short time. The final contraction of the inner layer induces compressive stresses in the surface of the glass balanced by tensile stresses in the body of the glass. This typical pattern of cooling can be observed under polarized light. Tempered glass exhibits typically a mechanical strength six times that of annealed glass and hence it is also called ***toughened glass***. However, this increased mechanical strength has a drawback. Due to the balanced stresses in the glass, any damage to the glass edges will result in the glass shattering into small sized pieces, and for that reason it is also called ***safety glass*** under the tradename ***Securit®***. Therefore, the glass must be cut to size before toughening and cannot be reworked once tempered. Moreover, the toughened glass surface is less hard than annealed glass and more prone to scratching.

Laminated glass. This multilayered composite material was first invented in 1903 by the French chemist Edouard Benedictus, who had been inspired by the breaking resistance of a glass flask coated with a layer of cellulose nitrate. Today, laminated glass is currently produced by bonding two or more layers of ordinary annealed glass together with a plastic interlayer of polyvinyl butyral (PVB). The polymer is sandwiched by the glass, which is then heated to around 70°C and passed through rollers to expel any air pockets and form the initial bond. A typical laminate has a 3-mm layer of glass, 0.38-mm interlayer, and another 3-mm layer of glass. This gives a final product that would be referred to as 6.38 laminated glass. The plastic interlayer keeps the two sheets of glass tightly bound even when broken, and its high strength prevents the glass from breaking up into large sharp pieces. Multiple laminates and thicker glass increase the strength. Bulletproof glass panels, made up of thick glass and several interlayers, can be as thick as 50 mm. The plastic interlayer also gives the glass a much higher acoustic insulation rating due to the damping effect.

10.9.4 Further Reading

BACH, H; NEUROTH, N. (1998) *The Properties of Optical Glass*. Springer, Berlin Heidelberg New York.
EITEL, W. (ed.) (1964–1973) *Silicate Science*, 6 volumes. Academic, New York.
FELTZ, A. (1993) *Amorphous Inorganic Materials and Glasses*. VCH, Weinheim.

Jones, G.O. (1956) *Glass.* Wiley, New York.
Morey, G.W. (1954) *The Properties of Glasses,* 2nd. ed. Reinhold-Van Nostrand, New York.
Shand, E.B. (1958) *Glass Engineering Handbook.* McGraw-Hill, New York.
Stanworth, J.E. (1950) *The Physical Properties of Glasses.* Clarendon, Oxford.
Zarzycky, J. (1981) *Les verres et l'état vitreux.* Masson, Paris.

10.10 Proppants

10.10.1 Fracturing Techniques in Oil-Well Production

Under the constant pressure of the market and the prediction of long-term depletion of oil resources, the oil and gas industry has constantly increased the productivity and injectivity of production wells. For instance, today, by increasing the drilling depth, it is possible to recover oil and natural gas from remote and shallow reservoirs. However, to recover fossil fuels more efficiently from existing or new oil fields exhibiting low original permeability, some additional techniques must be used for better results. Actually, in many cases including severe damage around well-bore, complex beds, layered unconnected reservoirs, or laminated reservoirs with low permeability, the best known techniques for the treatment of reservoir beds are the ***fracturing technologies***, which provide the only stimulation method possible. These techniques will be briefly described in the next several paragraphs. There are basically two methods in the industry to stimulate well production by extensive improvement of the inflow conditions in a reservoir bed[35]: ***hydraulic fracturing*** and, to a lesser extent, ***pressure acidizing***, which is restricted to carbonated rocks (e.g., limestones and dolomites). As mentioned previously, the aim of both types of stimulation is to improve the cumulative production versus time behavior of an oil well.

10.10.1.1 Hydraulic Fracturing

This technique was developed by the oil industry in the 1940s for opening up tight reservoir[36] rocks to improve product recovery. It consists in pumping and injecting a fluid into the production well until the hydrostatic pressure increases to a level sufficient to expand the strata and fracture the rock, which results in the creation of a network of cracks in the rock formation. The pumping rate is high enough to overcome the maximum rate of fluid loss in the medium to be fractured. The fractures produced are generally only a few millimeters wide, but they may be either horizontal or vertical depending on the path of least resistance. With a widening fracture, the oil increasingly migrates into the pore space of the rocks. Therefore, the presence of high-conductivity fractures affects the overall oil mass transfer efficiency, and they have a significant impact on reservoir performance. To insure the success of reservoir-bed treatment, a sufficient depth of penetration must be achieved originating at the well bore and extending into the rock mass. Usually, a penetration depth of 40 to 70 m is common. After the porous rock has been fractured, it is necessary to prop open the newly formed cracks to act against the closure stress in order to facilitate the continued flow of gas and oil and to avoid the catastrophic collapse of reservoir walls due to the elevated surrounding lithostatic pressure. If the cracks were not propped open, they would close under the overburden. The most common technique consist in pumping a slurry made of a mixture of viscous carrier fluid (i.e., frac fluid: water or brines) and solid particulate materials into the

[35] Rischmuller, H. (1993) *Resources of Oil and Gas.* In: *Ullmann's Encyclopedia of Industrial Chemistry,* Vol. A23. VCH, Weinheim, p. 183.

[36] The reservoir is the underground formation where oil and gas has accumulated. It consists of a porous rock to hold the oil or gas, and an impervious cap rock that prevents them to escape.

fractured formation. The particulate materials are called ***propping agent*** or, most commonly, ***proppants***. Note that at the end of the treatment, the proppant slurry is displaced from the well bore and tubing by a clean flush fluid. However, the volume of flush fluid must be accurately determined so as not to release the proppants into the reservoir. In conclusion, proppants are agents that keep cracks open against closure stresses, avoid the collapse of reservoir walls, and insure an efficient mass transfer for both oil and gas while the production well is moved to another location. Note that hydraulic fracturing is used not only in oil and gas industry but also in hydrogeology to improve the performance of aquifers.

10.10.1.2 Pressure Acidizing

This method uses an aqueous acidic solution instead of brines or water for creating unpropped fractures. To do this, a greater increase in permeability than with common hydraulic fracturing is achieved by chemical dissolution of a part of the carbonate matrix along the fracture walls. Obviously, this treatment can only be used in the case of carbonated formations, such as limestones or dolomites. The most common acidic agents are aqueous solutions of hydrochloric acid (5 to 15 wt.% HCl), hydrofluoric acid (1 to 6 wt.% HF), acetic acid, citric acid, and surfactants. The surfactants are used as dispersing agents to promote the dispersion of solids, to improve the wettability of the rock, and to prevent emulsification. To determine whether acidizing or fracturing will yield the greatest economic benefit, the current condition of the well must be known. This includes knowledge of the undamaged production capacity of the reservoir and the type of damage in existence including the severity and cause of the damage. Worldwide, acidizing has about a 50% success rate, which is believed to be due to a lack of knowledge about the true well condition. On the other hand, fracturing can result in substantially greater improvement in productivity than acidizing just to remove the skin damage, but at a much higher cost. Acidizing is a good stimulation candidate in moderate to high permeability reservoirs that show substantial damage after completion. If there is a damaged zone around the well bore where effective permeability is reduced, acidizing can increase productivity by as much as fivefold, depending on the degree and depth of damage. On the other hand, if acidizing is used to increase permeability above the average reservoir effective permeability, very little stimulation benefit results. Increasing near-well-bore permeability by even an order of magnitude results in a less than twofold productivity improvement.

In conclusion, these two techniques increase economically recoverable reserves and improve vertical communication in layered and unconnected reservoirs. Reserves can be increased either by increasing the flow capacity of an uneconomic well or by increasing the drainage radius of a well or by contacting producing layers that are not connected to the well through perforations. Economic benefit can also be obtained by accelerating production from low-permeability reservoirs. Sweep efficiency can be increased by forming line sources or sinks for injection or production.

10.10.2 Proppant and Frac Fluid Selection Criteria

10.10.2.1 Proppant Materials

As a general rule, not all materials can be used as efficient propping agents. Actually, due to the existing geothermal gradient (e.g., 30°C/km) and lithostatic pressure (e.g., 23 MPa/km) in sedimentary basins, a suitable proppant material must withstand both these harsh conditions (i.e., high bottom temperature and elevated pressure) encountered in deep wells (up to 6 km) and insure the good mass transfer of oil and gas. Hence, the critical characteristics and properties that must be taken into account for the proper selection of proppants are listed below:

- high crushing or compressive strength;
- both elevated hardness and fracture toughness;
- high gas and oil permeability;
- narrow particle-size distribution;
- low specific surface area;
- low bulk and tap densities;
- chemical inertness in hot acidic solutions and hot brines;
- good thermal stability;
- good flowability and rheology in frac fluids;
- low abrasiveness;
- low cost and large commercial availability.

10.10.2.2 Frac Fluids

The above-mentioned proppants can only be used successfully when mixed with a viscous fluid to form a pumpable slurry; hence the carrying fluid also plays an important rheological role in the final fracturing job. Many fluids have been used in fracturing operations, including lease crude oil, water, brines, linear gels, foams, cross-linked polymer gels, emulsions, and even carbon dioxide. Most fracturing fluids used today are either linear or cross-linked aqueous polymer gels or foams. These fluids are extremely complex and exhibit rheological properties that are sensitive to several critical operating parameters. Some of their properties, such as their shear rate, are also time and temperature dependent. The most common fracturing fluids used today include guar cross-linked with borate or zirconium, carboxymethyl-hydroxypropyl guar (CMHPG) cross-linked with zirconium, foams of guar-based fluids and nitrogen or carbon dioxide for gas assist, and gelled oils made of phosphate esters and sodium aluminate. When pH adjustment is required, this can be achieved by adding sodium hydroxide (NaOH) or magnesium oxide (MgO) directly to the frac fluid. Other additives include fluid loss additives, breakers, and surfactants. Breakers are typically enzymes or oxidizers for guar-based products. For gelled oil, acids and bases are used to break the association polymer structure including magnesium oxide.

10.10.2.3 Properties and Characterization of Proppants

To select a suitable propping agent, the properties of the material must be characterized according to well-known standards, e.g., the standards edited by the American Petroleum Institute (API) or other professional societies involved in the oil and gas industry (e.g., IoP, IFP, or ASTM). The critical properties that must be measured to qualify a suitable propping agent with the explanation of the critical values are listed in Table 10.24.

10.10.2.4 Classification of Proppant Materials

The materials commonly used as proppants can be grouped into three main categories, listed in Table 10.25. The first proppant material used was rounded silica sand mined from glacial deposits. This material was initially selected owing to both its wide availability near production wells and its low cost, but since the early days several other industrial materials have been selected and used as proppants, and today we observe the increased use of synthetic materials, especially sintered and fused ceramics. The main impetus in focusing on ceramics was driven by the fact that ceramic materials offer suitable properties for use in modern deep wells today.

Table 10.24. Critical properties for proppants

Critical properties		Description	Requirements	Benefits
Mechanical properties	High crushing strength (σ_c/MPa)	The crushing strength is the compressive strength of a material, i.e., its ability to resist compaction or compression under axial load.	$\sigma_c > 35$ MPa	Insures prop of cracks and avoids the collapse of cap rock if crushing strength is above the lithostatic pressure at the given depth.
	Crushing resistance (wt.%)	A series of crushing resistance tests consists in determining the stress at which the proppant material shows excessive generation of fines. Tests are conducted on samples that have been sieved. Four specific stress levels (i.e., 7.5, 10, 12.5, and 15 ksi) are used in the recommended practice (see API RP-61).	Suggested fine limit (API): 12/20 25 wt.% 16/20 25 wt.% 20/40 10 wt.% 40/70 8 wt.% Suggested fine limit (Stim. Lab.) All 5 wt.%	The lower the fine generation, the better the permeability of the reservoir; moreover; this avoids fooling of cavities and porosities.
	Pycnometer density (ρ_{pyc}/kg.m^{-3})	The pycnometer density measures the true skeletal density including closed internal porosity. Its knowledge is required for the determination of the specific surface area. It is usually measured with a helium pycnometer because helium gas will penetrate all open pores and intricate channels.	$\rho_{pyc} < 2800$ kg.m^{-3}	A low-pycnometer-density material allows the use of low-viscosity carrying fluids, leading to both lower power consumption and pumping rates during well injection.
	Bulk density (ρ_{bulk}/kg.m^{-3})	The bulk density corresponds to the mass of proppants that fill a unit volume and includes both proppant and porosity void volume.	$\rho_{bulk} < 1600$ kg.m^{-3}	A low-bulk-density material leads to the use of less mass of proppant for a given volume of reservoir bed or storage tank to fill.
	Tap density (ρ_{tap}/kg.m^{-3})	The tap density corresponds to the volume occupied by a weighed powder after a set number of taps, usually 10, have been applied to the bottom of its container. As such, the tap density provides a measure of the compactability of a powder.	$\rho_{tap} < 1800$ kg.m^{-3}	A low-tap-density material allows one to use less mass of proppant for a given volume of reservoir bed to fill.
	Permeability coefficient (darcy)	Characterizes the volume flow rate of a fluid into a porous medium exhibiting a cross-section area, *A*, and a thickness, *l*, under a given pressure differential ΔP (see standard API RP-61). Conductivity is *kA/l*.	k > 340 darcies[37] P > 6 darcy-ft	A high permeability coefficient insures a good mass transfer of oil and gas into the fractured rock.

[37] The darcy corresponds to the volume flow rate of one cubic centimeter per second of a liquid having a dynamic viscosity of 1 centipoise, which flows through an area of one square centimeter of a porous medium in 1 s when it undergoes a pressure gradient of one atmosphere per centimeter of length. Hence, 1 darcy = $9.869232266 \times 10^{-13}$ m^2.

Table 10.24. *(continued)*

Critical properties		Description	Requirements	Benefits
Size and dimensions[38]	Particle size distribution (PSD)	Distribution of the particle size is determined by sieving according to ASTM C429-82(1996).	425 < D < 850μm (40 < mesh < 20)	A narrow grain size distribution insures a good intercalation of grains into tight fractures and excellent packing ability. The sizes used in hydraulic fracturing are typically in US mesh: 20/40, 16/20, and 16/30. Some 30/50 and 12/20 is used in specialty applications.
	Sphericity and roundness indices (*S*, *RI*)	Dimensionless quantities that measure the ellipsoidal shape and particle smoothness, respectively. It gives the ratio of the smaller diameter to the larger diameter, and the ratio of actual area to theoretical area.	$S > 0.9$ $RI > 0.9$	Beadlike particles allow a good flowability of the slurry during pumping operation.
	Porosity (ε/vol.%)	Dimensionless quantity equal to the void volume fraction, i.e., ratio of volume of voids to the overall volume of material.	ε < 30 vol.%	The lower the porosity, the better the crushing strength, the lower the brittleness.
Chemical properties	Chemical composition	Chemical composition of the dried solid, expressed as oxides or elements, indicates also the oxidation degree of multivalent species (e.g., Fe, Mn, Cr, Co, and V).	No hazardous substances, low Fe(II) and Fe(III).	Soluble hazardous substances must be avoided so as not to contaminate aquifers, and release of iron cations is not recommended for well control.
	Weight loss in acidic media and boiling water	Weight loss during dissolution of a representative sample in an acidic mixture of 12 wt.% HCl and 3 wt.% HF at a given temperature of 100°C and/or in boiling water.	Less than 2 wt.% in acid mixture, and no dissolution in boiling water.	Chemical inertness allows it to resist the dissolution of beads into acidic frac fluids, brines, and corrosive agents used in chemical fracturing (e.g., HCl, HF, etc.).
Other	Maximum operating temperature (°C)	Capability to withstand maximum temperatures encountered in the deepest production wells owing to the usual geothermal gradient encountered in sedimentary basins (i.e., 30°C/km)	At least above 250°C with no phase changes	Helps to avoid creep phenomena and softening of the material under permanent load.
	Price (US$/tonne)	Specific cost, i.e., cost per unit mass of material including raw material cost, production cost, and distribution cost.	Less than 500 US$/tonne	Allows for competitive product on the market.

[38] ASTM F1877-98 – Standard Practice for Characterization of Particles.

Table 10.25. Major materials used as proppants

Material	Description	Characteristics
Rounded silica sand (e.g., Arizona Sand, Badger Mining, Borden, Colorado Silica, Hepworth, Oglebay Norton, Uninim)	Naturally occurring sand usually mined from glacial deposits, sometimes called Ottawa sands. Silica sand is essentially made of quartz (SiO_2) grains.	**Pressure range:** $28 < \sigma_c < 35$ MPa ($4 < \sigma_c < 5$ ksi) **Advantages:** Low cost, low density, wide availability, and excellent chemical resistance in acidic media except those containing free HF **Drawbacks:** Low permeability, low crushing strength, and poor resistance to flow back
Resin-coated sand (e.g., Borden, Santrol, Hepworth)	The first application of a phenolic resin coating was to particles such as silica sand, glass beads, and ceramics. The concept was to inject a partially cured resin-coated proppant into a well and let the elevated bottom hole temperature finish the resin polymerization and bond the coated particles together, forming a down-hole filter. When a proppant is coated with a phenolic formaldehyde resin that is securely attached to the proppant surface by a silane or other coupling agent, the former brittle material becomes crush resistant. Actually, the resin coating helps to provide a smooth and round substrate surface. In addition, it reduces stress between grains and maintains particle integrity. The coated grains are less sensitive to embedment and generate fewer fines. Finally, it increases the chemical resistance of particles.	**Pressure range:** $35 < \sigma_c < 69$ MPa ($7 < \sigma_c < 10$ ksi) **Advantages:** Better resistance to flow back and improved crushing strength **Drawbacks:** Higher tendency to produce dust under high shear conditions (dust explosions)
Ceramics (e.g., Carbo-Ceramics Inc., Norton Alcoa, Sintex Minerals Inc.)	Synthetic materials made by sintering of bauxite and kaolinite clay. After processing, the final material mineralogical composition consists of a mixture of mullite and corundum. Sometimes less common ceramics are also used, e.g., carborundum, stabilized cubic zirconia, other oxides, and silicates.	**Pressure range:** σ_c up to 140 MPa (σ_c up to 20 ksi) **Advantages:** High crushing strength **Drawbacks:** High density, high cost

10.10.2.5 Production of Synthetic Proppants

All ceramic proppant producers use as feedstock essentially bauxite and, to a lesser extent, other industrial minerals with a high alumina content such as kaolin, nepheline syenite, wollastonite, talc, and feldspars. The final spherical shape is obtained by several processing routes currently used in the ceramics industry for producing beads and other particulate materials. The most common of these processes are pelletizing and sintering, atomization, fire polishing, and flame spraying.

Pelletizing and sintering. In this process, which is the most common among ceramic proppant producers, raw material with a high alumina content (i.e., bauxite, or kaolin clay) is ground to a final particle size of several micrometers by ball milling prior to formation of the pellets. Then the material is calcined at 1000°C to drive off moisture and water from hydratation and reground to less than a micrometer to obtain through granulation as high a green density as possible. Afterwards, various pelletizing techniques can be used to agglomerate the material. Usually, ball forming consists in finely grounding the fired material and mixing it in a rotary dryer with water and a binder that gives them temporary cohesion and that does not affect the final strength of the material (e.g., molasses, starch, cellulose gum, polyvinyl alcohol, bentonite, sodium metasilicate, and sodium lignosulfonate). Until the final desired size of green pellets is obtained, the binder is continuously added. Finally, the green pellets are fired with a suitable parting agent (e.g., pure alumina powder) into a kiln between 1200 and 1650°C for a sufficient amount of time for vitrification to occur. After sintering the balls are made of alpha-alumina (i.e., corundum) and mullite grains with a diameter of 50 to 300 μm formed by crystalline growth during sintering together with a glassy phase. Hence sintered bauxite exhibits only a low residual porosity (<5%) because it has shrunk by 20%, and it is dense (>3550 $kg.m^{-3}$) and strong.

Atomization. This technique involves the melting at high temperatures (i.e., above 1800°C) of raw material particles together to obtain a molten bath of bulk liquid. Usually, the bulk liquid contains more than thousands of times the amount of raw material required to make a single product particle. A thin stream of molten material is atomized by dropping it into a disruptive air jet, subdividing the stream into fine, molten droplets. The droplets are then kept away from one another and from other objects until they have been cooled and solidified. Then they can be recovered as substantially discrete ellipsoidal glassy (i.e., amorphous) particles.

Fire polishing. In this techniques, discrete solid particles are heated to the softening or melting temperature of the material, usually between 1200 and 1650°C, while suspended and dispersed in a hot gaseous medium (e.g., fluidized bed). As particles become soft or molten, surface tension forms them into an ellipsoidal shape. If kept in suspension until cooled below softening temperature, the particles may be recovered as spherical grains.

Flame spraying. In this technique, finely ground raw particles are premixed with a combustible gas mixture, i.e., fuel and oxidant, and the mixture is then introduced into a burner. Hence, in the hot flame, the tiny particles melt or soften, and the surface tension leads them to exhibit an ellipsoidal shape. To prevent molten droplets or soften particles from contacting any surface before cooling, the flame must be allowed to move freely in a large combustion chamber. The droplets or softened particles are then kept away from one another and from the reactor walls until they have been cooled and solidified.

It is important to point out that, despite the wide availability of industrial minerals for preparing commercial ceramic proppants (e.g., bauxite, kaolin clay), all the major proppant producers must overcome technical issues for obtaining crush-resistant beads made of tough and hard corundum or mullite minerals. Actually all the above-mentioned processes require at least a comminution step and also a firing or sintering step conducted at high temperature. Hence, the overall process for obtaining suitable ceramic proppants is always energy demanding and accounts for 60% of the cost of the final product. Hence, the process represents a major pitfall for obtaining competitive ceramic proppants at low prices.

10.10.2.6 Properties of Commercial Proppants

The properties of most common commercial proppants measured according to standardized tests are summarized in Table 10.26.

Table 10.26. Properties of commercial proppants

	Proppant (trade name, producer)	Particle size distribution[39] (mesh)	Bulk and pycnometer densities (ρ/kg.m^{-3})		Crushed fraction (wt.%)	Weight loss HCl-HF 60°C (/wt.%)	Weight loss boiling water (wt.%)	Permeability (121°C) (k/D)	Conductivity (121°C) (k/D-ft)	Roundness index (RI)	Sphericity	Price (US$/tonne)
	Rounded silica sand (Oglebay Norton)	8/16, 12/20, 16/30, **20/40**	1540	2650	2.0 ksi 0.68 3.0 ksi 1.99 4.0 ksi 6.70	0.9	Traces	2 ksi 249 4 ksi 119 6 ksi 58 8 ksi 0	2 ksi 4.592 4 ksi 2.051 6 ksi 0.952 8 ksi 0.000	0.6	0.9	50
Resin-coated sand	Precured resin-coated sand (Tempered Econoflex®, Santrol)	12/20, 16/30, **20/40** (620μm)	1490	2590	2 ksi 0.00 4 ksi 0.10 6 ksi 0.10 8 ksi 0.20 10 ksi 0.50 12 ksi 1.40	<0.3	Traces	4 ksi 263 6 ksi 248 8 ksi 210 10 ksi 144 12 ksi 107	4 ksi 5.049 6 ksi 4.679 8 ksi 3.881 10 ksi 2.564 12 ksi 1.811	0.9	0.9	200
	Precured resin-coated sand (Tempered HS®, Santrol)	12/20, 16/30, **20/40** (730μm)	1530	2530	6 ksi 0.40 8 ksi 0.60 10 ksi 2.30	<0.3	Traces	2 ksi 328 4 ksi 249 6 ksi 178 8 ksi 103 10 ksi 61	2 ksi 6.743 4 ksi 4.302 6 ksi 3.011 8 ksi 1.753 10 ksi 995	0.8	0.8	200
	Precured resin-coated sand (Tempered DC®, Santrol)	12/20, 16/30, **20/40** (710μm)	1540	2570	2 ksi 0.10 4 ksi 0.30 6 ksi 0.70 8 ksi 1.20 10 ksi 5.10	<0.3	Traces	2 ksi 301 4 ksi 229 6 ksi 166 8 ksi 89 10 ksi 46	2 ksi 6.432 4 ksi 4.038 6 ksi 3.033 8 ksi 1.446 10 ksi 742	0.8	0.8	200
	Precured resin-coated sand (Tempered LC®, Santrol)	12/20, 16/30, **20/40** (720μm)	1530	2600	2 ksi 0.20 4 ksi 0.40 6 ksi 1.90 8 ksi 4.50 10 ksi 10.70	<0.3	Traces	2 ksi 272 4 ksi 201 6 ksi 115 8 ksi 59 10 ksi 31	2 ksi 5.034 4 ksi 3.629 6 ksi 2.014 8 ksi 987 10 ksi 503	0.8	0.8	200

Resin-coated sand	Precured resin-coated sand (Tempered TF®, Santrol)	12/20, 16/30, **20/40** (570µm)	1540	2600	2 ksi 0.40 4 ksi 1.30 6 ksi 2.50 8 ksi 5.30 10 ksi 13.00	<0.3	Traces	2 ksi 146 4 ksi 117 6 ksi 68 8 ksi 29 10 ksi 16	2 ksi 2.804 4 ksi 2.072 6 ksi 1.180 8 ksi 472 10 ksi 254	0.8	0.8	200
Sintered ceramics	Lightweight (Naplite®, Norton-Alcoa)	12/18, 16/20, **20/40** (721µm)	1560	2600	5.0 ksi 0.5 7.5 ksi 2.4 10 ksi 7.7	3.2	Traces	2 ksi 465 4 ksi 401 6 ksi 300 8 ksi 181 10 ksi 111	2 ksi 9.507 4 ksi 7.999 6 ksi 5.840 8 ksi 3.434 10 ksi 2.055	0.9	0.9	422–556
	Lightweight (Valuprop®, Norton-Alcoa)	**20/40** (657µm), 30/50	1560	2600	5.0 ksi 1.2 7.5 ksi 4.0	3.2	Traces	2 ksi 356 4 ksi 302 6 ksi 212 8 ksi 150 10 ksi 106	2 ksi 6.936 4 ksi 5.723 6 ksi 3.911 8 ksi 2.683 10 ksi 1.843	0.9	0.9	422–556
	Low-cost and lightweight ceramics (Carbo-Econoprop®, CCI)	**20/40** (635µm), 30/50	1560	2700	5.0 ksi 0.8 7.5 ksi 2.8	1.7	Traces	2 ksi 340 4 ksi 320 6 ksi 220 8 ksi 130 10 ksi 70	2 ksi 3.300 4 ksi 2.900 6 ksi 2.000 8 ksi 1.100 10 ksi 0.500	0.9	0.9	422–556
	Lightweight and intermediate-strength ceramics (CarboLite®, CCI)	12/18, 16/20, **20/40** (730µm)	1570	2710	10.0 ksi 5.2 12.5 ksi 8.3	1.7	Traces	2 ksi 570 4 ksi 480 6 ksi 340 8 ksi 210 10 ksi 120	2 ksi 10.70 4 ksi 8.900 6 ksi 6.000 8 ksi 3.700 10 ksi 2.000	0.9	0.9	333–480
	Medium-density (Interprop®, Norton-Alcoa)	16/30, **20/40** 30/50	1880	3200	5.0 ksi 0.3 7.5 ksi 0.8 10 ksi 2.2 12.5 ksi 4.0	2.5	Traces	2 ksi 485 4 ksi 415 6 ksi 335 8 ksi 235 10 ksi 160 12 ksi 110	2 ksi 7.830 4 ksi 6.585 6 ksi 5.230 8 ksi 3.615 10 ksi 2.375 12 ksi 1.720	0.9	0.9	422–556

[39] ASTM C429-96 – Standard Test Method for Sieve Analysis of Raw Materials for Glass Manufacture.

Table 10.26. *(continued)*

	Proppant (trade name, producer)	Particle size distribution[39] (mesh)	Bulk and pycnometer densities (ρ/kg.m^{-3})		Crushed fraction (wt.%)	Weight loss HCl-HF 60°C (/wt.%)	Weight loss boiling water (wt.%)	Permeability (121°C) (k/D)	Conductivity (121°C) (k/D-ft)	Roundness index (RI)	Sphericity	Price (US$/tonne)
Sintered ceramics	Intermediate-strength ceramics (CarboProp®, CCI)	16/30, **20/40** (658μm), 30/60	1880	3270	10.0 ksi 2.2 12.5 ksi 5.1	4.5	Traces	2 ksi 385 4 ksi 345 6 ksi 290 8 ksi 250 10 ksi 200 12 ksi 150 14 ksi 100	2 ksi 6.180 4 ksi 5.430 6 ksi 4.450 8 ksi 3.720 10 ksi 2.890 12 ksi 2.145 14 ksi 1.445	0.9	0.9	333–480
	Sintered bauxite (Norton-Alcoa)	16/30, **20/40** (662μm), 30/50	2020	3450	7.5 ksi 0.4 10 ksi 1.1 12.5 ksi 2.2 15 ksi 4.0	2.0	Traces	2 ksi 472 4 ksi 408 6 ksi 352 8 ksi 299 10 ksi 217 12 ksi 173 14 ksi 141	2 ksi 4 ksi 6 ksi 8 ksi 10 ksi 12 ksi 14 ksi	0.9	0.9	422–556
	High-strength sintered bauxite ceramics (CarboHSP® 2000, CCI)	12/18, 16/20, 16/30, **20/40** (717μm), 30/60	2040	3560	10.0 ksi 0.7 12.5 ksi 1.4 15.0 ksi 2.7	3.5	Traces	2 ksi 539 4 ksi 440 6 ksi 370 8 ksi 302 10 ksi 246 12 ksi 204 14 ksi 166	2 ksi 8.168 4 ksi 6.595 6 ksi 5.368 8 ksi 4.283 10 ksi 3.405 12 ksi 2.719 14 ksi 2.140	0.9	0.9	650

10.10.2.7 Proppant Market

The world annual consumption of proppants is estimated at about 2.2 million tonnes per year. The international market share has been roughly 65% North Sea, 15% Africa, 10% Middle East, 5% Far East, and 5% South America including Mexico. As a general rule, the majority of fracturing jobs use rounded silica sand (i.e., Ottawa sand), which is the most popular fracturing proppant used in oil-well production owing to its low price (68 US$/tonne) and because of its large geological availability. Nevertheless, newly developed sintered and fused synthetic ceramic proppants represent today roughly 22% of the previous market (ca. 196,000 tonnes/year). Actually, over the last 20 years, ceramic proppants, despite their high cost of 650–750 US$/tonne, have become increasingly popular due to their improved properties, especially their high crushing strength, near-spherical shape, and excellent chemical inertness. On the other hand, resin-coated sands, which are priced 360 US$/tonne, fill the gap between sand and ceramics. The leading producing countries of ceramic proppants are mainly the United States and, to a lesser extent, Brazil. The leading company is Carbo-Ceramics (CCI), which produces 60% of all the ceramic proppants consumed worldwide. CCI is followed by Norton Alcoa, also located in the USA, the third company being Sintex in Brazil, which sells a sintered bauxite under the trade name Sinterball.

Table 10.27. Worldwide production of proppants (2006)

Proppant material	Worldwide production (tonnes/annum)
Sand	2,000,000
Resin-coated sand	180,000
Synthetic ceramics	200,000
Total =	2,380,000

10.10.2.8 Proppant Producers

Table 10.28. Ceramic proppant producers

Producer	Proppant trade names
Carbo Ceramics 6565 MacArthur Blvd., Suite 1050, Irving, TX 75039, USA Telephone: +1 (972) 401-0090 Fax: +1 (972) 401-0705 URL: http://www.carboceramics.com/	- Carbo HSP - Carbo Prop - Carbo Lite - Ceramax P - Ceramax I+SI - Ceramax I - Ceramax E
Norton Alcoa 11300 North Central Expressway, Suite 402, Dallas, TX 75243, USA Telephone: +1 (972) 233-0661 Fax: +1 (972) 233-1409 URL: http://www.nortonalcoa.com/	- Sintered bauxite - Inter Prop - Nap Lite - Valu Prop
Sintex Minerals & Services (Brazil) 29810 Southwest Freeway, Rosenberg, TX 77471, USA URL: http://www.sintexminerals.com/	- SinterBall - SinterProp

Table 10.29. Resin-coated sand producers

Producer	Proppant trade names
Borden Chemicals 180 E. Broad Street , Columbus, OH 43215-3799, USA Telephone: +1 (614) 225-4127 E-mail: anschutzda@bordenchem.com URL: http://www.bordenchem.com/	– Diamond Flex – Econo Flex- AcFrac CR – AcPack – SB Prime – PR-6000 – SB-Excel – PR-6000 – CR-4000
Santrol P.O. Box 639, 2727 FM 521, Fresno, TX 77545, USA Telephone: +1 (281) 431-0670 Fax: +1 (281) 431-0044 E-mail: sales@santrol.com URL: http://www.santrol.com/	AcFrac Black – Tempered HS – Tempered DC – Tempered LC – Tempered TF – Super HS – Super LC – Super TF- Super DC

Table 10.30. Rounded silica sand producers

Producer	Sand trade name
Fairmount Minerals Chardon, OH, USA E-mail: sales@fairmountminerals.com URL: http://www.fairmount-minerals.com/	Ottawa sand
Oglebay Norton Industrial Sands Phoenix Office, Suite 320, 410 North 44th Street, Phoenix, AZ 85008, USA Telephone: +1 (602) 389-4399 Fax: +1 (602) 389-4389 URL: http://www.oglebaynorton.com/	Brady sand
Unimin Corp. 258 Elm Street, New Canaan, CT 06840-5300, USA Telephone: (800) 243-9004 Fax: (800) 243-9005	Jordan sand
Wedron Silica Division of Fairmount Minerals, South Olive Street, Wedron, IL 60557, USA Telephone: +1 (815) 433-2449 Fax: +1 (815) 433-9393 E-mail: sales@wedronsilica.com URL: http://www.wedronsilica.com/	Round silica sand

Table 10.31. Proppant testing laboratories

Stim-Lab P.O. Box 1644, Duncan, OK 73534-1644, USA Telephone: +1 (580) 252-4309 Fax: +1 (580) 252-6979 E-mail: stimlab@stimlab.com URL: http://www.stimlab.com/
FracTech Unit 204, Bedfont Lakes Industrial Park, Challenge Road, Ashford, Middlesex, TW15 1AX, UK Telephone: (+44) (0) 1784 244100 Fax: (+44) (0) 1784 250544 URL: http://www.fractech.co.uk/
Fracture Technologies Aaron House, 6 Bardolph Rd, Richmond, Surrey, TW9 2LS, UK Telephone: (+44) (0) 20 8288 7405 Fax: (+44) (0) 20 8287 1802 E-mail: sales@wellwhiz.com URL: http://www.wellwhiz.com/

10.10.3 Further Reading

API Recommended Practice RP-56 (1998) *Recommended Practices for Evaluating Sand used in Hydraulic Fracturing Operations*. American Institute of Petroleum Engineers, Washington, D.C.

API Recommended Practice RP-60 (1995) *Recommended Practices for Testing High-Strength Proppants Used in Hydraulic Fracturing Operations*. American Institute of Petroleum Engineers, Washington, D.C.

ECONOMIDES, M.J.; NOLTE, K.G. (2000) *Reservoir Stimulation*, 3rd ed. Wiley, New York.

ELY, J.W. (1994) *Stimulation Engineering Handbook*. Pennwell.

GIDLEY, HOLTDICTH, NIERODE, and VEATCH – Recent Advances in Hydraulic Fracturing. Society of Petroleum Engineers (SPE) Monograph Vol. 12.

GIDLEY, J.L. (1990) Recent Advances in Hydraulic Fracturing (No. 30412). Society of Petroleum Engineers.

MADER, D. (1989) *Hydraulic Proppant Fracturing and Gravel Packing*. In: Developments in Petroleum Science, Vol. 26. Elsevier, Amsterdam.

MCDANIEL, R.R. – *The effect of various proppants and proppants mixtures on fracture permeability*. – Society of Petroleum Engineers of AIME Paper No. SPE 7573, Houston, TX.

(1978)SMITH, S.A. (1989) *Manual of Hydraulic Fracturing for Well Stimulation and Geologic Studies*. National Groundwater Association.

VALKO, P.; ECONOMIDES, M.J. (1995) *Hydraulic Fracture Mechanics*. Wiley, London.

YEW, C.H. (1997) *Mechanics of Hydraulic Fracturing*. Gulf Professional Publishing.

11 Polymers and Elastomers

11.1 Fundamentals and Definitions

Historically, the plastics industry originated in 1869 with the first commercial production of celluloid and later in 1907 with the production of the first phenol-formaldehyde resins. Today polymers and elastomers are an important class of materials in both consumer goods and industrial applications.

11.1.1 Definitions

Macromolecules are long organic or inorganic molecules with a high relative molecular mass[1], the structure of which comprises the multiple repetition of building units of low relative molecular mass called ***monomers*** or ***mers*** (formerly ***residues***) and obtained from them by a particular chemical reaction called ***polymerization***. Polymers and elastomers are natural or synthetic organic ***macromolecules*** obtained either by biological or chemical processes. Polymers are usually made of a long linear chain of thousands of monomers but the monomers can also be linked together by cross linking or reticulation to form large bi- or tri-dimensional molecular structures. A general classification of polymers both natural and synthetic is given in Table 11.1. In this chapter, only natural and synthetic macromolecules having industrial applications will be described. Polymers can be classified into three main groups:

---

[1] As a rule of the thumb, a molecule can be regarded as having a high relative molecular mass if the addition or the removal of one or a few units has a negligible effect on the molecular properties.

Table 11.1. Classification of natural and synthethic polymers

Macromolecules	Origin	Groups
Polymers	Natural	Polypeptides, proteines, polynucleotides
		Polysacharides
		Gums and resins
		Elastomers
	Synthetic	Elastomers
		Thermoplastics
		Thermosets

(i) ***Thermoplastics*** are the simplest type of polymers. They consist of long, unconnected chains of monomers. Due to the high degree of freedom of the macromolecule structure, thermoplastics soften with increasing temperature but they recover their original strength when cooled. Most of thermoplastics are meltable and hence they can be easily extruded into long fibers or filaments from the molten state.

(ii) ***Thermosetting plastics*** or simply ***thermosets*** are particular polymers produced by polymerization of soft or viscous monomers that irreversibly set into an infusible and insoluble polymer 3D network by curing. Curing is usually obtained by heating or by electron-beam irradiation or both. Hence after curing, they set into a permanent shape irreversibly and they retain their strength and shape when cooled. Because reheating of these polymers leads to irreversible degradation they cannot be reworked from scrap.

(iii) ***Elastomers*** or ***rubbers*** consist of long coiled linear chains resulting from the introduction of a *cis*-bond in the chain inducing an asymmetry in the polymer sequence. Therefore, elastomers can withstand high elastic deformation without rupture.

11.1.2 Additives and Fillers

In addition to monomers, polymers can also incorporate in their structures ***additives*** and ***fillers*** that enhance some specific properties of the polymer. Usually, additives are chemical compounds that affect chemical properties. The major additives categories are:

(i) **Dyes** impart a given color to the polymer and also improve the resistance to sunlight. The dye can be soluble or insoluble in the polymer matrix or it can also take part entirely into the macromolecule (i.e., chromophoric monomer).

(ii) **Plasticizers** are small organic molecules that replace some monomer units in the macromolecule to allow a greater degree of freedom to the long chain, decreasing its mechanical rigidity and hence enhancing plasticity. Plasticizers are added to:

- improve workability during fabrication;
- extend the original properties of the resin;
- enhance new properties.

Therefore, plasticizers must exhibit a good chemical affinity to the polymer, and they should not be easily removable or volatile to insure good stability of the polymer. Usually, plasticizers are organic esters made from alcohol and ortho-phosphoric acid or organic acids (e.g., phthalic, stearic, adipic). Diisooctyl phthalate is a common plasticizer.

(iii) **Solvents** are organic liquids that dissolve the polymer to give a homogeneous solution for specific applications such as paints and varnishes.

(iv) **Lubricants** are chemicals such as alkali-metal stearates (e.g., lithium stearate) that enhance the good formability of the polymer, for instance during extruding or cold-molding. They can be incorporated into the polymer matrix or added by spraying during the process.

(v) **Stabilizers** are additives that prevent the chemical degradation of the polymer against several external factors (e.g., UV radiation, heat load, free radicals, oxidation, chemical attack, etc.). 2-Hydroxybenzophenones are the largest and most versatile class of ultraviolet stabilizers used in polymers.

(vi) **Antioxidants** are chemicals that slow down the autoxidation of the polymer. Antioxidants usually exhibit amine or hydroxyl functional group such as secondary aryl amines or phenols. They prevent oxidation by transferring hydrogen to free peroxy radicals. Butylated hydroxytoluene, tris(nonylphenyl) phosphite, distearyl pentaerythritol diphosphite, and dialkyl thiodipropionates are typical examples.

Fillers are different from additives, in the sense that they are only mixed with the polymer matrix to impart specific physical properties instead of chemical properties such as mechanical strength, electrical conductivity or dielectric properties, thermal insulation and they never take part in the macromolecule chain. Because fillers together with the polymer matrix are considered as composite materials they will be discussed in detail in the corresponding chapter. The major categories of filler, with some example, are however given briefly here:

(i) **conductors** that impart a good electrical conductivity and prevent electrostatic discharge (e.g., powders of graphite and copper);

(ii) **fire-retardants materials** impart a low flammability rating for fireproofing materials (e.g., aluminum trihydrate (ATH), antimony compounds, borates, and gypsum);

(iii) **reinforcement materials** enhance mechanical strength and stiffness to the polymer.

11.1.3 Polymerization and Polycondensation

Polymerization. A chemical reaction that consists of the formation of a macromolecule from individual unsaturated monomers containing a double bond. The polymerization can be activated either by pressure, temperature or by a catalyst – a compound, present in minute amounts, that accelerates a chemical reaction (i.e., it enhances the rate constant and hence the hourly yield) but does not participate in the overall reaction scheme (i.e., the equilibrium constant remains the same). Several mechanisms can be involved during polymerization.

Polymerization by addition. A reaction characterized by the stoichiometric combination of residues from one or several monomers each containing at least one unsaturated (i.e., double or triple) chemical bond in their molecules such as is found in hydrocarbons like the alkenes and alkynes. The overall chemical reaction can be sketched as follows:

$$(n/2)\ R = R \longrightarrow -(R-R-\ldots-R-\ldots-R-R)_n- = -[R]_n-$$

where the subscript n denotes the degree of polymerization. The intermediate products of the reaction are either free radicals or ions and hence this type of polymerization is referred to as polymerization by free radicals or by ions. **Polymerization by free-radicals** involves free radicals, that is, unstable chemical species R with an unpaired electron which are denoted by the species with a dot $R^{\cdot}$. Free radicals can be produced either by a chemical reaction such as the decomposition of organic peroxides (e.g., benzoyl peroxide) or by the action of UV light (i.e., photolysis) or by ionizing radiation (i.e., radiolysis) on molecules. Polymerization by free radicals usually proceeds in three steps.

(i) **Initiation** or **activation** which consists of the production of free radicals by reacting an organic peroxide and a monomer molecule:

$R^{\cdot} + M \longrightarrow RM^{\cdot}$

(ii) **Propagation** of the chain, where a chain reaction occurs between the free radical and monomers until the last molecule of monomer has reacted.

$RM^{\cdot} + M \longrightarrow RM^{\cdot}_{2}$

$RM^{\cdot}_{2} + M \longrightarrow RM^{\cdot}_{3}$

... ...—>........

$RM^{\cdot}_{n-2} + M \longrightarrow RM^{\cdot}_{n-1}$

(iii) **Termination**, which consists of the completion of the reaction by combination of last radical monomers to form the final macromolecule

$RM^{\cdot}_{n-1} + M \longrightarrow RM^{\cdot}_{n}$

Ionic polymerization involves ions.

Polycondensation. A chemical reaction (e.g., esterification) occurring between two monomers with complementary function, for instance alcohols and carboxylic acids, with the removal of a small molecule (e.g., water, hydroacid, ammonia, formol, etc.) as a by-product.

Esterification: $R_1—CO—OH + HO—R_2 \longrightarrow R_1—\mathbf{CO—O}—R_2 + H_2O$
$R_1—CO—OH + X—R_2 \longrightarrow R_1—\mathbf{CO—O}—R_2 + HX$

Peptide formation: $R_1—CO—OH + H_2N—R_2 \longrightarrow R_1—\mathbf{CO—NH}—R_2 + H_2O$

Copolymerization. Copolymerization involves two or more different monomers, and the resulting macromolecule is named a dipolymer, terpolymer, etc.

11.2 Properties and Characteristics of Polymers

11.2.1 Molar Mass and Relative Molar Mass

The ***molecular molar mass*** or simply the ***molar mass,*** denoted M, corresponds to the mass per amount of substance of the macromolecule. For a macromolecule containing a number n of monomer units or residues R its molar mass is simply given by:

$$M = n \times M_R$$

with M molecular molar mass of the polymer in $kg.mol^{-1}$,
n dimensionless number of monomers or residues,
M_R molecular molar mass of the monomer in $kg.mol^{-1}$.

Note that in some old textbooks dealing with macromolecular chemistry and polymer science, the molar masses were expressed in an old unit called the Dalton (Da) which is the old name for the atomic mass unit ($1 \text{ u} = 1.66054021 \times 10^{-27}$ kg).

The ***relative molecular molar mass*** or simply the ***relative molar mass***, denoted M_r (formerly the molecular weight, *MW*) is a dimensionless quantity that corresponds to the ratio of the mass of the macromolecule to 1/12 of the mass of an atom of the nuclide ^{12}C. It is the most important property of a macromolecule, because it is directly related to its physical properties such as mechanical, thermal and electrical properties. For instance, for a simple macromolecule —$[R]_n$— made of n monomer units or residues, R, its relative molar mass is simply given by:

$$M_r = n \times M_R$$

with
M_r relative molar mass of the polymer,
n degree of polymerization,
M_R relative molar mass of the monomer.

11.2.2 Average Degree of Polymerization

The ***average degree of polymerization***, denoted X_k, were k represent the type of average (i.e., arithmetic, geometric), represents the dimensionless number of monomer units or residues R constitutive of a macromolecular chain. It can be defined as the ratio of the relative molar mass of the macromolecule to the relative molar mass of the monomer unit:

$$X_k = M_r/M_R$$

Depending of the degree of polymerization, it is possible to distinguish several subgroups of polymers listed in Table 11.2.

Table 11.2. Degree of polymerization and polymer subgroups

Degree of polymerization	Subgroups	Examples
$2 < X_k < 10$	Oligomers	Polysacharides, polypeptides
$10 < X_k < 100$	Low-mass polymers	
$100 < X_k < 1000$	Medium-mass polymers	
$1000 < X_k$	High-mass polymers	Most of the commercial plastics

11.2.3 Number-, Mass- and Z-Average Molar Masses

During polymerization, the length of each macromolecule produced is determined entirely by random events. Therefore, the random nature of the growth process requires that the final polymer consists of a mixture of chains of macromolecules having a different length and hence a different relative molar mass. Consequently, a polymer is better defined by a statistical distribution that plots the number fraction of the polymer having a given molar mass rather than a definite molar mass. Therefore, polymer scientists soon introduced several quantities called molar-mass averages. In practice, the most important molar mass averages are defined by simple moments of the distribution functions.

The first quantity is a colligative property called the ***number-average relative molar mass*** or simply the ***number-average molar mass*** denoted M_n and defined as the weighed average of the number fraction of each macromolecule by the following equation.

$$M_n = \Sigma_i N_i M_i / \Sigma_i N_i = \Sigma_i n_i M_i = \Sigma_i w_i / \Sigma_i (w_i / M_i)$$

with
M_n averaged relative molar mass,
N_i number fraction of the macromolecule i,
M_i relative molar mass of the macromolecule i.

Because the number-average molar mass is a ***colligative property***, it can be determined experimentally by techniques which are able to count the number of macromolecules such as osmotic pressure measurements.

The second quantity is a constitutive property called the ***mass-average relative molar mass*** or simply the ***mass-average molar mass*** denoted M_w and defined as the weighed average of the mass fraction of each macromolecule having a given molecular mass.

$$M_w = \Sigma_i w_i M_i / \Sigma_i w_i = \Sigma_i N_i M_i^2 / \Sigma_i N_i M_i$$

with
- M_w weighed relative molar mass,
- w_i mass fraction of the macromolecule i,
- M_i relative molar mass of the macromolecule i.

The mass-average molar mass is determined by optical methods such as Rayleigh light scattering measurements.

The third quantity is called the ***z-average relative molar mass*** or simply the ***z-average molar mass*** denoted M_z and defined as the weighed average of the mass fraction of each macromolecule having a given molecular mass.

$$M_z = \Sigma_i w_i M_i^2 / \Sigma_i w_i M_i = \Sigma_i N_i M_i^3 / \Sigma_i N_i M_i^2$$

with
- M_z z-average relative molar mass,
- w_i mass fraction of the macromolecule i,
- M_i relative molar mass of the macromolecule i.

It can be measured by ultracentrifugation.

The fourth quantity is called the ***(z+1)-average relative molar mass*** or simply the ***(z+1)-average molar mass*** denoted M_z and defined as the weighed average of the mass fraction of each macromolecule having a given molecular mass.

$$M_z = \Sigma_i w_i M_i^3 / \Sigma_i w_i M_i^2 = \Sigma_i N_i M_i^4 / \Sigma_i N_i M_i^3$$

with
- M_z z-average relative molar mass,
- w_i mass fraction of the macromolecule i,
- M_i relative molar mass of the macromolecule i.

A schematic plot representing the above three quantities is presented in Figure 11.1. The width of the distribution can be measured by introducing the ***heterogeneity index*** which is the dimensionless ratio (M_w/M_N). For most polymers the most probable value is ca. 2.0.

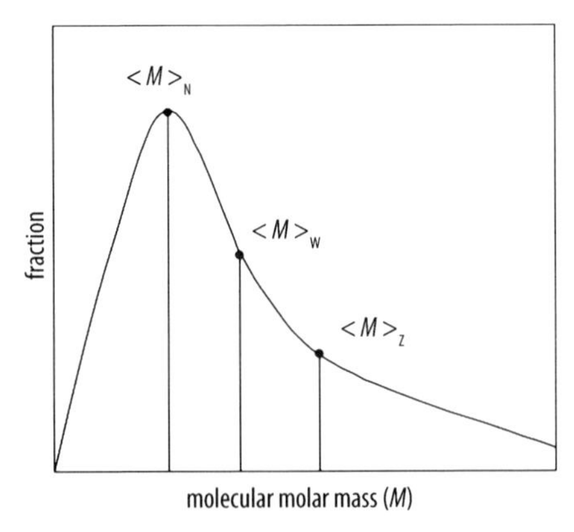

Figure 11.1. Schematic plot of molar masses of polymers

11.2.4 Glass Transition Temperature

Pure crystalline materials possess a well-defined temperature of fusion or melting point. By contrast, most polymers exhibit a temperature range over which the crystalline order progressively disappears. Actually, upon cooling, molten polymers begin to crystallize below their melting point with a contraction of volume. For amorphous polymers, the volume contraction continues until a given temperature, called the ***glass transition temperature*** denoted T_g, at which the supercooled liquid polymer exhibits enormous viscosity, is reached. Below T_g all polymers are rigid. Many physical properties of polymers such as the dynamic viscosity, the specific heat capacity, the coefficients of thermal expansion and elastic moduli, change abruptly at T_g.

11.2.5 Structure of Polymers

A homopolymer is a polymer made of the same monomer while a heteropolymer is made of at least two or more distinct monomers.

Copolymer. A copolymer is a heteropolymer formed when two or more different types of monomer are linked in the *same* polymer chain.

Graft polymer. A graft polymer is a heteropolymer with polymer chains of one monomeric composition branching out of the sides of a polymeric back-bone with a different chemical composition.

Block copolymer. A block copolymer denotes a heteropolymer having a great regularity of structure with large repeating unit containing dozens of monomers.

Tacticity. The orderliness of the succession of configurational repeating units in the main chain of a polymer molecule.

Tactic polymer. A regular polymer, the molecules of which can be described in terms of only one species of configurational repeating unit in a single sequential arrangement.

Isotactic polymer. A regular polymer, the molecules of which can be described in terms of only one species of configurational base unit having chiral or prochiral atoms in the main chain in a single sequential arrangement.

Syndiotactic polymer. A regular polymer, the molecules of which can be described in terms of alternation of configurational base units that are enantiomeric.

Atactic polymer. A regular polymer, the molecules of which have equal numbers of the possible configurational base units in a random sequence distribution.

11.3 Classification of Plastics and Elastomers

See Table 11.3, page 698.

11.4 Thermoplastics

11.4.1 Naturally Occurring Resins

11.4.1.1 Rosin

Rosin, formerly called *colophony*, is a solid form of resin obtained from several pine trees extensively found in Asia, Europe and North America and to a lesser extent from other conifers

Table 11.3. Classification of thermoplastics, thermosets and elastomers

Derivatives of natural products	Polyaddition resins	Polycondensation resins
1. Naturally occurring resins - Amber and succinite - Rosin - Shellac - Lignin	4. Polyolefins - Polyethylene (PE) - Polypropylene (PP) - Polybutylene (PB)	12. Phenolics - Phenol-formaldehyde - Resorcinol-formaldehyde
2. Derivative of cellulose: 2.1 Cellulose esters - Cellulose acetate (CA) - Cellulose propionate (CP) - Cellulose acetobutyrate (CAB) - Cellulose acetopropionate (CAP) - Cellulose nitrate (CN) 2.2 Cellulose ethers - Methyl cellulose (MC) - Ethyl cellulose (EC) - Carboxymethyl cellulose (CMC) 2.3 Regenerated cellulose - Viscose, rayon	5. Polyvinyls - Polyvinyl ethers - Dipolyvinyls - Polyvinyl chloride (PVC) - Polyvinyl fluoride (PVF) - Chlorinated polyvinyl chloride (CPVC)	13. Aminoplastics - Urea-formaldehyde - Melamine-formaldehyde - Melamine-phenolics 14. Furan resins - Phenol-furfural
	6. Polyvinylidenes - Polyvinylidene chloride (PVDC) - Polyvinylidene fluoride (PVDF)	15. Polyesters - Alkyd resins - Polycarbonates (PC)
	7. Polyvinyl derivatives - Polyvinyl alcohol (PVA) - Polyacetals (PAc)	16. Polyethers - Polyformaldehydes - Polyglycols
	8. Styrenics - Polystyrene (PS) - Acrylonitrile-butadiene-styrene (ABS) - Styrene-acrylonitrile (SAN) - Styrene-butadiene	17. Polyurethanes (PU)
	9. Fluorocarbons - Polytetrafluoroethylene (PTFE) - Polytrichlorofluoroethylene (PTCFE) - Fluorinated ethylene propylene (FEP) - Perfluoroalkoxy (PFA) - Ethylenetetrafluoroethylene (ETFE)	18. Polyamides (PA) 19. Polyimides (PI) 20. Polyaramides (PAR) 21. Sulfones - Polysulfones (PSF) - Polyethersulfone - Polyphenylsulfone
3. Derivatives of vegetal proteins - Caseine-formaldehyde - Zein-formaldehyde	10. Acrylics - Polymethylmethacrylate (PMMA)	22. Epoxy resins
	11. Coumarone-indenes	23. Polysiloxanes - Silicones

conifers. It is prepared by cutting a long slice in the tree to allow exudation and to collect the liquid resin in containers. Afterwards the liquid resin is steam heated to remove volatile terpene and ***turpentine*** and leaving ***gum rosin*** as residue. It is semi-transparent and varies in color from yellow to black. It chiefly consists of different organic acids, among those ***abietic acid*** ($C_{19}H_{29}COOH$) is the most important. Rosin is a brittle and friable resin, with a faint piney odor; the melting-point varies with different specimens, some being semi-fluid at the temperature of boiling water, while others melt between 100°C to 120°C. It is very flammable, burning with a smoky flame. It is soluble in alcohol, ether, benzene and chloroform. Rosin combines with caustic alkalis to yield salts called rosinates or pinates that are known as rosin soaps. In addition to its extensive use in soap making, rosin is largely employed in making inferior varnishes, sealing-wax and various adhesives.

11.4.1.2 Shellac

Shellac is a brittle or flaky secretion produced by the lac insect *Tachardia lacca*, commonly found in the rain-forests of Southern Asia (e.g., Thailand). Actually, the larvae of the insect settle on the branches, pierce the bark and feed from the sap. The female insect produces a protective coating over their bodies that produce a thick incrustation over the twig. When larvae emerge, the thick incrustation is scraped off and dried to yield the ***stick lac*** which still contains wood, lac resin, lac dye and various organic debris. After grinding, screening and washing the stick lac, the purified product called ***seed lac*** is obtained. Shellac is a naturally occurring polymer and it is chemically similar to synthetic polymers, thus it is considered as a natural plastic. It can be molded by heat and pressure methods, so it is classified as a thermoplastic. It is soluble in alkaline solutions such as ammonia, sodium borate, sodium carbonate, and sodium hydroxide, and also in various organic solvents. When dissolved in acetone or alcohol, shellac yields a coating of superior durability.

11.4.2 Cellulosics

Cellulosic materials are thermoplastics derivatives of ***cellulose***. The raw material for the industrial preparation of cellulosics is natural cellulose, itself a natural polymer, i.e., a polysaccharide chain $(C_6H_{10}O_5)_n$, with ca. 3500 glucosidic monomer units which is found extensively in plants and woods. In plants, cellulose acts as a structural reinforcement material. Actually, natural cellulose exhibits no plasticity because of the cross-linking existing between two polysaccharide chains. This reticulation is ensured by hydrogen bonds between two adjacent alcohol functions. Therefore, in order to become plastic, the alcohol functions of the cellulose must be converted either by esterification by an organic acid or by etherification with another alcohol. The most common esters of cellulose used commercially are the nitrate, acetate, butyrate, acetobutyrate, and propionate while major ethers are methylcellulose, ethylcellulose and benzylcellulose; finally xanthate of cellulose, called rayon, is also produced. Cellulose used for making cellulosics comes either from linters by-produced during the extraction of cotton or from wood pulp. However, before use, the raw cellulose must be carefully purified by removing pectine and fatty acids. Therefore, the raw cellulose is first treated by a strong caustic solution of NaOH followed by washing with water and finally bleaching is performed by sodium hypochlorite (NaOCl).

As a general rule, cellulosics do not constitute any major use but are encountered daily in a number of smaller items such as name plates, electrical component cases, high impact lenses, and other applications requiring a transparent plastic with good impact resistance. Weathering properties of the materials are good, particularly that of propionate, but overall chemical resistance is not comparable to other thermoplastics. Water and salt solutions are readily handled, but any appreciable quantity of acid, alkali, or solvent can have an adverse effect on the plastic.

11.4.2.1 Cellulose Nitrate

Cellulose nitrate or ***pyrolyxin*** was the first synthetic thermoplastic of industrial significance, and was first discovered in 1833 by H. Bracconot[2] after the nitration of wood flour and paper and later by Friedrich Schonbein in 1846. In 1870 the American chemist J.W. Hyatt[3] invented its gelification in order to find a substitute for ivory. It is obtained by nitration of purified

11 Polymers and Elastomers

[2] Braconnot, H. *Ann.*, **1**(1833)242–245.

[3] Hyatt, J.W.; Hyatt, I.S. US Patent 105,388, 1870.

cellulose with the sulfo-nitric mixture (i.e., conc. HNO_3 + fuming H_2SO_4). Actually, if concentrated nitric acid is used alone the by-produced water stops nitration by diluting the acid:

$$(C_6H_{10}O_5)_n + 2n\ HNO_3 \longrightarrow -[C_6H_8O_3(NO_3)_2]_n- + 2n\ H_2O$$

After nitration, the cellulose nitrate paste is carefully washed with water, bleached with sodium hypochlorite, washed with hydrochloric acid and washed again with water. Afterwards, the wet paste is progressively dried by mixing it with camphor or excess ethanol in order to replace the water by the alcohol which acts both as a platicizer and stabilizer. Actually, dry nitrate of cellulose is highly flammable, poorly stable towards heat and sunlight, and even choc sensitive. Addition of camphor up to 20–25 wt.% yields a safer material, called ***celluloid***, having good dimensional stability, low absorption of moisture and toughness. It can be laminated and colored providing a huge variety of material textures. The upper temperature of usefulness is 60°C. First uses of celluloid were to produce imitations of the expensive tortoise shell, the manufacture of photographic and cinematographic films and fabrication of varnishes by dissolving it into acetone or ethanol to yield ***collodions***. But today, due to its high flammability, it has been abandoned in favor of other esters of cellulose.

11.4.2.2 Cellulose Acetate (CA)

Industrially, cellulose acetate is made quite exclusively from pure cellulose obtained from linters of cotton because purified cellulose obtained from wood pulp still contains deleterious impurities. Cellulose is esterified by using a mixture of acetic anhydride and concentrated sulfuric acid, the latter acting as dehydrating agent for removing by-produced water, and also zinc (II) chloride acting as a catalyst.

$$(C_6H_{10}O_5)_n + 3n\ CH_3COOH \longrightarrow [C_6H_7O_2(OOCCH_3)_3]_n + 3n\ H_2O$$

Usually, glacial acetic acid is introduced gradually in the reactor to dissolve the newly formed cellulose acetate. The product obtained is the nonflammable ***cellulose triacetate*** which is insoluble in most organic solvents including acetone, and ether but it is soluble in chloroform, dichloromethane, glacial acetic acid and nitrobenzene. Cellulose triacetate exhibits good dimensional stability and heat resistance, possesses a dielectric constant, resistance to water and good optical transparency. Mechanically, cellulose triacetate has good folding endurance and burst strength. In order to render it soluble, a retrogadation (reverse) reaction, in which one ester functional group is partially saponified, is used to yield the ***cellulose diacetate*** or simply ***cellulose acetate*** which is soluble in acetone but not in chloroform. Afterwards, its dissolution into a solvent and addition of a plasticizer yields commercial cellulose acetate. Commercially, the product contains between 38 and 40% of acetyl groups. Cellulose acetate can be used up to 70°C, it is resistant to water and transparent to UV radiation. Dissolved in solvent it is used as a varnish. Cellulose acetate replace the highly hazardous cellulose nitrate for the manufacture of photographic and cinematographic films.

11.4.2.3 Cellulose Propionate (CP)

Cellulose propionate is the ester of cellulose with propionic acid and it is similar to cellulose acetate but with a higher plasticity. Cellulose propionate is a tough, strong, stiff, with a greater hardness that cellulose acetate and has excellent impact resistance. Articles made from it have a high gloss and are suitable for use in contact with food.

11.4.2.4 Cellulose Xanthate

Cellulose xanthate is obtained by reacting first pure cellulose with a strong caustic solution of NaOH to yield alkali-cellulose. Then, alkali-cellulose is reacted with carbon disulfide (CS_2). The commercial product is called rayon.

11.4.2.5 Alkylcelluloses

These ethers of cellulose are all prepared industrially in the same manner. The pure cellulose is digested into a strong caustic solution of 50 wt.% NaOH to yield the alkali-cellulose. Afterwards, the alkali-cellulose is alkylated by mixing it with an etherification agent such as methyl, ethyl or benzyl chloride or sulfate. The resulting products are alkylcelluloses in which the hydrogen in the hydroxyl groups is replaced by a methyl-, ethyl- or benzyl-group. Outstanding properties are unusually good low-temperature flexibility and toughness, wide range of compatibility, heat stability, and dielectric properties. The three ether groups are usually distinguished according to their solubilities.

Methylcelluloses are soluble in water, aqueous alkaline solutions and organic solvents. They are used for forming emulsions of given viscosities or as gelifying agent for electrolytes in batteries.

Ethylcelluloses are low density polymers (1070–1180 kg.m^{-3}) with solubilities depending on the degree of ethylation; usually commercial grade contains 44–48% ethoxyl functional groups. Solid masses of ethylcellulose exhibit low absorption of moisture, excellent dimensional stability and low temperature toughness and impact resistance. Chemically they are less resistant towards acids than cellulose esters but much more resistant to alkalis. They can be processed by injection molding. Because ethylcellulose is soluble in a wide variety of solvents, it provides a wide variety of varnish formulations. Benzylcelluloses yield plastics with excellent dielectric properties and chemical stability.

11.4.3 Casein Plastics

The casein plastics first produced in 1885 are a particular class of thermoplastic materials made from the rennet casein extracted from milk. With cellulosics they were the first synthetic thermoplastics produced industrially. The purified casein is reacted with formaldehyde to yield by condensation a ***casein-formaldehyde*** thermoplastics also known commercially as ***Galathite***®, meaning milkstone. By contrast with other condensation resins, casein-formaldehyde is obtained by impregnating the semi-products or preforms (e.g., rod, sheet, and plates) made by compressing an aqueous slurry of casein powder with formaldehyde until the compound penetrates inside the material. The raw material, that is, casein, can be prepared from milk by several routes. Actually, milk that contains between 2.7 and 3.5 wt.% casein also contains fatty acids that must be removed prior to precipitation. Therefore, the raw milk is strongly mixed to promote the coalescence of lipid droplets, that are removed by centrifugation. Then, the casein from the fat-free milk can be coagulated either by adding rennet, or by promoting the lactic fermentation using yeast or by adding a mineral acid. However, for the industrial preparation of casein-formaldehyde resins, only the precipitation by rennet provides a suitable product. Precipitation is performed in large jacketed vessels in order to maintain the temperature at 37°C at the beginning, ending at 65°C when coagulation is complete. The casein is removed by filtration, washed, dried, ground, and gelatinized with water before molding into semi-products. The complete reaction of semi-products with formaldehyde takes months but it yields an easily machinable and molded, nonflammable material. However, galathite exhibits a high water absorption and is attacked by alkalis. On the other hand, artificial wool can be obtained by dissolving casein into sodium hydroxide and then forcing the viscous solution through nozzles into a coagulating bath of acidified formaldehyde. The synthetic fiber produced resembles wool and was sold under the trade names ***Lanital*** and ***Aralac***.

11.4.4 Coumarone-Indene Plastics

Coumarone-indene plastics along with cellulose nitrate were the first synthetic resins developed commercially in the middle of the nineteenth century. Industrially, they are produced from coal-tar light oils by-produced either during coking or petroleum cracking operations. By treating the fraction distilling between 150 and 200°C that contains mainly indene and coumarone, with concentrated sulfuric or phosphoric acid, polymerization occurs readily. The synthetic resins obtained are mixtures of polyindene and polycoumarone, that are called ***cumar gum*** and commercialized under the trade name ***Nevindene*** with properties varying from a soft gum melting at 4°C to a hard brown solid with a melting point of ca. 150°C depending of the ratio of the two monomers. In all cases, the density is usually close to 1080 $kg.m^{-3}$. These resins are resistant towards alkalis but are easily soluble in organic solvents and hence are used as lacquers, varnishes, or waterproofing compounds.

11.4.5 Polyolefins or Ethenic Polymers

These thermoplastics have in common all the same basic monomer structure of the ethylene (i.e., $H_2C{=}CH_2$).

11.4.5.1 Polyethylene (PE)

Polyethylene also named polythene (PE) with the basic ethylene molecule $[—H_2C—CH_2—]_n$ as monomer was first produced on a commercial basis in 1934. Polyethylene is prepared directly from the polymerization of ethylene (C_2H_4). Ethylene is obtained from a refinery or from the steam cracking of naphtha or natural gas. Two industrial processes are today used:

(i) high pressure synthesis; and

(ii) low pressure synthesis.

In the **high pressure process,** the high purity ethylene is compressed under pressures ranging from 150 to 300 MPa at 300°C in the presence of traces of oxygen acting as a catalyst. The polymerization reaction is very simple and can be written as follows:

$$n\ CH_2{=}CH_2 \longrightarrow [—CH_2—CH_2—]_n$$

The majority of the polyethylene macromolecules produced are mostly linear and they have the same chain length. To a lesser extent however, some macromolecules have lateral chains (ramification). As a general rule, the longer the chain, the higher the mechanical strength and the lower the heterogeneity ratio. In the **low pressure process,** a stiffer product with a high softening point is obtained. Pure polyethylene solid crystallizes in the orthorhombic system with the crystal lattice parameters a = 741 pm, b = 494 pm and c = 255 pm respectively with 2 molecules (C_2H_4) per formula unit. Therefore the theoretical density of polyethylene is 996 $kg.m^{-3}$ which is only approached by the UHMW grade (see below). Hence polyethylene is a low density thermoplastics that floats on water.

Polyethylene is a thermoplastic material which varies from type to type according to the particular molecular structure of each type. Actually, several products can be made by varying the molecular weight (i.e., the chain length), the crystallinity (i.e., the chain orientation), and the branching characteristics (i.e., chemical bonds between adjacent chains). Polyethylene can be prepared in four commercial grades:

(i) **low density** (i.e., LDPE) and linear low density polyethylene (LldPE);

(ii) **medium density** (i.e., MDPE);

(iii) **high density** (i.e., HDPE); and
(iv) **ultra-high molecular weight** (i.e., UHMWPE) polyethylene.

Low density polyethylene (**LDPE**) exhibits a melting point of 105°C, toughness, stress cracking resistance, clarity, flexibility, and elongation. Hence, it is used extensively for piping and packaging, because of its ease of handling and fabrication. The chemical resistance of the product is outstanding, although not as good as high density polyethylene or polypropylene but it is resistant to many strong mineral acids (e.g., HCl, HF) and alkalis (e.g., NaOH, KOH, NH_4OH), it can be used for handling most organic chemicals but alkanes, aromatic hydrocarbons, chlorinated hydrocarbons, and strong oxidants (e.g., HNO_3) must be avoided. Assembly of parts made of PE can be achieved by fusion welding of the material which is readily accomplished with appropriate equipment. For instance, installations of piping made in this manner are the least expensive and most durable of any material available for waste lines, water lines, and other miscellaneous services not subjected to high pressures or temperatures. Nevertheless several limitations avoid its uses in some applications. These limitations are: a low modulus, a low strength, a low heat resistance, actually the upper temperature limit for the material is 60°C, combined with a tendency to degrade under UV irradiation (e.g., sunlight exposure). However, the polyethylene can be compounded with a wide variety of materials to increase strength, rigidity, and other suitable mechanical properties. It is now available in a fiber reinforced product to further increase the mechanical properties. Stress cracking can be a problem without careful selection of the basic resin used in the product or proper compounding to reduce this effect. Compounding of the product is also recommended to reduce the effect of atmospheric exposures over long periods. Linear low density polyethylene (LldPE) is produced by adding alpha-olefins (e.g., butene, 4-methyl-pentene-1, hexene, or octene) during ethylene polymerization to give a polymer with a similar density to LdPE but with the linearity of the HdPE. High density polyethylene (**HDPE**) has considerably improved mechanical properties, has better permeation barrier properties, and its chemical resistance is also greatly increased compared to the low density grade with a superior temperature limit of 75°C. Only strong oxidants will attack the material appreciably within the appropriate temperature range. Stress cracking of the HDPE can again be a problem if proper selection of the resin is not made. The better mechanical properties of this product extend their use into larger shapes, the application of sheet materials on the interior of appropriately designed vessels, such as packing in columns, and as solid containers to compete with glass and steel. Fusion welding can be achieved with a hot nitrogen gun. HdPE is produced by the catalytic polymerization of ethylene either in suspension, solution or gas phase reactors using traditional Ziegler–Natta, chromium or metallocene catalysts. Ultra-high molecular weight polyethylene (**UHMWPE**) is a linear polyethylene with an average relative molecular mass ranging from $3x10^6$ to $5\ x10^6$. Its long linear chains provide great impact strength, wear resistance, toughness, and freedom stress cracking in addition to the common properties of PE such as chemical inertness, self-lubricant, low coefficient of friction. Therefore, this thermoplastic is suitable for applications requiring high wear/abrasion resistance for components used in machinery. As a general rule, polyethylenes are highly sensitive to UV irradiation, especially sunlight exposure. Nevertheless, it is possible to avoid UV-light sensitivity by adding particular UV stabilizers.

11.4.5.2 Polypropylene (PP)

Polypropylene with the basic methyl substituted ethylene (i.e., propylene) as monomer is prepared industrially by the polymerization of propylene (C_3H_6) in a low pressure process using a mixture of aluminum triethyl [$(C_2H_5)_3Al$] and titanium tetrachloride ($TiCl_4$) as catalysts. The polymerization reaction is very simple and can be written as follows:

$$n\,(CH_2{=}CH_2{—}CH_2) \longrightarrow [{—}CH(CH_2){—}CH_2{—}]_n$$

In fact, the reaction mechanism induces a chain having a helical structure that exhibits the same asymmetrical stereochemical configuration of carbon atoms. This leads to a macromolecule having a high degree of crystallinity. Hence, polypropylene has considerably improved mechanical properties compared to polyethylene; actually it has a low density (900–915 $kg.m^{-3}$), it is stiffer, harder, and has a higher strength than many polyethylene grades. Moreover, due to its higher melting point (160°C), it can be used at higher temperatures than PE with a superior temperature limit of 100°C. Its chemical resistance is also greatly increased, and it is only attacked by strong oxidants. Stress cracking of the PP can be a problem if proper selection of the resin is not made. In comparison with PE it exists in few commercial grades but the plastic is stereospecific and can be isotactic and atactic. The better mechanical properties of this products extend their use into larger shapes, the application of the sheet materials on the interior of appropriately designed vessels, such as packing in columns, and as solid containers to compete with glass and steel. The modulus of the polypropylene is somewhat higher, which is beneficial in certain instances. The coefficient of thermal expansion is less for polypropylene than for the high density polyethylene. Fusion welding with a hot nitrogen gun is practical in the field for both materials when the technique is learned. The two main applications of polypropylene are injected molded parts and fibers and filaments. Polypropylene is prepared either by the Spherisol process licensed by Basell that combines both liquid and gas phase polymerization with a Ziegler–Natta catalyst or the Borstar process introduced by Borealis.

11.4.5.3 Polybutylene (PB)

The linear macromolecule of polybutylene is made of the following monomer unit $[—CH_2—CH(CH_2\text{-}CH_3)—]_n$. The ethyl groups are all located on the same side of the chain leading to an isotactic structure. Polybutylene is made from isobutylene, a distillation product of crude oil. Polybutylene exhibits high tear, impact, and puncture resistance. It also has low creep, excellent chemical resistance, and abrasion resistance.

11.4.6 Polymethylpentene (PMP)

Polymethylpentene (PMP) is a transparent thermplastic obtained industrially by means of a Ziegler-type catalytic polymerization of 4-methyl-1-pentene. Polymethylpentene exhibits both a high stiffness and impact resistance, good dielectric properties similar to those of fluorocarbons, and a good resistance towards chemicals and to high temperatures. Actually, it withstands repeated autoclaving, even at 150°C.

11.4.7 Polyvinyl Plastics

11.4.7.1 Polyvinyl Chlorides (PVCs)

Polyvinyl chloride $[—CH_2—CHCl—]_n$ was the first thermoplastic to be used in any quantity in industrial applications. It is prepared by reacting acetylene gas with hydrochloric acid in the presence of a suitable catalyst. PVC has grown steadily in favor over the years, primarily because of the ease of its fabrication. It is easily worked and can be solvent welded or machined to accommodate fittings. It is very resistant to strong mineral acid and bases, and as a consequence, the materials have been extensively used for over 40 years as piping for cold water and chemicals. However, in the design of a piping structure, the thermal coefficient of linear expansion must be taken into consideration, and the poor elastic modulus of the material must be considered. With these limitations, the product as a piping material can accommodate a wide range of products found in the chemical process industries. The industrial production of polyvinylchloride utilizes several polymerization processes depending on the final application. In the mass polymerization, liquid vinyl chloride monomer (VCM) is polymerized in a pressurized batch reactor at an operating temperature ranging from 40 to 70°C. This process yields PVC resins with a high clarity. On the other hand, suspension polymerization is the most common industrial process because of the versatility of PVC grades obtained while emulsion polymerization is conducted in aqueous solution and yields colloidal PVC more suitable for preparing paints and printing inks. Two grades of the primary PVC material are available: rigid PVC grade which accounts for 65% of demand is extensively used in the construction industry for piping used for water drainage, while flexible PVC grade that contains a plasticizer is used in calendered sheet, wire and cable coating.

11.4.7.2 Chlorinated Polyvinylchloride (CPVC)

Polyvinyl chloride can be modified through chlorination to obtain a vinyl chloride plastic with improved corrosion resistance and the ability to withstand operating temperatures that are 20–30°C higher. Hence, CPVC, which has about the same range of chemical resistance as rigid PVC, is extensively used as piping, fittings, ducts, tanks, and pumps for handling highly corrosive liquids and for hot water. For instance, it has been determined that the chemical resistance is satisfactory for CPVC in comparison with PVC on exposure for 30 days in such environments as 20wt.% acetic acid, 40–50 wt.% chromic acid, 60 to 70 wt.% nitric acid, at 30°C and 80 wt.% sulfuric acid, hexane, at 50°C and 80 wt.% sodium hydroxide until 80°C.

11.4.7.3 Polyvinyl Fluoride (PVF)

The linear macromolecule of ***polyvinyl fluoride*** (PVF) is based on the monomer unit: $[—CH_2—CHF—]_n$. PVF which is only used industrially as a thin film, exhibits good resistance to abrasion and resists staining. It also has outstanding weathering resistance and maintains useful properties from −100 to 150°C.

11.4.7.4 Polyvinyl Acetate (PVA)

The macromolecule of polyvinyl acetate (PVA) is based on a monomer where an acetate group replaces a hydrogen atom in the ethylene monomer. It is not used as a structural polymer because it is a relatively soft thermoplastic and hence it is only used for coatings and adhesives.

11.4.8 Polyvinylidene Plastics

11.4.8.1 Polyvinylidene Chloride (PVDC)

Polyvinylidene chloride identified as PVDC or polyvinyl dichloride is based on a dichloroethylene monomer $[—CH_2—CCl_2—CH_2—CHCl—]_n$. It has improved chemical resistance and mechanical properties. Actually, it has better strength than common PVC. The material has an upper temperature limitation of 65°C for the normal (Type I) and 60°C for the high impact (Type II) products. The chemical resistance is good in inorganic corrosive media with an outstanding resistance to oxidizing agents. However, contamination by solvents of almost all types must be avoided. This material has had great significance in chemical industry applications over the years. The product has been made into a number of specific items designed to serve the chemical process industry. Among these are valves, pumps, piping, and liners particularly on the inside of the pipe. The latter product was the first thermoplastic to be used for this purpose and found extensive and useful service as its trade-name ***Saran®***. Also, the material is available as a rigid or pliable sheet liner for application on the interior of vessels. It must be recognized that many modifications of the PVC can be made. Fiber reinforced products are also available.

11.4.8.2 Polyvinylidene Fluoride (PVDF)

The macromolecule of ***polyvinylidene fluoride*** (PVDF) consists of a linear chain in which the predominant monomer unit is $[—CH_2—CF_2—]_n$. PVDF has good weathering resistance and it is resistant to most chemicals and solvents but less inert than PTFE, PFA and FEP in the same conditions. PVDF is nonflammable and exhibits greater mechanical strength, wear and creep resistance than other fluorocarbons. PVDF is heat resistant up to 150°C. However, the material is much more workable and has been made into essentially any shape necessary for the chemical process industry. Complete pumps, valves, piping, smaller vessels, and other hardware have been made and have served successfully. The material may also be applied as a coating or as a liner.

11.4.9 Styrenics

11.4.9.1 Polystyrene (PS)

Polystyrene is based on the monomer of styrene (i.e., phenylethene). Polystyrene $[—CH(C_6H_5)—CH_2—]_n$ is essentially a light amorphous and atactic thermoplastic. The aromatic ring confers stiffness on the plastic and avoids chain displacement which would render the plastic brittle. The material is not recommended for applications handling corrosive chemicals because its chemical resistance by comparison with other available thermoplastics is poor and the material will stress crack in certain specific media. However, it has a high light transmission in the visible region, it has an excellent moldability rendering the ease of fabrication and possesses a low cost of the material so that it will always be considered if the properties are adequate for the use. Nevertheless, polystyrene is sensitive to UV irradiation (e.g., sunlight exposure) which gives a yellowish color to the material and the heat resistance of the material is only 65°C. The plastic will be encountered as casings for equipment and in various electrical applications. Fittings for piping have been made from the plastic, and many containers may be found made of the modified polystyrene. Joining can be achieved by solvent welding of the product to fabricate devices but restricts its use to waters and services not containing organic and inorganic chemicals. Polystyrene is prepared industrially by three different polymerization processes:

(i) the suspension polymerization produces polymers of different molecular weights and it can produce crystal polystyrene and high impact grades;
(ii) the solution polymerization that can be either a batch or a continuous process yields the purest polystyrene grades; while
(iii) bulk polymerization yields a high transparency and colorless polystyrene.

Polystyrene is available commercially in several grades: general purpose polystyrene (GPPS), medium impact (MIPS) and high impact (HIPS) polystyrenes and finally expandable polystyrene (EPS). Polystyrene is the third largest consumed thermoplastic in use today after PE and PP with 20% of the market.

11.4.9.2 Acrylonitrile Butadiene Styrene (ABS)

This is a terpolymer: the first monomer is butadiene; the second, acrylonitrile, consists of an ethylene molecule in which a hydrogen atom is replaced by a nitrile group (i.e., CN); and the third is styrene (i.e., an ethylene molecule with a phenyl group replacing a hydrogen atom).

The material can be varied considerably in properties by changing the ratio of acrylonitrile to the other two components of the terpolymer. This offshoot of the original styrene resins has achieved a place in industrial work of considerable importance. Actually, the strength, toughness, dimensional stability and other mechanical properties were improved at the sacrifice of other properties. Although the material has poor heat resistance (90°C), a relatively low strength, and a restricted chemical resistance, the low price, ease of joining, and ease of fabrication make the material most attractive for distribution piping for gas, water, waste and vent lines, automotive parts, and numerous consumer service items ranging from the telephones to automobile parts. Actually, the plastic withstands attack by very few organic compounds, but is readily attacked by oxidizing agents and strong mineral acids. Moreover, stress cracks can occur in the presence of certain organic products.

11.4.10 Fluorinated Polyolefins (Fluorocarbons)

Fluorocarbons represent certainly the most versatile and important group of thermoplastics for use in the chemical process industries (CPI). Most of these fluorinated polymers are able to handle, without any corrosion, extremely harsh environment and highly corrosive chemicals that only refractory, noble or precious metals or particular ceramics can tolerate. Nevertheless, owing to the presence of the most electronegative fluorine in the chain, the fluorocarbons are seldom attacked by molten alkali metals such as sodium and lithium forming graphite. As with other products, when such inertness is obtained in the material, certain other properties must be sacrificed. In this case, the fluorocarbon materials are more difficult to work in any manner and are much more limited in design and application than are other thermoplastic materials. The materials are porous and the permeation of specific chemicals must be considered to insure the proper selection of the proper fluorocarbon for the intended service. There are currently six types of commercial fluorocarbon thermoplastics. All have exceedingly good chemical stability, but there are differences which should be noted. These materials have been designed into a number of solid items for chemical service, such as impellers, mixers, spargers, packing, smaller containers, and a few more intricate shapes. However, the vast proportion of the use of fluorocarbons in the chemical process industry is as linings in steel or ductile iron. All shapes of lined pipe can be obtained. In addition, certain of the thermoplasts can be cut and jointed in the field using appropriate tools. Lined pumps and valves are available.

11.4.10.1 Polytetrafluoroethylene (PTFE)

The basic monomer unit is a totally fluorinated ethylene molecule ($—CF_2—CF_2—$). It is well known under its common trade name ***Teflon®***. It was discovered in 1938 by Roy J. Plunkett a DuPont scientists. Industrially, polytetrafluoroethylene is obtained from several consecutive of steps. First, chloroform reacts with hydrofluoric acid to yield chlorodifluoromethane. The chlorodifluoromethane is then pyrolized at 800–1000°C to yield the monomer, i.e., tetrafluoroethylene ($CF_2{=}CF_2$, TFE) which is purified and polymerized in aqueous emulsion or suspension using organic peroxides, persulfates or hydrogen peroxide as catalysts. The simple polymerization reaction is as follows.

$$n\,(CF_2{=}CF_2) \longrightarrow (—CF_2—CF_2—)_n$$

The macromolecule exhibits a structure with a high crystallinity that explains its unusually high melting point of 327°C. This characteristics ensures the highest useful temperature limit among plastics of 260°C. PTFE exhibits also good mechanical strength and an extremely low coefficient of friction that imparts to the material excellent self-lubricating abilities similar to those of graphite and molybdenum disilicide. PTFE exhibits the strongest anti-adhesive properties and no compound is known to adhere durably to PTFE. However, PTFE remains the most difficult of the fluorocarbons to work because of its high viscosity in the molten state; it cannot be cast and extrusion requires particular procedures. Therefore shapes and parts are usually made using powder metallurgy techniques such as sintering to produce it in usable forms. During sintering the PTFE powder or granules filling a mold of the desired shape are compressed under 14–70 MPa at a temperature well above 327°C and after cooling the PTFE exhibits the desired shape. From a chemical point of view, PTFE is one of the most chemically inert materials known apart from glass, tantalum, platinum and iridium for servicing (possessing a long service life) in various severely corrosive chemicals even at high temperatures. Moreover, no organic or mineral solvent dissolves PTFE. The only chemicals that are known to attack PTFE readily are molten alkali metals (e.g., Li, Na, K) and nascent fluorine gas. Finally, PTFE has excellent insulating properties with a low loss tangent factor at high frequencies. Nevertheless, permeation is an issue depending on the specific exposure but sometimes no better than many of the newer materials. Some problems associated with thermal cycling which can cause fatigue due to repeated expansion and contraction over a period of time when going through high temperatures were reported. Nevertheless, owing to their porosity, one particular mode of deterioration for fluorocarbons is the adsorption of a chemical, followed either by reaction with another component inside the thermoplastic or by polymerization of the product within the plastic. When this phenomena occurs, it leads to the surface degradation such as blistering. The material has also a definite heat limitation and overheating should be avoided. Cold flow of the resins is well known implies that the design and use of the fluorocarbon should be such that excessive compressive stresses are not imposed to create a cold flow condition. PTFE is now largely used in consumer goods as anti-stick materials. Industrially, its chemical inertness favors PTFE in applications involving harsh environments such as piping, pump parts and protective lining in the chemical process industry, while its self-lubricating properties allow its use in moving parts such as braking-pads on machinery, rotors and shafts packing etc., where lubrication is an issue.

11.4.10.2 Fluorinated Ethylene Propylene (FEP)

The ***fluorinated ethylene propylene*** (FEP) macromolecule consist mainly of a linear chain with the basic monomer unit $[—(CF_2)_3—CF(CF_3)—]_n$.This translucent fluorocarbon is flexible and more workable than PTFE, and like PTFE it resists to all known chemicals except molten alkali metals, elemental fluorine, fluorine precursors, and concentrated perchloric acid. It withstands temperatures up to 200°C and may be sterilized by all known chemical and thermal methods. Certain carefully prepared films of FEP can be used as windows in equipment when necessary. The product has found extensive use as a pipe fitting liner as well as a liner in small vessels.

11.4.10.3 Perfluorinated Alkoxy (PFA)

The macromolecule of ***perfluorinated alkoxy*** (PFA) or simply ***perfluoroalkoxy*** is based on the monomer unit: $[—(CF_2)_2—CF(O—C_nF_{2n+1})—(CF_2)_2—]_n$. Perfluoroalkoxy is similar to other fluorocarbons such as polytetrafluoroethylene and fluorinated ethylene propylene regarding its chemical resistance, dielectric properties, and coefficient of friction. Its mechanical strength, Shore hardness, and wear resistance are similar to PTFE and superior to that of FEP at temperatures above 150°C. PFA has a good heat resistance from -200°C up to 260°C near to that of PTFE but having a better creep resistance.

11.4.10.4 Polychlorotrifluoroethylene (PCTFE)

Polychlorotrifluoroethylene (PCTFE) consists of a linear macromolecule with the following monomer unit $[—CF_2—CF(Cl)—]_n$. PCTFE possesses outstanding barrier properties to gases, especially water vapor. Its chemical resistance is surpassed only by PTFE and few solvents dissolve it at temperatures above 100°C, and it swells in chlorinated solvents. It is harder and stronger than perfluorinated polymers but its impact strength is much lower. PCTFE has heat resistance up to 175°C. It is commercialized under the common trade name ***Kel-F®***. The working properties of PCTFE are relatively good, and it can be formed by injection molding, and hence the material is used as a coating as well as a prefabricated liner for severe chemical applications.

11.4.10.5 Ethylene-Chlorotrifluoroethylene Copolymer (ECTFE)

Ethylene-chlorotrifluoroethylene copolymer (ECTFE) has a linear macromolecule in which the predominant alternating copolymer is $[—(CH_2)_2–CF_2—CFCl—]_n$. This copolymer exhibits useful properties for a wide range of temperatures, that is, from cryogenic temperatures up to 180°C. Its permittivity is low but stable over a broad temperature and frequency range.

11.4.10.6 Ethylene-Tetrafluoroethylene Copolymer (ETFE)

Ethylene-tetrafluoroethylene copolymer (ETFE) is a linear macromolecule with the monomer unit: $[—(CH_2)_2—(CF_2)_2—]_n$. ETFE exhibits physical properties similar to those of ethylene-chlorotrifluoroethylene copolymer.

11.4.11 Acrylics and Polymethyl Methacrylate (PMMA)

The ***polymethyl methacrylate*** (PMMA) macromolecule is based on a monomer that corresponds to an ethylene molecule with one hydrogen atom substituted by a methyl group (i.e., CH_3—) while the second hydrogen atom on the same carbon is replaced by an acetyl group (i.e., CH_3COO—) giving the basic monomer unit $[—CH_3—C(CH_3)(COOCH_3)—]_n$. The raw chemical intermediate used for making polymethyl methacrylate is 2-hydroxy-2-methylpropanenitrile, which is prepared by reacting acetone with hydrocyanic acid according to the following reaction:

$$CH_3COCH_3 + HCN —> (CH_3)_2C(OH)CN$$

Afterwards, the 2-hydroxy-2-methylpropanenitrile produced is reacted with methanol (CH_3OH) to yield the methacrylate ester. Another former route consisted of reacting methylacrylic acid $[CH_2{=}C(CH_3)COOCH_3]$ directly with methanol. The polymerization reaction is initiated either by organic peroxides or azo catalysts to finally produce the polymethyl methacrylate macromolecule. The polymerization is highly exothermic as indicated by the elevate enthalpy of the reaction (58 $kJ.mol^{-1}$). Therefore, the process requires fast heat removal from the reactor vessel by efficient cooling. Continuous or batch processes are used. Batch process yield higher molecular weight PMMA while continuous process leads to copolymers. The commercial product is well known under the common trade name ***Plexiglas®***. PMMA is a clear and rigid thermoplastic and in addition it is readily formed by injection molding. Main applications are guards and covers. As a general rule, the good atmospheric stability and clarity of acrylics have made them useful as high impact window panes and other see-through barriers important in industry. Various modifications are being made to alter the properties of the basic acrylic resins for specific services. However, these are not found in any great use in the industrial area to date. The upper temperature of usefulness is approximately 90°C. The loss in light transmission is only 1% after five years' exposure in locations with high sunlight exposure.

11.4.12 Polyamides (PA)

Polyamides thermoplastics are prepared by condensation by reacting a carboxylic acid (i.e., RCOOH) and an amine (i.e., $R'NH_2$) giving off water. Hence, the basic monomer unit in polyamides is $[—NH—(CH_2)_2—CO—]_n$. These resins are well known under the common trade name ***Nylon***®. Nylon was one of the first resinous products to be used as an engineering material. Actually, their excellent mechanical properties combined with their ease of fabrication have assured their continued growth for mechanical applications. Excellent strength, toughness, abrasion/wear resistance, and a high Young's modulus are the chief valuable properties of nylons and explained the important applications as mechanical parts in various operating equipment such as gears, electrical fittings, valves, fasteners, tubing, and wire coatings. Actually, some nylon grades have tensile properties comparable to that of the softer aluminum alloys. In addition, coatings and structural items can be obtained. The heat resistance of the nylon can be varied, but must be considered in the range of 100°C. The chemical resistance is remarkably good for a thermoplastic, the most notable exception being the poor resistance to strong mineral acids. Moreover, stress corrosion cracking of nylon parts can occur, particularly when in contact with acids and alkaline solutions. Owing to the wide diversity of different additives or copolymer as starting materials, there are several commercial grades of nylon resins available. each of them with particular properties. The main grades are nylon®6 and nylon®66, these being the two grades having the highest strength. Industrially, nylon 6 is obtained in a batch process by mixing caprolactam, water and ethanoic acid in a reaction vessel heated under inert nitrogen atmosphere at 230°C, while nylon 66 is prepared from adiponitrile, itself obtained from butadiene or propylene, which is converted into hexamethylene diamine (HMD). HMD is then reacted with adipic acid to yield nylon by a condensation reaction.

11.4.13 Polyaramides (PAR)

More recently, new commercial grades of nylon resins were developed in order to overcome the limitation of the previously discussed nylons grades. These products consist of polyamides that contain an aromatic functional group in their monomer, and hence are called aramid resins, aramid nylons after the acronym of **ar**omatic and **amides**. The basic monomer units in polyaramides is $[—NH—C_6H_4—CO—]_n$.They are well know under the trade names ***Kevlar***® and ***Nomex***® from E.I. DuPont de Nemours or ***Twaron***® and ***Technora***® from Teijin, the two companies that have 50% of the market. In practice, the basic difference between Kevlar and Nomex is the orientation of the aromatic rings, Kevlar being para-oriented while Nomex is meta-oriented. In both cases, this leads to a typical rod-like structure resulting in a high temperature of glass transition, and poor solubility in organic solvents along with an improved clarity.

11.4.14 Polyimides (PI)

These plastics offer the most unique combinations of properties available for use in industrial service. The plastics are usable from –190°C to 370°C. These excellent low temperature properties are often overlooked where a plastic is required to retain some ductility and toughness at such low temperatures. Some combinations of the resins can be taken to 510°C for short periods without destroying the parts. The plastic has excellent creep and abrasion resistance, excellent elastic modulus for a thermoplastic, and good tensile strength that does not drop off rapidly with temperature. The chemical resistance must be rated as good.

11.4.15 Polyacetals (PAc)

Polyacetals or simply ***acetals*** under the common trade name ***Delrin***® differ from other polymers due to the presence of an oxygen heteroatom in their monomer giving an heterochain polymer. The basic polymer unit is usually formaldehyde $[—CH—O—]_n$.Polyacetals resins are obtained by the polymerization of formaldehyde using trioxanes to give the homopolymer while copolymers are usually prepared by incorporating other monomers. The main properties of acetals include high melting point, elevate strength and stiffness, low friction coefficient, and high resistance to fatigue. Higher molecular weight increases toughness but reduces melt flow. The excellent dimensional stability and toughness of the acetal resins recommends their use for gears, pump impellers, other types of threaded connections such as plugs, mechanical uses. The material has an upper useful temperature limitation of ca. 105°C. The chemical resistance indicated in the literature shows a wide range of tolerance for various inorganic and organic products. As with many other resins, this formaldehyde polymer will not withstand strong acids, strong alkalis, or oxidizing media.

11.4.16 Polycarbonates (PC)

Polycarbonate (PC) is prepared by reacting bisphenol A and phosgene or by reacting a polyphenol with dichloromethane and phosgene. The basic monomer unit is $[—OC_6H_4C(CH_3)_2C_6H_4COO—]_n$. Polycarbonate is a linear, low crystalline, transparent, high molecular mass thermoplastic commonly know under the commercial trade name ***Lexan***®. It exhibits a good chemical resistance to greases, and oils but has a poor organic solvent resistance. Moreover, it is greatly restricted in its resistance by a severe propensity to stress crack. This property can be modified greatly by proper compounding but remains the most serious problem when considering the polycarbonate for chemical exposures. The exceedingly high impact resistance of this thermoplastic (30 times that of safety glass) combined with high electrical resistivity, ease of fabrication, fire resistance, and light transmission (90%) has promoted its use into a wide range of industrial applications. The most notable of these for industrial applications is the use of the sheet material as a glazing product. Where a high impact, durable, transparent shield is required, the polycarbonate material is used extensively. In addition, many smaller mechanical parts for machinery, particularly those with very intricate molding requirements, impellers in pumps, safety helmets, and other applications requiring light weight and high impact resistance have been satisfied by the use of polycarbonate plastics. The material can be used from –170°C up to a temperature of 121°C.

11.4.17 Polysulfone (PSU)

Polysulfone plastics comprise ***polysulfone*** $[—Ph—C(CH_3)_2—Ph—O—Ph—SO_2—O—]_n$, ***polyester sulfone*** $[—O—Ph—SO_2—O—]_n$ and ***polyphenylsulfone*** $[—O—Ph—SO_2—Ph—O—Ph—Ph—]_n$ with the aryl radical Ph = C_6H_4. The isopropylidene linkage imparts chemical resistance, the ether bond imparts temperature resistance, while the sulfone linkage imparts impact strength. The brittleness temperature of polysulfones is –100°C. Polysulfones are clear, strong, nontoxic, and virtually unbreakable. However, stress cracking can be a problem and should be considered before use. PSU do not hydrolyze during autoclaving and they are resistant to acids, alkalis, aqueous solutions, aliphatic hydrocarbons, and alcohols. Hence they have added another dimension to thermoplastics in heat resistance and strength at high temperature. The ease of molding the material, and its retention of properties as temperatures increase, has made it one of the faster growing resins in the market place. Use of the product for chemical equipment applications has not been noted to date but is anticipated.

11.4.18 Polyphenylene Oxide (PPO)

PPO is a thermoplastic with excellent heat and dimensional stability, and satisfactory chemical resistance. The material may be found primarily in pump parts and certain other applications where impact strength, good modulus, and reasonable abrasion resistance are required. The chemical resistance of the material is good and the allowable temperature limit of 120°C, under appropriate conditions, extends the attractiveness for use of the product. The cost requires specific need for an identifiable application before choosing the product over many less costly thermoplastics.

11.4.19 Polyphenylene Sulfide (PPS)

The macromolecule of ***polyphenylene sulfide*** (PPS) consists of the basic monomer of para-substituted benzene rings $[—C_6H_5—S—]_n$. The high crystallinity and thermal stability of the chemical bond existing between the aromatic ring and the sulfur atom are responsible for the high melting point, thermal stability, inherent flame-retardance, and good chemical resistance of polyphenylene sulfide. There are no known solvents of polyphenylene sulfide below 205°C. Chemical resistance is outstanding and the temperature usefulness ranges from −170°C to 190°C. Coatings prepared from the resin are available. Considerable strength with high elastic modulus can be obtained by the addition of glass or other mineral fillers to the material.

11.4.20 Polybutylene Terephthalate (PBT)

Polybutylene terphthalate (PBT) is a semicrystalline thermoplastic polyester considered as a medium performance engineering polymer. It is produced industrially in a two-step batch or continuous process. The first step involves the transesterification of dimethyl terephthalate (DMT) with 1,4-butanediol (BDO) to produce hydrobutyl terephtlate (bis-HBT) at a temperature of 200°C. The second step consists to the polycondensation of bis-HBT at 250°C to yield PBT. It exhibits both excellent electrical properties and chemical resistance. When reinforced with glass fibers, it has improved stiffness and mechanical strength. Typical uses include connectors, capacitors and cable enclosures. PBT is also used in hot appliances such as iron and kettles.

11.4.21 Polyethylene Terephthalate (PET)

Polyethylene terephthalate (PET) is obtained by reacting purified terephthalic acid (PTA) and monoethylene glycol (MEG) and melting the reaction product to initiate the polycondensation. The molten polymer is then extruded, cut into chips and cooled. PET main use is in the soft drinks and water bottles, other applications include thick-walled containers for cosmetics and pharmaceuticals.

11.4.22 Polydiallyl Phthalate (PDP)

Polydiallyphthalate (PDP) is prepared from the diallyl 1,3-phthalate [$C_6H_4(CH_2—CH=CH_2)$]. The linear polymer obtained is a solid thermoplastic still containing unreacted allylic groups spaced at regular intervals along the polymer chain. When mixed with various fillers such as mineral, glass, or synthetic fibers it exhibits good electrical properties under high humidity and high temperature conditions, stable low-loss factors, high surface and volume resistivity, and high arc resistance.

11.5 Thermosets

11.5.1 Aminoplastics

These thermosetting polymers are synthetic resins containing the amine group ($-NH_2$) in their macromolecules. The major commercial resins in this group are:

(i) ***urea-formaldehyde***;
(ii) ***melamine-formaldehyde***; and
(iii) ***aniline-formaldehyde.***

Urea-formaldehyde first appeared in 1929 in the USA as a substitute of glass for windows while melamine-formaldehyde was first commercialized by the American Cyanamid Co. in 1939.

Urea-formaldehyde. This thermoset is obtained by the condensation of urea [$(NH_2)_2CO$] or derivatives such as hydroxymethylurea and formaldehyde (HCHO) in the presence of a proper catalyst either basic or acid. The general equation of the first reaction is given below:

$$NH_2\text{-CO-}NH_2 + HCHO \longrightarrow NH_2\text{-CO-NH-}CH_2OH + OHCH_2\text{-NH-CO-NH-}CH_2OH$$

A wide variety of urea-formaldehyde resins can be obtained by careful selection of the pH, reaction temperature, reactant ratio, amino monomer, and degree of polymerization. If the reaction is carried far enough, an infusible polymer network is produced. The condensation proceeds in several consecutive stages: First it yields a liquid and transparent resin easily soluble in organic solvents. Secondly, the condensation continues and the resin becomes easy to mold. Thirdly, upon heating, the resin hardens yielding a hard solid, non fusible and insoluble in organic solvents. Industrially, the condensation is maintained in the second state in order to be able to mold the resin easily. The major fillers consist of pure cellulose, caseine or cotton flocks, in order to not alter the whiteness. Once cured, urea-formaldehyde exhibits good mechanical and dielectric properties and good chemical resistance.

Melamine-formaldehyde. The monomer used for preparing melamine formaldehyde is formed by reacting melamine with formaldehyde to yield hexamethylolmelamine. The monomer can further condense in the presence of an acid catalyst; ether linkages can also form. A wide variety of resins can be obtained by careful selection of pH, reaction temperature, reactant ratio, amino monomer, and extend of condensation. Liquid coating resins are prepared by reacting methanol or butanol with the initial methylolated products. These can be used to produce extreme surface hardness, discoloration and solvent-resistant coatings by heating with a variety of hydroxy, carboxyl, and amide functional polymers to produce a cross-linked film.

11.5.2 Phenolics

Phenol-formaldehyde resins or simply ***phenolics*** are prepared from the condensation of a mixture of phenol (i.e., carbolic acid) and cresols with formaldehyde as follows:

$$n\ C_6H_5OH + n\ HCHO \longrightarrow [-C_6H_2(OH)CH_2-]_n + nH_2O$$

Addition of formaldehyde ensures the formation of di- and trimethylolphenol, which later condense and polymerize rapidly.

Industrial preparation. Two processes are currently used:

(i) In the one-stage process, formaldehyde, phenol and an alkaline catalyst are introduced into a stainless steel vessel and reacted together. The elevated ratio of formaldehyde to phenol allows the thermosetting process to take place without any addition of another cross-linking agent. After discharge, further heating terminates the polymerization yielding an insoluble and non fusible resin.

(ii) In the two-stage process, formaldehyde, phenol and concentrated sulfuric acid are introduced in a stainless steel vessel with a low ratio of formaldehyde to phenol in order to prevent the thermosetting reaction from occurring during manufacture of the resin. After 4 hours at 150°C, separation of condensation water and resin occurs and overlying water is simply removed by vacuum pumping. At this point the resin is a viscous liquid termed ***novolac resin***. Subsequently, an activator, hexamethylenetetramine, is incorporated into the material to complete the polymerization and yields the final thermoset in the cured state. Ground phenolic resin can be mixed with a plasticizer and fillers such as asbestos, graphite, or silica to give materials with desired properties.

Properties. Phenolics are cheap thermosets that exhibit good strength, heat stability, and impact resistance along with a good machinability. Chemically speaking, phenolics demonstrate high chemical resistance and moisture penetration, except towards strong alkalis.

Industrial applications. Due to their chemical resistance, phenolics are widely used as linings and impregnating resins for chemical process equipment for handling strong acids. Other uses include brake linings, electrical components, laminates, glues, adhesives, molds and binders.

11.5.3 Acrylonitrile-Butadiene-Styrene (ABS)

Acrylonitrile-butadiene-styrene (ABS) is the largest volume engineering resin mainly used in the automotive industry and electronic appliances. ABS is made by the polymerization of styrene with acrylonitrile and butadiene. Three main processes are used industrially:

(i) the oldest process which is performed by emulsion polymerization is the more polluting;

(ii) suspension polymerization consists of blending together a rich-rubber medium with styrene-acrylonitrile; and finally

(iii) the continuous mass polymerization which does not use an aqueous medium is the preferred route because it generates less waste.

ABS is usually sold as odorless solid pellets. From a health and safety point of view, when burning ABS produces dense fumes containing noxious gases such as carbon monoxide (CO) and hydrogen cyanide (HCN).

11.5.4 Polyurethanes (PUR)

Polyurethanes are thermosets prepared by a condensation reaction involving diisocyanate (e.g., toluene diisocyanate, polymethylene diphenylene diisocyanate) with an appropriate polyol. They were first discovered by Wurtz in 1848. The polymers can be used in several forms such as flexible and rigid foams, elastomers, and liquid resin. The polyurethanes exhibit low corrosion resistance to strong acids and alkalis, and to organic solvents. Flexible foams are extensively used for domestic applications (e.g., bedding, and packaging), while rigid foams are used as thermal insulation material for transportation of cryogenic fluids, and frozen food products.

11.5.5 Furan Plastics

Furan plastics or simply ***furans*** are the collective names for a wide range of thermosetting resins made either from furfuraldehyde (i.e., furol, furfural) or furfuryl alcohol. All these raw materials can be prepared from agricultural wastes (e.g., cornstalks, corncobs, oats husks, bagasse, and rice). Furfuryl alcohol resins are made by reacting furfuryl alcohol with an acid catalyst. They form low cost liquid resins having a good chemical resistance. They are used for making corrosion resistant coatings, industrial tank linings for protecting against various corrosives and finally mixed with silica sand they provide acid-proof cements (e.g., Alkor® cement). On the other hand, furfuraldehyde condenses with phenol to yield self-curing furalphenol resins. These resins possess excellent heat resistance up to 177°C and chemical resistance (e.g., acids, alkalis, alcohol, hydrocarbons) together with good dielectric properties. Moreover, their excellent adhesion capabilities onto metals and other materials make them highly suitable for making protective coatings for the CPI such as tank linings. As a general rule, they are more expensive than the phenolic resins but also offer somewhat higher tensile strengths. Furan plastics, filled with asbestos, have much better alkali resistance than phenolic asbestos. Some special materials in this class, based on bisphenol, are more alkali resistant.

11.5.6 Epoxy Resins (EP)

Glycidal ether-based epoxies represent perhaps the best combination of corrosion resistance and mechanical properties. Epoxy novolac resins are produced by glycidation of the low-molecular-weight reaction products of phenol or cresol with formaldehyde. Highly cross-linked systems are formed that have superior performance at elevated temperatures. Epoxies reinforced with fiberglass have very high strengths and excellent resistance to heat. Chemical resistance of the epoxy resin is excellent in non-oxidizing and weak acids but not good against strong acids. Alkali resistance is excellent in weak solutions. Chemical resistance of epoxy-glass laminates may be affected by any exposed glass in the laminate. Epoxies are available as castings, extrusions, sheet, adhesives, and coatings. They are used as pipes, valves, pumps, small tanks, containers, sinks, bench tops, linings, protective coatings, insulation, adhesives, dies for forming metal. When epoxies are used as adhesive, the epoxy resin and the aliphatic polyamine are packaged separately and mixed just before use.

11.6 Rubbers and Elastomers

Rubber and elastomers are widely used as lining materials for columns, vessels, tanks, piping. The chemical resistance depends on the type of rubber and its compounding. A number of synthetic rubbers have been developed to meet the demands of the chemical industry. Despite the fact that none of these has all the properties of natural rubber, they are superior in one or more ways. (trans-) polyisoprene and (cis-) Polybutadiene synthetic rubbers are close duplicates of natural rubber. A variety of rubbers and elastomers has been developed for specific uses.

11.6.1 Natural Rubber (NR)

Natural rubber (NR) or cis-1,4-polyisoprene has as basic monomer unit a cis-1,4-isoprene (it is sometimes called *caoutchouc*). Natural rubber is made by processing the sap of the rubber tree (i.e., *Hevea brasiliensis*) with steam, and compounding it with vulcanizing agents, antioxidants, and fillers. If a color is desired, it can be obtained by incorporation of suitable pigments (e.g., red: iron oxide, Fe_2O_3, black: carbon black and white: zinc oxide, ZnO). Natural rubber have good dielectric properties, an excellent resilience, an elevate damping capacity and a good tear resistance. As a general rule, natural rubbers are chemically resistant to non-oxidizing dilute mineral acids, alkalis, and salts. However, they are readily attacked by oxidizing chemicals, atmospheric oxygen, ozone, oils, benzene, and ketones and as a general rule they have also poor chemical resistance to petroleum and its derivatives and many organic chemicals in which the material soften. Moreover, natural rubbers are highly sensitive to UV-irradiation (e.g., sunlight exposure). Hence, natural rubber is a general-purpose material for applications requiring abrasion/wear resistance, electric resistance, and damping or shock absorbing properties. Nevertheless, owing to their mechanical limitations, natural as well as many synthetic rubbers are converted into a harder and more stable product by vulcanization and compounding with additives. The **vulcanization process** consists of mixing crude natural or synthetic rubber with 25 wt.% sulfur and to heat the blend at 150°C in a steel mold. The resulting rubber material is harder and stronger than the previous raw material due to the cross-linking reaction between adjacent carbon chains. Therefore, industrial applications of natural rubber include components such as internal lining for pumps, valves, piping, hoses, and for machined components when hardened by vulcanization. However, because natural rubber has a low chemical resistance and is sensitive to exposure to sunlight, unsuitable properties in many industrial applications, it is today replaced by newer improved elastomers.

11.6.2 Trans-Polyisoprene Rubber (PIR)

Trans-1,4-polyisoprene rubber (i.e., PIR, sometimes called *Gutta Percha* in the past) is a synthetic rubber with properties similar to those of its natural counterpart. It was first industrially prepared during World War II because of a lack of supply of natural rubber but despite containing fewer impurities than natural rubbers and having a simpler preparation process it is not widely used because it is also more expensive. Mechanical properties and chemical resistance is identical to that of natural rubber. As with many other rubbers its mechanical properties can be also improved by the vulcanization process.

11.6.3 Polybutadiene Rubber (BR)

Polybutadiene rubber (BR) is similar to natural rubber in its properties but it is more costly to process into intricate shapes than rubbers such as styrene butadiene rubber. Hence, it is essentially used as an additive in order to increase the tear resistance of other rubbers.

11.6.4 Styrene Butadiene Rubber (SBR)

Styrene butadiene rubber (SBR) is obtained by copolymerization of styrene and butadiene as basic monomer units usually mixed in the 3:1 mass ratio. It is well known commercially under the common trade name *Buna*®*S*. SBR exhibits a superior abrasion resistance than polybutadiene and natural rubber that explains its extensive use in automobile tires. Its chemical resistance is similar to that of natural rubber, that is, a poor resistance to oxidizing media, hydrocarbons and mineral oils. Hence, it offers no particular advantages in chemical service in comparison with other rubbers. Two main industrial processes are used for producing SBR: in the emulsion process, feedstocks are suspended in water together with a catalyst and a stabilizer; in the continuous-solution process, feedstocks are solubilized in a hydrocarbon solvent with an organometallic complex acting as a catalyst. SBR is the largest volume synthetic rubber used extensively in automobile tires, belts, gaskets, hoses, and other miscellaneous products.

11.6.5 Nitrile Rubber (NR)

Nitrile rubber (NR) is a copolymer of butadiene and acrylonitrile. It is produced in different ratios varying from 25:75 to 75:25. The manufacturer's designation should identify the percentage of acrylonitrile. Nitrile rubber under the common trade name *Buna*® *N* is well known for its excellent resistance to oils and solvents owing to its resistance to swelling when immersed in mineral oils. Moreover, its chemical resistance to oils is proportional to the acrylonitrile content. However, it is not resistant to strong oxidizing chemicals such as nitric acid, and it exhibits fair resistance to ozone and to UV irradiation which severely embrittles it at low temperatures. Nitrile rubber is used for gasoline hoses, fuel pumps diaphragm, gaskets, seals and packings (e.g., o-rings) and finally oil-resistant soles for safety work shoes.

11.6.6 Butyl Rubber (IIR)

Butyl rubber (IIR) is a copolymer of isobutylene and isoprene as basic monomer units. Butyl rubber is chemically resistant to non-oxidizing dilute mineral acids, salts and alkalis, and a good chemical resistance to concentrated acids, except sulfuric and nitric acids. Moreover, it has a low permeability to air and an excellent resistance to aging and ozone. However, it is readily attacked by oxidizing chemicals, oils, benzene, and ketones and as a general rule it has also poor chemical resistance to petroleum and its derivatives and many organic chemicals. Moreover, butyl rubbers are sensitive to UV-irradiation (e.g., sunlight exposure). As with other rubbers, its mechanical properties can be largely improved by the vulcanization process. Industrial applications are the same as for natural rubber. Butyl rubber is used for tire inner tubes and hoses.

11.6.7 Chloroprene Rubber (CPR)

Polycholoroprene is a chlorinated rubber material well-known under its common tradename ***Neoprene***® or grade M. This elastomer is an extremely versatile synthetic rubber with nearly 70 years of proven performance in a broad spectrum of industry. It was the first commercial synthetic rubber originally developed in 1930s as an oil resistant substitute for natural rubber. The polymer structure can be modified by copolymerizing chloroprene with sulfur and/or 2,3 dichloro-1,3-butadiene to yield a family of materials with a broad range of chemical and physical properties. By proper raw material selection and formulation of these polymers, the compounder can achieve optimum performance for a given end-use. Initially developed for resistance to oils and solvents it may resist various organic chemicals including mineral oils, gasoline, and some aromatic or halogenated solvents. It also exhibits good chemical resistance to aging and attack by ozone, and good resistance to UV irradiation (e.g., exposure to sunlight), until moderately elevated temperatures. Moreover, it has outstanding resistance to damage caused by flexing and twisting, an elevated toughness, and it resists burning but its electrical properties are inferior to that of natural rubber. Therefore, neoprene is noted for a unique combination of properties which has led to its use in thousands of applications throughout industry. It is extensively used as wire and cable jacketing, hose, tubes and covers. In the automotive industry, neoprene serves as gaskets, seals, boots, air springs, and power transmission belts, molded and extruded goods, cellular products adhesives and sealants, both solvent- and water-based foamed wet suits, latex dipped goods (e.g., gloves, balloons), paper, and industrial binders (e.g., shoe board). In civil engineering and construction applications, neoprene is used for bridge pads/seals, soil pipe gaskets, waterproof membranes, and asphalt modification.

11.6.8 Chlorosulfonated Polyethylene (CSM)

Chlorosulfonated polyethylene (CSM) is well known under its common trade name ***Hypalon®***. It is prepared by reacting polyethylene with sulfur dioxide and chlorine. This elastomer has outstanding chemical resistance to oxidizing environments including ozone, but it is readily attacked by fuming nitric and sulfuric acids. It is oil-resistant but it has poor resistance to aromatic solvents and most fuels. Except for its excellent resistance to oxidizing media, its physical and chemical properties are similar to that of neoprene with however improved resistance to abrasion, heat and weathering.

11.6.9 Polysulfide Rubber (PSR)

Polysulfide rubbers are usually prepared by reacting dichloroalkyls with sodium polysulfide as follows:

$$2nCl—R—Cl + nNa_2S_4 —> [—R—S_4—R—]_n + 2nNaCl$$

Polysulfide rubbers exhibit excellent chemical resistance towards oils and greases and they have very good dielectric properties. They are commercialized under the trade name ***Thiokol®***.

11.6.10 Ethylene Propylene Rubbers

Ethylene propylene rubber (EPR, or EPDM,) is a copolymer of ethylene and propylene. It has much of the chemical resistance of the related plastics: excellent resistance to heat and oxidation; good resistance to steam and hot water. It is used as a standard lining material for steam hoses; widely used in chemical services as well, having a broad spectrum of resistance.

11.6.11 Silicone Rubber

Polysiloxanes are inorganic polymeric materials well known under the common name of silicone rubbers or simply ***silicones.*** Instead of the classic carbon chain skeleton, these particular class of polymers are based on a chemical bond occurring between silicon and oxygen (Si—O) similar to that found in silicates. Silicones are usually prepared from the hydrolysis of chlorosilanes such as dimethyl dichlorosilane that yields a silanol according to the chemical reaction listed below:

$$(CH_3)_2SiCl_2 + 2H_2O \longrightarrow (CH_3)_2Si(OH)_2 + 2HCl$$

Afterwards, in a second step, the unstable silanol formed yields by condensation a polysiloxane:

$$n\,(CH_3)_2Si(OH)_2 \longrightarrow [\text{—}Si(CH_3)_2\text{—}O\text{—}]_n$$

Polysiloxanes are characterized by a three-dimensional branched-chain structure. Various organic groups introduced within the polysiloxane chain impart peculiar characteristics and properties to silicones. For instance, methyl groups impart water repellency, surface hardness, and nonflammability, while aromatic functions impart heat and wear resistance, and compatibility with organic chemicals. On the other hand, vinyl groups improve the stiffness of the macromolecule by reticulation. Finally, methoxy and alkoxy groups facilitate crosslinking at low temperatures. As a general rule, silicones have outstanding temperature resistance over an unusually wide temperature range (e.g., –75°C to +200°C). Silicones have relatively poor abrasion resistances and fair chemical resistance towards aromatic hydrocarbons (e.g., benzene, toluene), and to high-pressure steam, but withstand aging and ozone, as well as aliphatic solvents, oils and greases.

11.6.12 Fluoroelastomers

Fluoroelastomers combine excellent chemical resistance (e.g. oxidizing acids, and alkalis) and high-temperature resistance (i.e., up to 275–300°C for short periods of time); excellent oxidation resistance; good resistance to fuels containing up to 30% aromatics; mostly poor resistance in solvents or organic media by contrast with fluorinated plastics.

Viton® fluoroelastomers. There are three major general use families of Viton fluoroelastomer: A, B, and F. They differ primarily in their resistance to fluids, and in particular aggressive lubricating oils and oxygenated fuels, such as methanol and ethanol automotive fuel blends. There is a full range of Viton® grades that accommodates various manufacturing processes including transfer and injection molding, extrusion, compression molding, and calendering. There is also a class of high performance Viton® grades such as GB, GBL, GF, GLT, and GFLT.

Viton®A is a family of fluoroelastomer dipolymers, that is they are polymerized from two monomers, vinylidene fluoride (VF2) and hexafluoropropylene (HFP). Viton®A fluoroelastomers are general purpose types that are suited for general molded goods such as o-rings and v-rings, gaskets, and other simple and complex shapes.

Viton®B is a family of fluoroelastomer terpolymers, that is they are polymerized from three monomers, vinylidene (VF2), hexafluoropropylene (HFP), and tetrafluoroethylene (TFE). Viton®B fluoroelastomers offer better fluid resistance than A type fluoroelastomer.

Viton®F is a family of fluoroelastomer terpolymers, that is they are polymerized from three monomers, vinyl fluoride (VF2), hexafluoropropylene (HFP), and tetrafluoroethylene (TFE). Viton®F fluoroelastomers offer the best fluid resistance of all Viton types. F types are particularly useful in applications requiring resistance to fuel permeation.

Viton®GBL is a family of fluoroelastomer terpolymers, that is they are polymerized from three monomers, vinyl fluoride (VF2), hexafluoropropylene (HFP), and tetrafluoroethylene (TFE). Viton GBL uses peroxide cure chemistry that results in superior resistance to steam, acid, and aggressive engine oils.

Viton®GLT is a fluoroelastomer designed to retain the high heat and the chemical resistance of general use grades of Viton fluoroelastomer, while improving the low temperature flexibility of the material. Viton GLT shows a glass transition temperature 8–12°C lower than general use Viton grades.

Viton®GFLT is a fluoroelastomer designed to retain the high heat and the superior chemical resistance of the GF high performance types, while improving the low temperature performance of the material. Viton GFLT shows a glass transition temperatures 6–10°C lower than general use Viton grades.

11.7 Physical Properties of Polymers

Physical properties of common polymers and elastomers are reported in Tables 11.4 and 11.5, while physical quantities commonly used in the previous table to describe polymers characteristics are listed in Table 11.6 with the corresponding ASTM standards. On the other hand, particular mechanical properties are briefly described below.

Shore hardness. Durometer hardness is a property that, as applied to elastomers, measures resistance to indentation. **Shore A** scale is used for soft elastomers, with **shore D** scale for harder materials.

Compression modulus. Compression modulus is the stress required to achieve a specific deflection, typically 50% deflection. This test measures the polymer rigidity or toughness.

Flexural or tear strength. Tear strength measures the resistance to growth of a nick or cut when tension is applied to a test specimen. Tear strength is critical in predicting an elastomer's working life in demanding and abusive applications.

Tensile strength. Tensile strength describes the ultimate strength of a material when enough stress is applied to cause it to break. In combination with elongation and modulus, tensile strength can predict a material's toughness.

Elongation at break. Elongation relates to the ability of an elastomer to stretch without breaking. Ultimate elongation is the percentage of the original length of the sample and is measured at the point of rupture. This property is useful in identifying the appropriate elastomer for stress or stretching applications.

Table 11.4. Polymers Physical Properties 1

Usual chemical name	Trade Names	Acronym, Abreviation or Symbol	Category	Density (ρ/kg.m^{-3})	Elastic or Young's modulus (E/GPa)	Flexural modulus (G/GPa)	Compressive modulus (K/GPa)	Poisson's ratio (ν/nil)	Yield tensile strenght (σ_{YS}/MPa)	Ultimate tensile strenght (σ_{UTS}/MPa)	Elongation at break (Z/%)	Ultimate compressive strenght (σ_{UCS}/MPa)	Flexural yield strenght (/MPa)	Notched Izod impact energy per unit width (/J.m^{-1})	Hardness Rockwell (or Shore SHD)	Static friction coefficient (μ/nil)	Wear resistance (i.e., weight loss per 1000 cycles) (/mg)
Acrylonitrile butadiene styrene	Cycolac®	ABS	TP	1040–1180	1.7–2.6	0.92–3.03	1.03–2.90	n.a.	32–45	41–62	20–100	36–69	28–97	105–440	R75-115	0.5	n.a.
Butyl rubber	Kalar®, GR-1	IIR	EM	917	0.3–3.4	n.a.	n.a.	n.a.	n.a.	17	700–950	n.a.	n.a.	n.a.	SHA30-100	n.a.	n.a.
Casein-formaldehyde	Galathite®, Ameroid®	GAT	TP	1340–1350	3.5–3.9					52–69	2.5	186–365		4.4			
Cellulose acetate	Protectoid	CA	TP	1270–1340	1.0–4.0	8.3–27.6	n.a.	n.a.	17–43	12–110	6–70	20–55	14–110	100–450	R34-125	n.a.	65
Cellulose acetobutyrate	Tenite®	CAB	TP	1150–1220	0.3–2.0	0.62–4.14	n.a.	n.a.	10.3–48.3	20–60	38.74	14.5–52	12.4–110	260	R31-99	n.a.	n.a.
Cellulose acetopropionate		CAP	TP	1150–1220	0.34–1.38	0.69–1.93	n.a.	n.a.	10.3–48.3	13.8–51.7	35–60	21–79	21–75	182	R20-R120	n.a.	n.a.
Cellulose nitrate	Celluloid®	CN	TP	1350–1600	1.03–2.76					35–83	4–60	138–207	62–75	2.7–11.6	R95-R115		
Chlorinated polyvinyl chloride		CPVC	TP	1490–1500	2.48–3.30	2.6–3.15	2.3–4.14	n.a.	n.a.	n.a.	n.a.	42–75	53–299	n.a.	R112-117	n.a.	n.a.
Chlorofluorinated polyethylene	Hypalon®	CSM	EM	n.a.	n.a.	n.a.	n.a.	n.a.	n.a.	21	600	n.a.	n.a.	n.a.	SHA40-90	n.a.	n.a.
Epichloridrin rubber		ECO	EM	1270	n.a.	n.a.	n.a.	n.a.	n.a.	17	400	n.a.	n.a.	n.a.	SHA60-90	n.a.	n.a.
Epoxy resin	Novalac®	n.a.	TS	1120–1180	1.5–3.6	n.a.	n.a.	n.a.	n.a.	69–121	n.a.	n.a.	n.a.	0.4–0.9	n.a.	n.a.	n.a.

11 Polymers and Elastomers

Table 11.4. *(continued)*

Usual chemical name	Trade Names	Acronym, Abreviation or Symbol	Category	Density (ρ/kg.m^{-3})	Elastic or Young's modulus (E/GPa)	Flexural modulus (G/GPa)	Compressive modulus (K/GPa)	Poisson's ratio (ν/nil)	Yield tensile strenght (σ_{YS}/MPa)	Ultimate tensile strenght (σ_{UTS}/MPa)	Elongation at break (Z/%)	Ultimate compressive strenght (σ_{UCS}/MPa)	Flexural yield strenght (/MPa)	Notched Izod impact energy per unit width (/J.m^{-1})	Hardness Rockwell (or Shore SHD)	Static friction coefficient (μ/nil)	Wear resistance (i.e., weight loss per 1000 cycles) (/mg)
Ethylene propylene diene rubber	Dutral®, Nordel®	EPDM	EM	850	n.a.	n.a.	n.a.	n.a.	n.a.	21	100–300	n.a.	n.a.	n.a.	SHA30-90	n.a.	n.a.
Ethylene tetrafluoroethylene	Tefzel®, Halon®	ETFE	TP	1700	1.4	1.2–1.4	n.a.	n.a.	45	44.85	300	n.a.	38	1000	SHD67-75	0.4	n.a.
Ethylene-propylene rubber		EBR	EM	n.a.	n.a.	n.a.	n.a.	n.a.	n.a.	n.a.	n.a.	n.a.	n.a.	n.a.	n.a.	n.a.	n.a.
Ethylene chlorotrifluoroethylene	Halar®	ECTFE	TP	1680	1.7	1.7	1.7	n.a.	n.a.	31–48	200–300	n.a.	48	nil	R93	n.a.	n.a.
Fluorinated ethylene propylene	Neoflon®	FEP	TP	2150	n.a.	0.62	n.a.	n.a.	n.a.	23	325	21	18	nil	SHD50-65	0.27	n.a.
Melamine formaldehyde		MF	TS	1500	7.6–10	n.a.	n.a.	n.a.	n.a.	n.a.	n.a.	36–90	n.a.	11–21	M115-125	n.a.	n.a.
Natural rubber (cis-1,4-polyisoprene)	Caoutchouc	NR	EM	920–1037	3.3–5.9	n.a.	2.2	0.50	17.1–31.7	29	660–850	n.a.	n.a.	n.a.	SHA30-95	n.a.	n.a.
Butadiene acrylonitrile rubber	Buna®N, Nytek®	NBR	EM	1000	n.a.	n.a.	n.a.	n.a.	n.a.	21	510	100–600	n.a.	n.a.	SHA30-90	n.a.	n.a.
Perfluorinated alkoxy		PFA	TP	2140–2150	0.66	0.66	n.a.	n.a.	n.a.	21–29	300	n.a.	n.a.	nil	SHD60	0.2	n.a.
Phenol formaldehyde		PF	TS	1360	n.a.	n.a.	n.a.	n.a.	n.a.	n.a.	n.a.	n.a.	n.a.	n.a.	n.a.	n.a.	n.a.
Polyacrylic butadiene rubber		ABR	EM	n.a.	n.a.	n.a.	n.a.	n.a.	n.a.	n.a.	n.a.	n.a.	n.a.	n.a.	n.a.	n.a.	n.a.
Polyamide-imide	Torlon®, Ultem®	PAI	TP	1420–1460	4.5–6.8	2	n.a.	0.38	n.a.	110–190	7–15	170–220	76–200	60–140	E72-86	n.a.	n.a.
Polyamide nylon 11	Nylon®11	PA	TP	1040	1.5	n.a.	n.a.	n.a.	38	54	320	n.a.	n.a.	96	M60	n.a.	n.a.
Polyamide nylon 12	Nylon®12	PA	TP	1010	2.0	n.a.	n.a.	n.a.	45	50–55	290–300	n.a.	n.a.	n.a.	R84-107	n.a.	n.a.
Polyamide nylon 4,6	Nylon®46	PA	TP	1180	3.1–3.3	3.1	n.a.	n.a.	95	55–100	50	n.a.	n.a.	80	M92	n.a.	n.a.
Polyamide nylon 6	Nylon®6	PA	TP	1130	2.6–3.0	0.97	n.a.	n.a.	44	78	300	n.a.	n.a.	30–250	M82	0.2–0.3	5

Polyamide nylon 6,10	Nylon®610	PA	TP	n.a.	n.a.	n.a.	n.a.	n.a.	55	n.a.	n.a.	n.a.	n.a.	n.a.	R107	n.a.	n.a.
Polyamide nylon 6,12	Nylon®612	PA	TP	1060	2.1	1.241	n.a.	n.a.	51	52–61	100–250	n.a.	n.a.	5–70	R95-120	n.a.	n.a.
Polyamide nylon 6,6	Nylon®66	PA	TP	1140	3.3	1.207	n.a.	n.a.	59	82	300	103	n.a.	40–110	M89	0.2–0.3	3–5
Polyaramide	Kevlar®	PAR	TP	1440	59–124	n.a.	n.a.	n.a.	n.a.	2760	n.a.	n.a.	n.a.	n.a.	n.a.	n.a.	n.a.
Polyaramide	Nomex®	PAR	TP	n.a.	n.a.	n.a.	n.a.	n.a.	n.a.	n.a.	n.a.	n.a.	n.a.	n.a.	n.a.	n.a.	n.a.
Polyarylate resins	Durel®	PAR	TP	1210	16.60	13.80	n.a.	n.a.	68.9–75.8	138	50	n.a.	130.90	117–294	n.a.	n.a.	n.a.
Polybenzene-imidazole		PBI	TP	1300	6	7	6	n.a.	n.a.	156	3	344	n.a.	26	E105	0	n.a.
Polybutadiene rubber		BR	EM	910	2.1–10.3	n.a.	n.a.	n.a.	13.8–17.2	n.a.	450	n.a.	n.a.	n.a.	SHA45-80	n.a.	n.a.
Polybutadiene terephtalate		PBT	TP	1310	2.6	2.5	n.a.	n.a.	52	50	250	n.a.	n.a.	60	M70	n.a.	n.a.
Polybutylene		PB	TP	935	0.3	0.75	n.a.	n.a.	16–18	n.a.	n.a.	n.a.	n.a.	640–800	SHD60	n.a.	n.a.
Polycarbonate	Lexan®, Macrolon®	PC	TP	1200	2.3–2.4	2.2	n.a.	n.a.	62	55–75	100–150	80	n.a.	600–850	M70	0.31	10–15
Polychloroprene rubber	Neoprene®	CPR	EM	1230–1250	0.7–20.1	n.a.	n.a.	n.a.	3.4–24.1	n.a.	100–800	n.a.	n.a.	n.a.	SHA30-95	n.a.	n.a.
Polyether ether ketone	Victrex®	PEEK	TP	1320	3.7–4.0	n.a.	n.a.	0.4	n.a.	70–100	50	n.a.	n.a.	85	M99	0.18	n.a.
Polyether imide	Ultem®	PEI	TP	1270	2.9	n.a.	2.9	n.a.	n.a.	85	60	140	n.a.	50	R125	n.a.	10
Polyether sulfone		PESV	TP	1370	2.4–2.6	n.a.	n.a.	n.a.	n.a.	70–95	40–80	n.a.	n.a.	85	M88	n.a.	6
Polyethylene (high density)		HDPE	TP	950–968	0.414–1.24	0.062–0.105	n.a.	n.a.	25–40	16–40	5–12	n.a.	20–38	20–210	SHD60-73	0.29	n.a.
Polyethylene (low density)		LDPE	TP	912–925	0.14–1.86	0.069–0.207	n.a.	n.a.	0.2–11.5	5–26	20–40	n.a.	n.a.	1000	SHD41-46	n.a.	n.a.
Polyethylene (medium density)		MDPE	TP	926–940	0.17–0.38	0.240–0.790	n.a.	n.a.	10–19	10–19	10–20	n.a.	n.a.	1000	SHD45-60	n.a.	n.a.
Polyethylene (ultra-high molecular weight)	Lennite®	UHMW	TP	940	0.135–6.90	0.520–0.970	n.a.	n.a.	n.a.	20–40	500	n.a.	n.a.	1000	R50-70	0.1–0.2	n.a.
Polyethylene naphtalate		PEN	TP	1360	5.0–5.5	n.a.	5.5	n.a.	n.a.	200	60	n.a.	n.a.	n.a.	n.a.	0.27	n.a.

Table 11.4. *(continued)*

Usual chemical name	Trade Names	Acronym, Abreviation or Symbol	Category	Density (ρ/kg.m^{-3})	Elastic or Young's modulus (E/GPa)	Flexural modulus (G/GPa)	Compressive modulus (K/GPa)	Poisson's ratio (ν/nil)	Yield tensile strenght (σ_{YS}/MPa)	Ultimate tensile strenght (σ_{UTS}/MPa)	Elongation at break (Z/%)	Ultimate compressive strenght (σ_{UCS}/MPa)	Flexural yield strenght (/MPa)	Notched Izod impact energy per unit width (/J.m^{-1})	Hardness Rockwell (or Shore SHD)	Static friction coefficient (μ/nil)	Wear resistance (i.e., weight loss per 1000 cycles) (/mg)
Polyethylene oxide		PEO	TP	n.a.	n.a.	n.a.	n.a.	n.a.	n.a.	n.a.	n.a.	n.a.	n.a.	n.a.	n.a.	n.a.	n.a.
Polyethylene terephtalate	Mylar®	PET	TP	1560	2.0–4.0	3.0	n.a.	n.a.	81	80	70	n.a.	n.a.	13–35	M94-101	0.2–0.4	n.a.
Polyhydroxybutyrate (biopoloymer)		PHB	TP	1250	3.5	n.a.	n.a.	n.a.	n.a.	40	n.a.	n.a.	n.a.	35–60	n.a.	n.a.	n.a.
Polyimide	Vespel®	PI	TP	1420	2.0–3.0	3.1–3.45	n.a.	n.a.	n.a.	70–150	8–70	n.a.	n.a.	80	E52-99	0.42	n.a.
Polyisoprene (trans-1,4-polyisoprene)	Gutta Percha	PIP	EM	n.a.	n.a.	n.a.	n.a.	n.a.	n.a.	n.a.	n.a.	n.a.	n.a.	n.a.	n.a.	n.a.	n.a.
Polylactic acid		PLA	TP	1250		3.7–3.83				48.3–145	2.5–100			12.8–29			
Polymethyl methacrylate	Plexiglas®	PMMA	TP	1180–1190	3.036	2.24–3.17	2.55–3.17	n.a.	54–73	72.4	2.5–4	72–124	72–131	16–32	M92-100	n.a.	n.a.
Polymethyl pentene	TPX®	PMP	TP	835	1.5	n.a.	n.a.	n.a.	n.a.	25.5	15	n.a.	n.a.	49	R85	n.a.	n.a.
Polyoxymethylene (Heteropolymer)	Acetal®	POM	TP	1400	2.9–3.2	2.41–3.10	4.62	n.a.	65–69	69–83	40–75	110	90	53–80	R120-M78	n.a.	n.a.
Polyoxymethylene (Homopolymer)	Delrin®500	POMH	TP	1420	3.6	2.62–3.585	3.11	n.a.	57–70	72	15–75	107–124	94–110	75–130	M94-101	0.20–0.35	n.a.
Polyphenylene Oxide	Noryl®	PPO	TP	1090	2.5	2.59	n.a.	n.a.	n.a.	55–65	50	110	n.a.	200	M78-R115	0.35	20
Polyphenylene sulfide	Milkon®, Ryton®	PPS	TP	1350	1350	3.8	n.a.	n.a.	n.a.	65.5	1.6	110	96	16	R120	n.a.	n.a.
Polypropylene (atactic)	Propylux®	PP	TP	850–900	0.689–1.520	0.9	n.a.	n.a.	n.a.	21.4	300	n.a.	n.a.	763	R95	0.10–0.30	13–16

Polypropylene (isotactic)	Propylux®	PP	TP	920–940	0.689–1.520	1.2–1.7	n.a.	n.a.	n.a.	31–41	100–600	n.a.	n.a.	20–53	R95	n.a.	n.a.
Polypropylene (syndiotactic)	Propylux®	PP	TP	890–915	0.689–1.520	n.a.	n.a.	n.a.	n.a.	35	n.a.	n.a.	n.a.	n.a.	R95	n.a.	n.a.
Polystyrene (high-impact)	Propylux®	HIPS	TP	1040	1.6–2.4	2.07	n.a.	0.34	24.8	35–100	36–50	n.a.	35–39.3	133	L73	n.a.	n.a.
Polystyrene (normal)	Crystal®	PS	TP	1054–1070	2.3–4.1	3.17	n.a.	n.a.	n.a.	27–69	1.6–3	83–117	90	19–24	M60-90	n.a.	n.a.
Polysulfide rubber	Thiokol®	PSR	EM	1340	n.a.	n.a.	n.a.	n.a.	n.a.	4.83–8.63	100–400	n.a.	n.a.	n.a.	n.a.	n.a.	n.a.
Polysulfone	Udel®, Thermalux®	PSU	TP	1240	2.48	2.69	2.58	0.37	70.3	n.a.	75	96	106	69	M69	n.a.	n.a.
Polytetrafluoroethylene	Teflon®	PTFE	TP	2130–2220	0.48–0.76	0.19–0.55	0.41	n.a.	17–27	10–40	200–400	11.7	nil	160	R45(SHD50)	0.05–0.20	n.a.
Polytrifluorochloroethylene	Kel-F®	PTFCE	TP	2100	1.3	1.17–1.38	1.03–2.07	n.a.	37	32–35	80–250	32–51	51–76	65	R75-112	n.a.	n.a.
Polyurethane		PUR	TS	1050–1250	n.a.	n.a.	n.a.	n.a.	n.a.	29–49	10–21	n.a.	n.a.	102	n.a.	n.a.	n.a.
Polyvinyl acetate		PVA	TP	1191	0.6	n.a.	n.a.	n.a.	n.a.	n.a.	n.a.	n.a.	n.a.	n.a.	n.a.	n.a.	n.a.
Polyvinyl alcohol		PVAL	TP	n.a.	n.a.	n.a.	n.a.	n.a.	n.a.	n.a.	n.a.	n.a.	n.a.	n.a.	n.a.	n.a.	n.a.
Polyvinylidene chloride	Saran®	PVDC	TP	1630	0.3–0.55	n.a.	n.a.	n.a.	n.a.	48	200	n.a.	n.a.	16–53	R98-106	0.24	n.a.
Polyvinylidene fluoride	Kynar®, Foraflon®	PVDF	TP	1760	1.0–3.0	1.17–8.3	2.09–2.90	n.a.	20–57	5–25	50–300	55–110	67–94	120–320	R77-83	0.20–0.40	24
Polyvinyl fluoride		PVF	TP	1380–1720	4.4–11	n.a.	n.a.	0.4	33–41	n.a.	n.a.	n.a.	n.a.	n.a.	n.a.	n.a.	n.a.
Polyvinyl chloride	Vinyl®	PVC	TP	1160–1550	2.1–2.7	1.0	n.a.	n.a.	55	7–27	4.5–65	n.a.	n.a.	21	SHD65-85	n.a.	n.a.
Propylene-vynilidene hexafluoride	Viton®, Fluorel®	PVHF	EM	1800–1860	2.07–15.17	n.a.	n.a.	n.a.	8.96–18.62	4.8–11.0	100–700	n.a.	n.a.	n.a.	SHA50-95	n.a.	n.a.

Table 11.4. *(continued)*

Usual chemical name	Trade Names	Acronym, Abreviation or Symbol	Category	Density (ρ/kg.m^{-3})	Elastic or Young's modulus (E/GPa)	Flexural modulus (G/GPa)	Compressive modulus (K/GPa)	Poisson's ratio (ν/nil)	Yield tensile strenght (σ_{YS}/MPa)	Ultimate tensile strenght (σ_{UTS}/MPa)	Elongation at break (Z/%)	Ultimate compressive strenght (σ_{UCS}/MPa)	Flexural yield strenght (/MPa)	Notched Izod impact energy per unit width (/J.m^{-1})	Hardness Rockwell (or Shore SHD)	Static friction coefficient (μ/nil)	Wear resistance (i.e., weight loss per 1000 cycles) (/mg)
Silicone rubber (polysiloxane)	Rhodorsil®	SR	EM	n.a.	n.a.	n.a.	n.a.	n.a.	n.a.	6.5	n.a.	n.a.	n.a.	n.a.	SHD60A	n.a.	n.a.
Styrene-butadiene styrene rubber	Kraton-D®	SBS	EM	n.a.	n.a.	n.a.	n.a.	n.a.	n.a.	n.a.		n.a.	n.a.	n.a.	n.a.	n.a.	n.a.
Styrene-butadiene rubber	Buna®S, GR-S	SBR	EM	940	2.1–10.3	n.a.	n.a.	n.a.	12.4–20.7	21	450–500	n.a.	n.a.	n.a.	SHA30-90	n.a.	n.a.
Synthetic isoprene rubber		IR	EM	940	n.a.	n.a.	n.a.	n.a.	n.a.	25	750	n.a.	n.a.	n.a.	SHA30-95	n.a.	n.a.
Unplastified polyvinyl chloride		UPVC	TP	1300–1450	24–40	n.a.	n.a.	n.a.	n.a.	38–62	2–40	n.a.	n.a.	21	n.a.	n.a.	n.a.
Unsaturated polyester		UP	TS	1780	5.5	n.a.	n.a.	n.a.	n.a.	41	n.a.	n.a.	n.a.	32	M88	n.a.	n.a.
Urea formaldehyde		UF	TS	1470–1520	n.a.	n.a.	n.a.	n.a.	n.a.	n.a.	n.a.	n.a.	n.a.	n.a.	n.a.	n.a.	n.a.

Table 11.5. Polymers Physical Properties 2

Usual chemical name	Minimum operating temperature range (/°C)	Maximum operating temperature range (/°C)	Vicat softening temperature (/°C)	Glass transition temperature (T_g/°C)	Melting point or range (*m.p.*/°C)	Specific heat capacity (c_p/J.kg^{-1}.K^{-1})	Thermal conductivity (k/W.m^{-1}.K^{-1})	Coefficient of linear thermal expansion (α/10^{-6} K^{-1})	Deflection temperature under 0.455 MP a flexural load (T/°C)	Heat deflection temperature under 1.82 MP a flexural load (T/°C)	Electrical resistivity (ρ/ohm.cm)	Relative electric permittivity (@1MHz) ($\varepsilon\rho$/nil)	Dielectric field strenght (E_d/kV.cm^{-1})	Loss factor	Refractive index (n_D/nil)	Transmittance (T/%)	Water absorption per 24 hours (/%wt.day^{-1})	Water absorption at saturation (/%wt.)	Flame rating ASTM UL94
Acrylonitrile butadiene styrene	n.a.	70–110	n.a.	n.a.	88–120	1506	0.17–0.34	53–110	77–98	99–112	1,E+15	2.4–3.3	140–250	0.0200	1.49	92	0.3–0.7	n.a.	HB
Butyl rubber	-45	150	n.a.	-75 to -67	n.a.	1950	0.13–0.23	n.a.	n.a.	n.a.	n.a.	n.a.	n.a.	n.a.	1.5081	n.a.	n.a.	n.a.	n.a.
Casein-formaldehyde			94					50				6.1–6.8	157–276	0.052			7–14		
Cellulose acetate	-20	55–95	n.a.	n.a.	230	1200–1900	0.16–0.36	80–180	52–105	73	1,E+12	5	110	0.0600	1.49	n.a.	1.9–7.0	n.a.	Comb.
Cellulose acetobutyrate	-40	60–100	n.a.	n.a.	140	1464	0.16–0.32	140	73	62	1,E+11	2.5–6.2	100	0.0400	1.478	n.a.	0.9–2.2	n.a.	Comb.
Cellulose acetopropionate	n.a.	60–105	n.a.	n.a.	n.a.	1200–1600	0.16–0.33	120–160	n.a.	n.a.	1,E+13	n.a.	n.a.	n.a.	1.478	n.a.	1.0–3.0	n.a.	Comb.
Cellulose nitrate		60	90			1422–1590	0.180	65–160	43	66	10^{10}–10^{11}	6.7–8.8	118–590	0.0700	1.46–1.58				
Chlorinated polyvinyl chloride	n.a.	110	n.a.	n.a.	110	n.a.	0.140	68–78	n.a.	n.a.	1,E+15	3.3–3.8	480–590	0.0019	n.a.	n.a.	0.1	n.a.	n.a.
Chlorofluorinated polyethylene	n.a.	n.a.	n.a.	n.a.	n.a.	n.a.	n.a.	n.a.	n.a.	n.a.	n.a.	n.a.	n.a.	n.a.	n.a.	n.a.	n.a.	n.a.	n.a.
Epichloridrin rubber	-46	121	n.a.	n.a.	n.a.	n.a.	n.a.	n.a.	n.a.	n.a.	n.a.	n.a.	n.a.	n.a.	n.a.	n.a.	n.a.	n.a.	n.a.
Epoxy resin	n.a.	200–260	n.a.	n.a.	n.a.	n.a.	n.a.	n.a.	n.a.	230–260	n.a.	n.a.	n.a.	n.a.	n.a.	n.a.	n.a.	n.a.	n.a.

Table 11.5. *(continued)*

Usual chemical name	Minimum operating temperature range (/°C)	Maximum operating temperature range (/°C)	Vicat softening temperature (/°C)	Glass transition temperature (T_g/°C)	Melting point or range (*m.p.*/°C)	Specific heat capacity (c_P/J.kg^{-1}.K^{-1})	Thermal conductivity (k/W.m^{-1}.K^{-1})	Coefficient of linear thermal expansion (α/10^{-6} K^{-1})	Deflection temperature under 0.455 MP a flexural load (T/°C)	Heat deflection temperature under 1.82 MP a flexural load (T/°C)	Electrical resistivity (ρ/ohm.cm)	Relative electric permittivity (@1MHz) ($\varepsilon\rho$/nil)	Dielectric field strenght (E_d/kV.cm^{-1})	Loss factor	Refractive index (n_D/nil)	Transmittance (T/%)	Water absorption per 24 hours (/%wt.day^{-1})	Water absorption at saturation (/%wt.)	Flame rating ASTM UL94
Ethylene propylene diene rubber	–51	150	n.a.	n.a.	n.a.	n.a.	2.22		n.a.	n.a.	n.a.	2.5	n.a.	n.a.	1.474	n.a.	n.a.	n.a.	n.a.
Ethylene tetrafluoroethylene	0	150	n.a.	n.a.	270	n.a.	n.a.	90	105	70	1,E+17	2.6	800	0.0050	1.4028	n.a.	0.03	n.a.	V0
Ethylene-propylene rubber	n.a.	n.a.	n.a.	n.a.	n.a.	n.a.	n.a.	n.a.	n.a.	n.a.	n.a.	n.a.	n.a.	n.a.	n.a.	n.a.	n.a.	n.a.	n.a.
Ethylene chlorotrifluoroethylene	n.a.	150	n.a.	n.a.	245	n.a.	0.16	80	115	77	n.a.	2.6	190	n.a.	n.a.	n.a.	0.01	n.a.	V0
Fluorinated ethylene propylene	n.a.	204	n.a.	n.a.	260	n.a.	n.a.	135	n.a.	n.a.	1,E+18	2.1	n.a.	0.0007	n.a.	n.a.	0.01	n.a.	V0
Melamine formaldehyde	n.a.	120–200	n.a.	n.a.	n.a.	1674	0.167	22	n.a.	183	n.a.	n.a.	110–160	n.a.	n.a.	n.a.	0.1	n.a.	n.a.
Natural rubber (cis-1,4-polyisoprene)	–56	82	n.a.	–70	n.a.	1830	0.15	n.a.	n.a.	n.a.	n.a.	2.6	n.a.	n.a.	n.a.	n.a.	n.a.	n.a.	n.a.
Butadiene acrylonitrile rubber	–40	121	n.a.	n.a.	n.a.	n.a.	n.a.	n.a.	n.a.	n.a.	n.a.	n.a.	n.a.	n.a.	n.a.	n.a.	n.a.	n.a.	n.a.
Perfluorinated alkoxy	n.a.	260	n.a.	n.a.	305	n.a.	0.25	n.a.	n.a.	n.a.	1,E+18	2.1	800	0.0001	n.a.	n.a.	0.03	n.a.	V0
Phenol formaldehyde	n.a.	n.a.	n.a.	n.a.	n.a.	n.a.	0.25	16	n.a.	163	1,E+12	5.0–6.5	120–160	0.0060	n.a.	n.a.	0.2	n.a.	n.a.
Polyacrylic butadiene rubber	n.a.	n.a.	n.a.	104	317	n.a.	n.a.	n.a.	n.a.	n.a.	n.a.	n.a.	n.a.	n.a.	n.a.	n.a.	n.a.	n.a.	n.a.
Polyamide-imide	–200	200–260	n.a.	280	n.a.	1000	0.26–0.54	n.a.	n.a.	278	1,E+17	3.9–5.4	230	0.0420	1.42–1.46	n.a.	0.3	n.a.	V0
Polyamide nylon 11	–50	70–130	n.a.	n.a.	n.a.	1226	0.3	125	150	55	1,E+13	3	160–200	0.0500	n.a.	n.a.	1.1	n.a.	V2
Polyamide nylon 12	n.a.	n.a.	n.a.	n.a.	n.a.	1226	0.19	100–120	130–135	48–55	1,E+13	3.5	260–300	0.0600	n.a.	n.a.	1.1	n.a.	V2

Polyamide nylon 4,6	−40	100–200	n.a.	42.85	290	n.a.	0.3	25–50	220	160	1,E+13	3.8–4.3	200	0.3500	n.a.	n.a.	1.3	n.a.	V2
Polyamide nylon 6	−40	80–160	n.a.	53	223	1600	0.23	45	200	65–80	5,E+12	3.6	250	0.2000	1.53	n.a.	2.7	n.a.	Self-E
Polyamide nylon 6,10	n.a.	n.a.	n.a.	49.85	n.a.	n.a.	n.a.	n.a.	n.a.	n.a.	n.a.	n.a.	n.a.	n.a.	n.a.	n.a.	n.a.	n.a.	V2
Polyamide nylon 6,12	n.a.	n.a.	n.a.	45.85	212	1670	0.22	120–130	130–180	55–90		3.1–3.8	270	0.0280	n.a.	n.a.	3.0	n.a.	HB-V2
Polyamide nylon 6,6	−30	80–180	n.a.	49.85	255	1670	0.25	40	235	90–100	1,E+13	3.4	250	0.2000	n.a.	n.a.	2.3	n.a.	Self-E
Polyaramide	−200	180–245	n.a.	375	640	1400	0.04	2//	n.a.	n.a.	n.a.	n.a.	n.a.	n.a.	n.a.	n.a.	n.a.	n.a.	n.a.
Polyaramide	n.a.	n.a.	n.a.	375	640	n.a.	n.a.	n.a.	n.a.	n.a.	n.a.	n.a.	n.a.	n.a.	n.a.	n.a.	n.a.	n.a.	n.a.
Polyarylate resins	−150	100	n.a.	n.a.	n.a.	n.a.	0.178	50.4–72.0	355	174	1,E+12	2.93–3.30	134–183.1	0.0220	n.a.	n.a.	n.a.	n.a.	V0
Polybenzene-imidazole	n.a.	n.a.	n.a.	n.a.	n.a.	n.a.	n.a.	23	n.a.	426	1,E+13	3.20	n.a.	n.a.	n.a.	n.a.	0.4	5.0	V0
Polybutadiene rubber	−100	95	n.a.	n.a.	n.a.	n.a.	n.a.	n.a.	n.a.	n.a.	1,E+15	2.5	n.a.	n.a.	n.a.	n.a.	0.01	n.a.	n.a.
Polybutadiene terephtalate	n.a.	120	210	50	223	1350	0.21	45	150	60	1,E+15	3.2	200	0.0020	n.a.	n.a.	n.a.	n.a.	HB
Polybutylene	n.a.	n.a.	113	n.a.	127	n.a.	0.22	13	102–113	54–60	n.a.	2.53	n.a.	0.0005	n.a.	n.a.	0.03	0.6	n.a.
Polycarbonate	−135	115–130	n.a.	n.a.	n.a.	1200	0.19–0.22	38–70	140	128–138	1,E+16	2.92	150–670	0.0010	1.585	n.a.	0.1	n.a.	V0–V2
Polychloroprene rubber	−43	107	n.a.	−50	80	2170	0.192	n.a.	n.a.	n.a.	1,E+11	2.0–6.3	n.a.	n.a.	n.a.	n.a.	n.a.	n.a.	n.a.
Polyether ether ketone	n.a.	250	n.a.	143	334	320	0.25	26–108	147	142	1,E+16	3.1	190	0.0030	n.a.	n.a.	0.3	n.a.	V0
Polyether imide	n.a.	n.a.	n.a.	215	n.a.	n.a.	0.22	31–56	210	190–210	1,E+15	3.1	280–330	0.0013	n.a.	n.a.	0.25	n.a.	V0
Polyether sulfone	−110	180–200	n.a.	225	n.a.	n.a.	0.13–0.18	55	260	203	1,E+17	3.7	160	0.0030	1.65	n.a.	1.0	n.a.	V0
Polyethylene (high density)	n.a.	55–120	n.a.	−90 to −200	125 to 137	1900	0.42–0.52	100–200	75	46	1,E+15	2.30–2.40	420–520	0.0010	1.53–1.54	n.a.	0.01	n.a.	Comb.
Polyethylene (low density)	−60	50–70	n.a.	−110 to −20	102 to 112	1900	0.33	100–200	50	35	1,E+15	2.25–2.35	270–390	0.0005	1.51–1.52	n.a.	0.015	n.a.	Comb.
Polyethylene (medium density)	n.a.	93	n.a.	−118	110 to 120	1900	n.a.	100–200	n.a.	n.a.	1,E+17	2.25–2.35	180–390	0.0010	1.52–1.53	n.a.	0.01	n.a.	Comb.

Table 11.5. *(continued)*

Usual chemical name	Minimum operating temperature range (/°C)	Maximum operating temperature range (/°C)	Vicat softening temperature (/°C)	Glass transition temperature (T_g/°C)	Melting point or range (*m.p.*/°C)	Specific heat capacity (c_P/J.kg^{-1}.K^{-1})	Thermal conductivity (k/W.m^{-1}.K^{-1})	Coefficient of linear thermal expansion (α/10^{-6} K^{-1})	Deflection temperature under 0.455 MP a flexural load (T/°C)	Heat deflection temperature under 1.82 MP a flexural load (T/°C)	Electrical resistivity (ρ/ohm.cm)	Relative electric permittivity (@1MHz) ($\varepsilon\rho$/nil)	Dielectric field strenght (E_d/kV.cm^{-1})	Loss factor	Refractive index (n_D/nil)	Transmittance (T/%)	Water absorption per 24 hours (/%wt.day^{-1})	Water absorption at saturation (/%wt.)	Flame rating ASTM UL94
Polyethylene (ultra-high molecular weight)	n.a.	55–95	n.a.	n.a.	125 to 135	1900	0.45–0.52	130–200	69	42	1,E+18	2.3	190–280	0.0010	n.a.	n.a.	0.01	n.a.	Comb.
Polyethylene naphtalate	n.a.	155	n.a.	n.a.	n.a.	n.a.	n.a.	20–21	n.a.	n.a.	1,E+15	3.2	1600	0.0048	n.a.	84	n.a.	n.a.	n.a.
Polyethylene oxide	n.a.	n.a.	n.a.	n.a.	n.a.	n.a.	n.a.	n.a.	n.a.	n.a.	n.a.	n.a.	n.a.	n.a.	n.a.	n.a.	n.a.	n.a.	n.a.
Polyethylene terephtalate	−40 to −60	115–170	235	68.85	255	1200	0.17–0.40	15–65	115	80	1,E+14	3.0	170	0.0020	1.58–1.64	n.a.	0.1	n.a.	HB
Polyhydroxybutyrate (biopoloymer)	n.a.	95	n.a.	n.a.	n.a.	n.a.	n.a.	n.a.	n.a.	n.a.	1,E+16	3.0	n.a.	n.a.	n.a.	n.a.	n.a.	n.a.	n.a.
Polyimide	−270	250–320	n.a.	280 to 330	365	1090	0.10–0.36	30–60	n.a.	360	1,E+18	3.4	220	0.0018	1.42	n.a.	0.2–2.9	n.a.	V0
Polyisoprene (trans-1,4-polyisoprene)	n.a.	n.a.	n.a.	n.a.	n.a.	n.a.	n.a.	n.a.	n.a.	n.a.	1,E+15	2.5	n.a.	n.a.	n.a.	n.a.	n.a.	n.a.	n.a.
Polylactic acid									55										
Polymethyl methacrylate	−40	50–90	n.a.	104.85	45	1450	0.17–0.19	34–77	105	74–95	1,E+15	2.76	150	0.0140	1.49	92	0.2–0.3	n.a.	HB
Polymethyl pentene	−20 to −40	75–115	n.a.	29	250	2000	0.17	117	100	40	1,E+16	2.12	n.a.	0.0020	n.a.	n.a.	0.01	n.a.	Comb.
Polyoxymethylene (Heteropolymer)	n.a.	105	n.a.	n.a.	175	1464	0.37	85	n.a.	136	1,E+15	3.8	200	0.0048			0.2	n.a.	HB
Polyoxymethylene (Homopolymer)	n.a.	80–120	n.a.	n.a.	175	1464	0.22–0.24	122	170	136	1,E+15	3.7	200	0.0050	n.a.	n.a.	0.25	n.a.	HB

Polyphenylene Oxide	-40	80–120	n.a.	84.9	267	n.a.	0.22	38–60	137–179	100–125	2,E+17	2.59	160–200	0.0040	n.a.	n.a.	0.1–0.5	n.a.	V0
Polyphenylene sulfide	-170	190	n.a.	85	285	1090	0.17–0.28	30–49	260	135	1,E+14	3.8	177–240	0.0014	n.a.	n.a.	0.05	n.a.	V0
Polypropylene (atactic)	n.a.	240	n.a.	-18	176	1960	0.12	68–95	85	43	1,E+16	2.2–2.3	200–260	0.0005	n.a.	n.a.	0.01	n.a.	Comb.
Polypropylene (isotactic)	n.a.	160	n.a.	-1.5 to -10	165	1960	0.154	81–100	110–120	50–60	1,E+16	2.2–2.3	200–260	0.0005	n.a.	n.a.	0.01	n.a.	Comb.
Polypropylene (syndiotactic)	n.a.	140	n.a.	-10 to -8.2	135	1960	0.154	60–90	n.a.	n.a.	1,E+16	2.2–2.3	200–260	0.0005			0.01	n.a.	Comb.
Polystyrene (high-impact)	n.a.	n.a.	98	100	n.a.	1250	0.124	90	91	82	1,E+16	2.3–2.5	177–240	0.0004	1.59–1.60	n.a.	0.1	n.a.	V0
Polystyrene (normal)	n.a.	50–95	109	100	115	1250	0.10–0.13	30–210	90	80	1,E+16	2.4–3.1	180–240	0.0002	1.59–1.60	n.a.	0.4	n.a.	V0
Polysulfide rubber	-54	100–400	n.a.	n.a.	n.a.	n.a.	n.a.	n.a.	n.a.	n.a.	1,E+08	1.3	n.a.	n.a.	n.a.	n.a.	n.a.	n.a.	n.a.
Polysulfone	n.a.	n.a.	n.a.	193	n.a.	1255	0.259	31–51	n.a.	174	1,E+16	3.14	166	0.0050	1.63	99	0.22	n.a.	V0
Polytetrafluoroethylene	-260	180–260	n.a.	-97	327	1000	0.25	100–160	120	54	1,E+18	2.0–2.1	400–800	0.0001	1.38	nil	0.01	n.a.	V0
Polytrifluorochloroethylene	n.a.	175	n.a.	45	215	920	0.19–0.22	126–216	130	75	n.a.	2.46	197–230	n.a.	n.a.	n.a.	nil	n.a.	V0
Polyurethane	n.a.	n.a.	n.a.	n.a.	n.a.	1800	0.21	100–200	n.a.	n.a.	1,E+12	n.a.	n.a.	n.a.	n.a.	n.a.	1	n.a.	n.a.
Polyvinyl acetate	n.a.	n.a.	n.a.	29	n.a.	n.a.	0.159	100–200	n.a.	n.a.	n.a.	3.5	394	n.a.	1.4669	n.a.	3.6	n.a.	n.a.
Polyvinyl alcohol	n.a.	n.a.	n.a.	84.85	258	1255	0.795	n.a.	n.a.	n.a.	n.a.	n.a.	n.a.	n.a.	n.a.	n.a.	n.a.	n.a.	n.a.
Polyvinylidene chloride	n.a.	80–100	n.a.	-18.15	198	1339	0.13	190	n.a.	n.a.	1,E+12	3.2–6.0	160–240	0.0450	1.63	n.a.	0.1	n.a.	Self-E
Polyvinylidene fluoride	-40	135–150	n.a.	-40 to -35	141–178	1381	0.10–0.25	80–140	120–150	80–115	1,E+14	6.4–8.9	100–130	0.0490	1.42	n.a.	0.04–0.06	n.a.	V0
Polyvinyl fluoride	-70	175	n.a.	-20 to 41	200	n.a.	0.17	50	n.a.	n.a.	1,E+09	6.2–7.7	80–130	n.a.	1.46	n.a.	n.a.	n.a.	Self-E

Table 11.5. *(continued)*

Usual chemical name	Minimum operating temperature range (/°C)	Maximum operating temperature range (/°C)	Vicat softening temperature (/°C)	Glass transition temperature (T_g/°C)	Melting point or range (*m.p.*/°C)	Specific heat capacity (c_p/J.kg^{-1}.K^{-1})	Thermal conductivity (k/W.m^{-1}.K^{-1})	Coefficient of linear thermal expansion (α/10^{-6} K^{-1})	Deflection temperature under 0.455 MP a flexural load (T/°C)	Heat deflection temperature under 1.82 MP a flexural load (T/°C)	Electrical resistivity (ρ/ohm.cm)	Relative electric permittivity (@1MHz) ($\varepsilon\rho$/nil)	Dielectric field strenght (E_d/kV.cm^{-1})	Loss factor	Refractive index (n_D/nil)	Transmittance (T/%)	Water absorption per 24 hours (/%wt.day^{-1})	Water absorption at saturation (/%wt.)	Flame rating ASTM UL94
Polyvinyl chloride	n.a.	60–105	n.a.	81 to 87	212	1674	0.167	60–70	n.a.	n.a.	1,E+16	2.9–3.6	160–590	0.0070	1.54	n.a.	0.15–1	n.a.	Self-E
Propylene-vynilidene hexafluoride	–29	204	n.a.	n.a.	n.a.	n.a.	n.a.	n.a.	n.a.	n.a.	n.a.	n.a.	n.a.	n.a.	n.a.	n.a.	n.a.	n.a.	Self-E
Silicone rubber (polysiloxane)	–60	232	n.a.	n.a.	n.a.	n.a.	n.a.	n.a.	n.a.	n.a.	n.a.	n.a.	n.a.	n.a.	n.a.	n.a.	0.1	n.a.	n.a.
Styrene-butadiene styrene rubber	n.a.	n.a.	n.a.	n.a.	n.a.	n.a.	n.a.	n.a.	n.a.	n.a.	n.a.	n.a.	n.a.	n.a.	n.a.	n.a.	n.a.	n.a.	n.a.
Styrene-butadiene rubber	–60	120	n.a.	n.a.	n.a.	n.a.	n.a.	n.a.	n.a.	n.a.	1,E+14	2.4	n.a.	n.a.	n.a.	n.a.	n.a.	n.a.	n.a.
Synthetic isoprene rubber	–54	82	n.a.	n.a.	n.a.	n.a.	n.a.	n.a.	n.a.	n.a.	1,E+15	2.4	n.a.	n.a.	n.a.	n.a.	n.a.	n.a.	n.a.
Unplastified polyvinyl chloride	n.a.	n.a.	n.a.	n.a.	n.a.	n.a.	n.a.	n.a.	n.a.	n.a.	n.a.	n.a.	n.a.	n.a.	n.a.	n.a.	n.a.	n.a.	Self-E
Unsaturated polyester	n.a.	n.a.	n.a.	n.a.	n.a.	n.a.	n.a.	16	n.a.	n.a.	n.a.	n.a.	n.a.	n.a.	n.a.	n.a.	n.a.	n.a.	HB
Urea formaldehyde	n.a.	n.a.	n.a.	n.a.	n.a.	250	0.30–0.42	22–96	n.a.	n.a.	n.a.	n.a.	120–160	0.0350	n.a.	n.a.	0.5	n.a.	n.a.

Table 11.6. Polymers physical quantities and ASTM standards

	Physical quantities	ASTM standard	SI unit	U.S. customary unit
Processing	Processing temperature range		°C	°F
	Molding pressure range		Pa	psi
	Compression ratio		nil	nil
	Melt mass flow rate	ASTM D1238	$kg.s^{-1}$	lb./h
	Mold linear shrinkage	ASTM D955	nil	in/in
Mechanical	Density (ρ)	ASTM D792	$kg.m^{-3}$	$lb.ft^{-3}$
	Specific gravity (d)	ASTM D792	nil	nil
	Poisson's coefficient (ν)	n.a.	nil	nil
	Yield tensile strength (σ_{YS})	ASTM D638	Pa	psi
	Ultimate tensile strength (σ_{UTS})	ASTM D638	Pa	psi
	Tensile strength at break	ASTM D412	Pa	psi
	Elongation at yield (Z)	ASTM D638	nil	%
	Elongation at break (Z)	ASTM D638	nil	%
	Compressive strength	ASTM D695	Pa	psi
	Flexural yield strength	ASTM D790	Pa	psi
	Tensile or elastic modulus (E)	ASTM D638	Pa	psi
	Compressive or bulk modulus (K)	ASTM D695	Pa	psi
	Flexural or shear modulus (G)	ASTM D790	Pa	psi
	Unotched Izod impact strength (i.e., impact energy per unit width)	ASTM D256A	$J.m^{-1}$	ft.lb/in
	Abrasion resistance per 1000 cycles	ASTM D1044	kg.Hz	$lb.cycles^{-1}$
	Hardness Rockwell (HR) scale M and R	ASTM D785	nil	nil
	Hardness Durometer Shore (SH) scale A and D	ASTM D2240	nil	nil
	Hardness Durometer Barcol	ASTM D2583	nil	nil
Thermal properties	Minimum operating temperature (T_{min})	n.a.	°C	°F
	Maximum operating temperature (T_{max})	n.a.	°C	°F
	Brittle temperature (T_{brit})	ASTM D746	°C	°F
	Glass transition temperature (T_g)	ASTM D3418	°C	°F
	Vicat softening point (T_{vicat})	ASTM D1525	°C	°F
	Melting point ($m.p.$)	ASTM D3418	°C	°F
	Thermal conductivity (k)	ASTM C177	$Wm^{-1}K^{-1}$	$Btu.ft^{-1}.h^{-1}.°F^{-1}$
	Specific heat capacity (c_p)	n.a.	$J.kg^{-1}.K^{-1}$	$Btu.lb^{-1}.°F^{-1}$
	Coefficient of linear thermal expansion (a)	ASTM E831	K^{-1}	°F-1
	Deflection temperature under flexural load (0.455MPa)	ASTM D648	°C	°F
	Deflection temperature under flexural load (1.82MPa)	ASTM D648	°C	°F
Electrical	Dielectric permitivity (e_r) (1MHz)	ASTM D150	nil	nil
	Dielectric field strength (E_d)	ASTM D149	$V.m^{-1}$	$V.mil^{-1}$
	Dissipation or loss factor (δ)	ASTM D149	nil	nil
	Electrical volume resistivity (ρ)	ASTM D257	Ω.m	Ω.cirft/in
	Surface resistivity	ASTM D257	Ω.m	Ω.cirft/in

Table 11.6. *(continued)*

	Physical quantities	ASTM standard	SI unit	U.S. customary unit
Miscellaneous	Refractive index (n_D) (589 nm)	n.a.	nil	nil
	Optical transmission (T) (visible light)	n.a.	nil	%
	Nuclear radiation resistance (e.g., α, β. γ, and X-rays)	n.a.	nil	nil
Chemical	Water absorption in 24 hours	ASTM D570	s^{-1}	wt.%day^{-1}
	Water absorption at saturation	ASTM D570	nil	wt.%
	Chemical resistance	n.a.	nil	nil
	Flammability rating index	ANSI/UL-94	nil	nil
	Oxygen permeability	ASTM D3985	$m^2s^{-1}Pa^{-1}$	barrers
	Water vapor transmission	ASTM E96	s	perm-inch
	Water vapor transmission rate	ASTM F1249	s	perm-inch
Molten state	Apparent viscosity	ASTM D3835	nil	
	Melt specific heat	ASTM C351	mPa.s	
	Melt thermal conductivity	ASTM C177	mPa.s	
	Melt viscosity	ASTM D3835	mPa.s	

11.8 Gas Permeability of Polymers

Table 11.7. Gas permeability coefficients of most common polymers (in barrers)

Polymer	O_2	N_2	H_2	He	CO_2	H_2O
PAN	0.0002	–	–	–	0.0008	300
Cellophane	0.0021	0.0032	0.0065	0.005	0.005	1900
PVDC	0.0053	0.00094	–	0.31	0.03	0.5
PVA	0.0089	0.001	0.009	0.001	0.001	–
TFE	0.025	0.003	0.94	6.8	0.048	0.29
PETP	0.035	0.0065	3.70	1.32	0.17	130
PA (Nylon 6)	0.038	0.0095	–	0.53	0.10	177
PVC	0.0453	0.0118	1.70	2.05	0.157	275
HDPE	0.403	0.143	3.0	1.14	0.36	12.0
CA	0.78	0.28	3.5	13.6	23	5500
PP	2.3	0.44	41	38	9.2	51
PTFE (Teflon)	2.63	0.788	23.3	18.7	10.5	1200
PS	2.63	23.2	23.2	18.7	10.5	1200
LDPE	2.88	0.969	12.0	4.9	12.6	90

Conversion factors: 1 barrers = 10^{-10} cm^3(STP).cm /(cm^2.s.cmHg) (E)

11.9 Chemical Resistance of Polymers

See Table 11.8, pages 735–744.

Table 11.8. Chemical resistance of polymers. A ≡ satisfactory; B ≡ fairly; C ≡ poor and NR ≡ nonresistant

Chemical	LDPE		HDPE		PP		PS		PMP	PC	PEEK	PSU	Ulltem	Radel	Acetal	ABS	PVC	FEP	PFA	PVDF	ETEE	ECTFE	PCTFE
	20°C	60°C	20°C	60°C	20°C	60°C	20°C	60°C	20°C	20°C	20°C/60°C												
Acetaldehyde	B	NR	B	C	B	NR	NR	NR	B	A	A/A				B	NR	B	A/A	A/A	NR	A/A	B	B
Acetamide (std.)	A		A		A		A		A	NR							NR						
Acetic acid (5 wt.%)	A	A	A	A	A	A	A	B			A/A	B/C	A/B	A/B	B/NR	B		A/A	A/A	A/A	A/A	B	B
Acetic acid (50 wt.%)	A	A	A	A	A	A	B	B	A	A	A/A	C/C	B/C	B/C	NR	NR	A	A/A	A/A	A/A	A/A	B	B
Acetic anhydride	NR		C		B		NR		A	NR							NR						
Acetone	A	B	A	A	A	B	NR	NR	A	NR	A/A	NR	B	B	C/C	NR	NR	A/A	A/A	NR	A/C	A/B	A/B
Acetonitrile	A		A		C		NR		C	NR							NR						
Acrylonitrile	A		A		C		NR		A	NR						A	NR						
Adipic acid	A		A		A		A		A	A							A						
Alanine	A	A	A				A		A	NR						A	NR						
Allyl alcohol	A	A	A	A	A	A	B	C	A	B							B						
Aluminum chloride											A/A					NR		A/A	A/A	A/A	A/A	A/A	A/A
Aluminum hydroxide	A		A		A		B		A	C	A/A				B		A	A/A	A/A	A/A	A/A	A/A	A/A
Aluminum sulfate											A/A	A	A	A	A	B		A/A	A/A	A/A		A/A	A/A
Amino acids	A	A	A	A	A	A	A	A	A	A							A						
Ammonia (anhydrous)	A	A	A	A	A	A	B	C	A	NR	A/A	C	C	C		B	A	A/A	A/A	NR	A/A	A/A	A/A
Ammonium acetate (std.)	A		A		A		A		A	A							A						
Ammonium carbonate	A	A	A	A	A	A	B	C															
Ammonium chloride											A/A	A	A	A	A/B			A/A	A/A	A/A	A/A	A/A	A/A
Ammonium glycolate	A	A	A				A		A	B							A						
Ammonium hydroxide (28 wt.%)	A		A		A		B		A	NR	A/A	B	B	B	C/NR	B	A	A/A	A/A	A/A	A/A	A/A	A/A
Ammonium nitrate											A/A	A/A	A/A	A/A		NR		A/A	A/A	A/A	A/A	A/A	A/A
Ammonium phosphate	A	A	A	A	A	A	B	B			A/A							A/A	A/A	A/A	A/A	A/A	A/A

Table 11.8. *(continued)*

Chemical	LDPE		HDPE		PP		PS		PMP	PC	PEEK	PSU	Ulltem	Radel	Acetal	ABS	PVC	FEP	PFA	PVDF	ETEE	ECTFE	PCTFE
	20°C	60°C	20°C	60°C	20°C	60°C	20°C	60°C	20°C	20°C	20°C/60°C												
Ammonium oxalate	A	A	A				A		A	A							A						
Ammonium sulfate	A	A	A	A	A	A	B	B															
n-Amyl acetate	B	A	B				NR		B	NR	A/A	NR	NR	NR	B	NR	NR	A/A	A/A	A/C	A/A	A/NR	A/NR
Amyl alcohol												A/A	A/A	A/A	B			A/A	A/A	A/A	A/A	A/A	A/A
Amyl chloride	NR	NR	C	NR	NR	NR	NR	NR	NR	NR							NR	A/A	A/A	A/A	A/A	A/A	A/A
Aniline	B	B	B	B	B	C	NR	NR	B	C							NR						
Antifreeze (Prestone)												C/NR	B/B	B/B	B	NR		A/A	A/A	A/A		A/A	A/A
Aqua regia	NR		NR		NR		NR		NR	NR							NR						
Barium carbonate																NR		A/A	A/A	A/A	A/A	A/A	A/A
Barium chloride											A	A/A				NR		A/A	A/A	A/A	A/A	A/A	A/A
Benzaldehyde	A		A		A		NR		A	C	A	NR/NR	C/C	C/C	B	NR	NR	A/A	A/A	C	A/A	A/A	A/A
Benzene	NR	NR	C	C	C	NR	NR	NR	B	NR	A/A	NR/NR	C/C	C/C	B/B	C	NR	A/A	A/A	A/A	A/A	B/NR	B/NR
Benzoic acid (std.)	A		A		A		B		A	A	A/A				C	B	A	A/A	A/A	A/A	A/A	A/A	A/A
Benzine	C	N	B	B	B	C	NR	NR															
Benzyl acetate	A		A		A		NR		A	C							NR						
Benzyl alcohol	NR	NR	C	NR	NR	NR	NR	NR	NR	B	A				B	NR	B	A/A	A/A	A/A	A/A	B/NR	B/NR
Borax																NR		A/A	A/A	A/A	A/A	A/A	A/A
Boric acid	A	A	A	A	A	A	A	B			A/A				B	NR		A/A	A/A	A/A	A/A	A/A	A/A
Bromic acid																NR		A/A	A/A	A/A	A/A	A/A	A/A
Bromine	NR	NR	NR	NR	NR	NR	NR	NR	NR	C	NR	B	B	B	NR	NR	B	A/A	A/A	A/A	A/C	A/A	A/A
Bromobenzene	NR		C		NR		NR		NR	NR							NR						
Bromoform	NR		NR		NR		NR		NR	NR							NR						
Butadiene	NR		C		NR		NR		NR	NR							C						
n-Butyl acetate	B		A		B		NR		B	NR		NR	C	C	B	B	NR	A/A	A/A	B	A/A	A/A	B/NR
Butyl alcohol	A		A		A		A		A	B	A	C	B	B	B	C	B	A/A	A/A	A/A	A/A	A/A	A/A
Butyl chloride	NR		NR		NR		NR		C	NR							C						

Butyric acid	C	NR	B	C	C	NR	NR	NR	NR	C							B						
Calcium chloride	A	A	A	A	A	A	A	A			A/A	A/A			B/B			A/A	A/A	A/A	A/A	A/A	A/A
Calcium hydroxide (std.)	A	A	A	A	A	A	B	B	A	NR					B/B	NR	A	A/A	A/A	A/A	A/A	A/A	A/A
Calcium hypochlorite (std.)	A		A		A		B		A	C							B						
Calcium nitrate											A					B		A/A	A/A	A/A	A/A	A/A	A/A
Calcium sulfate	A	A	A	A	A	A	B	B															
Carbazole	A		A		A		A		A	NR							NR						
Carbon disulfide	NR		NR		NR		NR		NR	NR							NR						
Carbon tetrachloride	C	NR	C	C	C	NR	NR	NR	NR	NR							B						
Caustic soda																B		A/A	A/A	N/T	A/A	A/A	A/A
Cellosolve acetate	A		A		A		NR		A	C							C						
Cetyl alcohol												NR	NR	NR		NR		A/A	A/A	N/T		A/A	A/A
Chlorobenzene	NR	NR	NR	NR	NR	NR	NR	NR	C	NR							NR						
Chloroform	C	N	B	C	B	C	NR	NR	NR	NR							NR						
Chlorine gas											NR				NR	NR		A/A	A/A	A/A	A/A	A/A	A/A
Chlorine 10 wt.% in water	B	NR	B	C	C	NR	NR	NR	B	B							A						
Chloroacetic acid	A		A		A		B		A	C							C						
p-Chloroacetophenone	A		A		A		NR		A	NR							NR						
Chromic acid (10 wt.%)	A	A	A	A	A	A	A	A			A	C/C	B	B	NR	B		A/A	A/A	A/A	A/A	A/A	A/A
Chromic acid (50 wt.%)	A	A	A	A	B	C	C	C	B	C	B	NR	NR	NR	NR	B	A	A/A	A/A	A/C	A/C	A/A	A/A
Chromic acid (80 wt.%)											C	NR	NR	NR	NR			A/A	A/A			A/A	A/A
Citric acid (10 wt.%)	A	A	A	A	A	A	A	B	A	A	A/A	B	B	B	B/NR		B	A/A	A/A	A/A	A/A	A/A	A/A
Copper (II) chloride											A/A	A/A	A/A	A/A	A	B		A/A	A/A	A/A	A/A	A/A	A/A
Copper (II) fluoride											A/A					NR		A/A	A/A	A/A	A/A	A/A	A/A
Copper (II) nitrate											A/A					A		A/A	A/A	A/A	A/A	A/A	A/A
Copper (II) sulfate											A/A					NR		A/A	A/A	A/A	A/A	A/A	A/A
Cresol	NR	NR	C	NR	B	C	NR	NR	NR	NR							NR						
Cyclohexane	C	C	B	C	C	C	C	NR	NR	A	A/A	A/A	A/A	A/A	B	NR	B	A/A	A/A	A/A	A/A	A/A	A/A
Cyclohexanone	NR		C		C		NR		B	NR							NR						

Table 11.8. *(continued)*

Chemical	LDPE		HDPE		PP		PS		PMP	PC	PEEK	PSU	Ulltem	Radel	Acetal	ABS	PVC	FEP	PFA	PVDF	ETEE	ECTFE	PCTFE
	20°C	60°C	20°C	60°C	20°C	60°C	20°C	60°C	20°C	20°C	20°C/60°C												
Cyclopentane	NR		C		C		NR		C	NR							C						
n-Decane	C		C		C		C		C	C							A						
Diacetone alcohol	C		A		A		B		A	NR							NR						
o-Dichlorobenzene	C		C		C		NR		C	NR							NR						
p-Dichlorobenzene	C		B		B		NR		B	NR							NR						
1,2-Dichloroethane	NR		NR		NR		NR		NR	NR							C						
2,4-Dichlorophenol	NR		NR		NR		NR		C	NR							NR						
Diethyl benzene	NR		C		NR		NR		NR	C							NR						
Diethyl ether	NR		C		NR		NR		NR	NR						NR	C	A/A	A/A	A/C		A/A	NR
Diethyl ketone	B	C	B	B	B	B	NR	NR	B	NR							NR						
Diethyl malonate	A		A		A		NR		A	C							B						
Diethylamine	NR		C		B		B		C	NR							NR						
Diethylene glycol	A		A		A		B		A	B							C						
Diethylene glycol ethyl ether	A		A		A		NR		A	C							C						
Dimethyl acetamide	C		A		A		NR		C	NR							NR						
Dimethyl formamide	A		A		A		NR		A	NR							C						
Dimethylsulfoxide	A	A	A	A	A	A	A	B	A	NR							NR						
1.4-Dioxane	B	C	B	B	C	C	NR	NR	B	B							C						
Dipropylene glycol	A		A		A		A		A	B							B						
Ether	NR		C		NR		NR		NR	NR							C						
Ethyl acetate	A	C	A	B	A	C	NR	NR	C	NR		NR	NR	NR	B/C	NR	NR	A/A	A/A	NR	A/C	A/A	C/C
Ethanol (95 wt.%)	A	A	A	A	A	A	A	A	A	A	A/A	NR	B/NR	B/NR	C/C	NR	A	A/A	A/A	A/A	A/A	A/A	A/A
Ethyl benzene	C	NR	B	C	C	NR	NR	NR	C	NR							NR						
Ethyl benzoate	C		B		B		NR		B	NR							NR						
Ethyl butyrate	B		B		B		NR		C	NR							NR						
Ethyl chloride	C		C		C		NR		C	NR		NR	NR	NR	B	NR	NR	A/A	A/A	A/A	A/A	A/A	A/A

Ethyl cyanoacetate	A		A		A		B		A	C							C						
Ethyl lactate	A		A		A		C		A	C							C						
Ethylene chloride	B		B		C		NR		NR	NR		NR	C	C	C		NR	A/A	A/A	A/A	A/A	A/A	A/A
Ethylene glycol	A	A	A	A	A	A	A	A	A	B	A/A	C/C	A/B	A/B	B	B	A	A/A	A/A	A/A	A/A	A/A	A/A
Ethylene glycol methyl ether	A		A		A		NR		A	C							C						
Ethylene oxide	C	C	B	C	C	C	NR	NR	C	C	A				B		C	A/A	A/A	A/A	A/A	A/A	NR
Fatty acids	A		A		A		A		A	B	A/A				B	B	A	A/A	A/A	A/A	A/A	A/A	A/A
Ferric chloride	A	A	A	A	A	A	A	A			C/C	A/A	A/A	A/A	B			A/A	A/A	A/A	A/A	A/A	A/A
Ferric nitrate											A							A/A	A/A	A/A	A/A	A/A	A/A
Ferric sulfate											A/A					C		A/A	A/A	A/A	A/A	A/A	A/A
Fluorides	A	A	A	A	A	A	B	B	A	A							A						
Fluorine gas	C	NR	B	NR	C	NR	NR	NR	C	B	NR				NR	C	A	NR	NR	C	C	NR	NR
Fluosilicic acid																NR		A/A	A/A	A/A	A/A	A/A	A/A
Formaldehyde (10 wt.%)	A	A	A	A	A	A	C	NR			A/A	A/A	A	A	B	C		A/A	A/A	A/C	A/A	A/A	A/A
Formaldehyde (40 wt.%)	B	C	A	B	A	B	NR	NR	A	A							B						
Formic acid (98–100 wt.%)	A		A		A		C		A	A	A/A	B/B	B	B	NR	B/C	C	A/A	A/A	A/A	A/A	A/A	A/A
Freon TF	A		A		A		C		C	B	A	NR	B	B	B	NR	B	C	C		A/A	NR	NR
Fuel oil	C		B		A		NR		B	A							A						
Gasoline	C		B		B		NR		B	C	A/A	NR	C	C	B/C		B	A/A	A/A	A/A	A/A	A/A	A/A
Glacial acetic acid	A	B	A	A	A	B	NR	NR	A	NR	A/A	NR	NR	NR	NR	C	A	A/A	A/A	A/A	A/A	B	B
Glutaraldehyde	A		A		A		A		C	A							A						
Glycerol	A	A	A	A	A	A	A	A	A	A	A/A	B/C	B/C	B/C	B	A	A	A/A	A/A	A/A	A/A	A/A	A/A
Heating oil	C	NR	B	C	A	B	NR	NR															
n-Heptane	C		B		C		NR		C	A	A	NR	NR	NR	A/A	B	B	A/A	A/A	A/A	A/A	A/A	A/A
Hexane	C	C	B	C	B	C	NR	NR	C	C	A	NR	B	B	B/C	NR	B	A/A	A/A	A/A	A/A	A/A	NR
Hydrazine	NR		NR		NR		NR		NR	NR							NR						
hydrochloric acid (5 wt.%)	A	A	A	A	A	A	A	A			A/A	C	A	A	NR	C		A/A	A/A	A/A	A/A	A/A	A/A
hydrochloric acid (20 wt.%)	A	A	A	A	A	A	A	A			A/A	NR	NR	NR	NR	C		A/A	A/A	A/A	A/A	A/A	A/A
hydrochloric acid (35 wt.%)	A	A	A	A	A	B	C	C	A	NR	A/A	NR	NR	NR	NR	B	B	A/A	A/A	A/A	A/A	A/A	A/A

Table 11.8. *(continued)*

Chemical	LDPE		HDPE		PP		PS		PMP	PC	PEEK	PSU	Ulltem	Radel	Acetal	ABS	PVC	FEP	PFA	PVDF	ETEE	ECTFE	PCTFE
	20°C	60°C	20°C	60°C	20°C	60°C	20°C	60°C	20°C	20°C	20°C/60°C												
Hydrocyanic acid	A	A	A	A	A	A	B	B															
Hydrofluoric acid	C	NR	B	NR	C	NR	NR	NR			NR	NR	C	C	NR	NR		A/A	A/A	A/A	A/A	A/A	A/A
Hydrofluoric acid (4 wt.%)	A	B	A	A	A	B	B	C															
Hydrofluoric acid (48 wt.%)	A	B	A	A	A	B	NR	NR	A	NR							B						
Hydrogen peroxide (3 wt.%)	A	A	A	A	A	A	A	B															
Hydrogen peroxide (10 wt.%)											A/A	C	B	B	C	C		A/A	A/A	A/A	A/A	A/A	A/A
Hydrogen peroxide (30 wt.%)	A	A	A	A	A	B	A	B			A/A	NR	NR	NR	NR	NR		A/A	A/A	A/A	A/A	A/A	A/A
Hydrogen peroxide (90 wt.%)	A		A		A		A		A	A		NR	NR	NR	NR	NR	A	A/A	A/A	C	A/A	A/A	A/A
Hydrogen sulfide											A/A				B	C		A/A	A/A	A/A	A/A	A/A	A/A
Iodine crystals	NR		NR		C		NR		B	NR	B/C					NR	NR	A/A	A/A	A/B	A/A	A/A	A/A
Isobutanol	A		A		A		B		A	A							A						
Isopropyl acetate	B	C	A	B	B	C	NR	NR	B	NR							NR						
Isopropyl alcohol	A	A	A	A	A	A	B	B	A	A	A	NR	B	B	B/B	B	A	A/A	A/A	A/B		A/A	A/A
Isopropyl benzene	C		B		C		NR		NR	NR							NR						
Isopropyl ether	NR		NR		NR		NR		A	NR							NR						
Jet fuel	C		C		C		B		C	NR							A						
Kerosene	C	NR	B	B	B	C	NR	NR	B	A	A	NR	B	B	A/A	B	A	A/A	A/A	A/A		B	B
Lacquer thinner	NR		C		C		NR		C	NR							NR						
Lactic acid (10 wt.%)	A	A	A	A	A	A	B	B			A/A	B/B	B/B	B/B	B			A/A	A/A	A/C	A/A	A/A	A/A
Lactic acid (85 wt.%)	A	A	A				B		A	A							B						
Lactic acid (90 wt.%)	A	A	A	A	A	A	B	B															
Lead (II) acetate	A	A	A	A	A	A	A	A															
Magnesium chloride											A/A							A/A	A/A	A/A	A/A	A/A	A/A
Magnesium hydroxide																		A/A	A/A	A/A	A/A	A/A	A/A
Magnesium nitrate																		A/A	A/A	A/A	A/A	A/A	A/A
Magnesium sulfate											A/A							A/A	A/A	A/A	A/A	A/A	A/A

Malic acid											A/A							A/A	A/A	A/A	A/A	A/A	A/A
Mercury	A		A		A		NR		A	NR						NR	A	A/A	A/A	A/A	A/A	A/A	A/A
Mercury (II) chloride											A/A					B		A/A	A/A	A/A	A/A	A/A	A/A
Mercury (II) nitrate																C		A/A	A/A	A/A	A/A	A/A	A/A
Methanoic acid (100 wt.%)	A	B	A	A	A	B	C	C															
2-Methoxyethanol	A		A		A		NR		A	NR							B						
Methoxyethyl oleate	A		A		A		NR		A	C							NR						
Methyl acetate	C		C		B		NR		A	NR							NR						
Methanol	A	A	A	A	A	A	C	NR	A	B	A/A	NR	B	B	B/NR	NR	A	A/A	A/A	A/A	A/A	A/A	A/A
Methyl chloride											A				B	C		A/A	A/A	A/A	A/A	B	B
Methyl ethyl ketone	B	C	B	B	B	C	NR	NR	NR	NR	A/A	NR	NR	NR	C	C	NR	A/A	A/A	NR	A/A	B/NR	B/NR
Methyl isobutyl ketone	B		A		B		NR		C	NR							NR						
Methyl propyl ketone	B	C	A	B	B	C	NR	NR	C	NR							NR						
Methyl-t-butyl ether	NR		C	C			NR		A	NR							NR						
Methylene chloride	C	NR	B	C	C	NR	NR	NR	C	NR	A/A	NR	NR	NR	B	C	NR	A/A	A/A	A/C	A/A	B/NR	B/NR
Mineral oil	B	C	A	A	A	A	A	A	A	A	A/A	C/C	C	C	A/B		A	A/A	A/A	A/A	A/A	A/A	A/A
Mineral spirits	C		C	C			C		A	NR							B						
Naphtha											A/A	NR	B	B	B			A/A	A/A	A/A	A/A	A/A	A/A
Nickel (II) chloride											A/A					C		A/A	A/A	A/A	A/A	A/A	A/A
Nickel (II) nitrate											A/A					NR		A/A	A/A	A/A	A/A	A/A	A/A
Nickel (II) sulfate											A/A					NR		A/A	A/A	A/A	A/A	A/A	A/A
n-amyl acetate	C	C	B	C	C	C	NR	NR															
n-butyl alcohol	A	B	A	A	A	B	B	B															
Nitric acid (10 wt.%)	A	A	A	A	A	A	B	NR	A	A	A/A	NR	NR	NR	NR	NR	A	A/A	A/A	A/A	A/A	A/A	A/A
Nitric acid (50 wt.%)	B	C	B	B	C	NR	NR	NR	B	B	NR	NR	NR	NR	NR	B	B	A/A	A/A	A/C	A/C	B/NR	B/NR
Nitric acid (70 wt.%)	C	NR	B	NR	NR	NR	NR	NR	B	NR	A	NR	NR	NR	NR		C	A/A	A/A	NR	NR	A/A	A/A
Nitric acid (fuming)												NR	NR	NR	NR	B		A/A	A/A	NR	NR	A/A	A/A
Nitrobenzene	NR		C		NR		NR		NR	NR							NR						
Nitromethane	NR		C		C		NR		A	NR							NR						

Table 11.8. *(continued)*

Chemical	LDPE		HDPE		PP		PS		PMP	PC	PEEK	PSU	Ulltem	Radel	Acetal	ABS	PVC	FEP	PFA	PVDF	ETEE	ECTFE	PCTFE
	20°C	60°C	20°C	60°C	20°C	60°C	20°C	60°C	20°C	20°C	20°C/60°C												
n-Octane	A	A	A	A	A	A	NR	NR	A	B							C						
Oleic acid	C	NR	B	B	B	C	B	B			A	C	C	C	B/B	NR		A/A	A/A	A/A	A/A	A/A	A/A
Oxalic acid	A	A	A	A	A	A	A	B															
Ozone	C	C	B	C	C	C	NR	NR	A	A	A/A	B	B	B	NR	B	A	A/A	A/A	A/A	A/A	A/A	A/A
Perchloric acid	B	NR	B	NR	B	NR	B	C	B	NR	A/A	NR	C	C		C	B	A/A	A/A	A/A	A/B	A/A	A/A
Perchloroethylene	NR	NR	NR	NR	NR	NR	NR	NR	NR	NR							NR						
Phenol	C	NR	B	B	B	C	NR	NR	A	NR		NR	C	C	NR	C/NR	C	A/A	A/A	A/C	A/A	A/A	A/A
Phosphoric acid (10 wt.%)	A	A	A	A	A	A	B	B			A/A	A/A	B	B	NR			A/A	A/A	A/A	A/A	A/A	A/A
Phosphoric acid (85 wt.%)	A	A	A	A	A	B	A	B	A	A	A/A	B/C	B/C	B/C	NR	C/NR	A	A/A	A/A	A/A	A/A	A/A	A/A
Phosphorus trichloride	B	C	B	B	B	C	NR	NR															
Picric acid	NR		NR		NR		B		A	NR							NR						
Pine oil	B		A		A		NR		B	B							C						
Potassium acetate	A	A	A	A	A	A	A	A															
Potassium bromide	A	A	A	A	A	A	A	A			A/A				B	B		A/A	A/A	A/A	A/A	A/A	A/A
Potassium carbonate	A	A	A	A	A	A	A	A			A					B		A/A	A/A	A/A	A/A	A/A	A/A
Potassium hydroxide (5 wt.%)	A	A	A	A	A	A	B	B															
Potassium hydroxide, concentrated	A	A	A	A	A	A	B	B	A	NR	A/A	B/B	B	B	B	B	A	A/A	A/A	A/C	A/A	A/A	A/A
Potassium sulfide											A/A							A/A	A/A	A/A	A/A	A/A	A/A
Potassium permanganate	A	A	A	A	A	A	C	C															
Propanol											A				B			A/A	A/A	A/B	A/A	A/A	A/A
Propionic acid	C		A		A		B		A	NR							B						
Propylene glycol	A	A	A	A	A	A	A	A	A	B							C						
Propylene oxide	A		A	A			NR		A	B							C						
Pyridine	C	NR	B	C	C	NR	NR	NR															
Resorcinol, Sat.	A		A		A		B		A	B							C						

Salicylaldehyde	A		A		A		NR		A	B							C						
Salicylic acid (std.)	A	A	A	A	A	A	A	B	A	A							B						
Silicone oil	A		A		A		A		A	A	A/A	C	B	B	B/B	C/NR	A	A/A	A/A	A/A	A/A	A/A	A/A
Silver acetate	A	A	A	A	A	A	B	B	A	A							B						
Silver nitrate	A	B	A	A	A	B	B	C	A	A	A/A					C/NR	A	A/A	A/A	A/A	A/A	A/A	A/A
Sodium acetate (std.)	A		A		A		B		A	A	A						B	A/A	A/A	A/A	A/A	A/A	A/A
Sodium carbonate	A	A	A	A	A	A	A	A			A	B	B	B		B		A/A	A/A	A/A	A/A	A/A	A/A
Sodium Bisulfite															NR			A/A	A/A	A/A	A/A	A/A	A/A
Sodium Bromide															A	B		A/A	A/A	A/A	A/A	A/A	A/A
Sodium chloride (std.)	A	A	A	A	A	A	A	A			A/A	B	B	B	C	C		A/A	A/A	A/A	A/A	A/A	A/A
Sodium dichromate	A	A	A	A	A	A	A	A															
Sodium Flouride																B		A/A	A/A	A/A	A/A	A/A	A/A
Sodium hydroxide (1 wt.%)	A	A	A	A	A	A	B	B	A	C							A						
Sodium hydroxide (50 wt.%)	A	A	A	A	A	A	B	B	A	NR	A/A	A/B	B	B	A/B		NR	A/A	A/A	A/C	A/A	A/A	A/A
Sodium hypochlorite (15 wt.%)	A	A	A	A	A	A	B	B	A	B	A/A	C	B	B	NR		A	A/A	A/A	A/A	A/A	A/A	A/A
Sodium nitrate	A	A	A	A	A	A	A	A			A/A					C		A/A	A/A	A/A	A/A	A/A	A/A
Sodium phosphate																NR		A/A	A/A	A/A	A/A	A/A	A/A
Sodium sulfate	A	A	A	A	A	A	A	A															
Stearic acid (crystals)	A		A		A		A		A	A					B	NR	A	A/A	A/A	A/A	A/A	A/A	A/A
Sucrose	A	A	A	A	A	A	A	A															
Sulfuric acid (6 wt.%)	A	A	A	A	A	A	A	B															
Sulfuric acid (20 wt.%)	A	A	A	A	A	B	A	B			A/A	B/C	B/C	B/C	NR	B		A/A	A/A	A/A	A/A	A/A	A/A
Sulfuric acid (60 wt.%)	A	B	A	A	A	B	B	NR	A	B	NR	C/C	C/C	C/C	NR	NR	A	A/A	A/A	A/A	A/A	A/A	A/A
Sulfuric acid (98 wt.%)	B	B	B	B	C	NR	NR	NR	B	NR	NR	NR	NR	NR	NR	NR	B	A/A	A/A	A/B	A/A	A/A	A/A
Sulfur dioxide	A		A		A		C		A	A							A						
Tannic acid	A	A	A	A	A	A	B	B			A/A					NR		A/A	A/A	A/A	A/A	A/A	A/A
Tartaric acid	A		A		A		B		A	A							A						
Tetrahydrofuran	C	NR	C	C	C	C	NR	NR	C	NR							NR						

Table 11.8. *(continued)*

Chemical	LDPE		HDPE		PP		PS		PMP	PC	PEEK	PSU	Ulltem	Radel	Acetal	ABS	PVC	FEP	PFA	PVDF	ETEE	ECTFE	PCTFE
	20°C	60°C	20°C	60°C	20°C	60°C	20°C	60°C	20°C	20°C	20°C/60°C												
Thionyl chloride	NR	NR	NR				NR		NR	NR							NR						
Toluene	C	NR	B	B	C	NR	NR	NR	C	C	A	NR	NR	NR	B/C	NR	NR	A/A	A/A	A/B	A/A	B/NR	B/NR
Tributyl citrate	B		A		B		NR		B	NR							C						
Trichloracetic acid	C	NR	B	C	C	NR	NR	NR	A	C							C						
1,2,4-Trichlorobenzene	NR		NR		NR		NR		B	NR							NR						
Trichloroethane	NR	NR	C	NR	NR	NR	NR	NR	NR	NR							NR						
Trichloroethylene	NR		C		NR		NR		NR	NR	A/A	NR	B	B	NR	B	NR	A/A	A/A	A/A	A/A	A/A	NR
2,2,4-Trimethylpentane	C		C		C		NR		C	NR							NR						
Tris buffer solution	A		A		A		B		A	B							B						
Turpentine oil	C	NR	B	C	B	C	NR	NR	C	C	A	NR	NR	NR	B		B	A/A	A/A	A/A	A/A	A/A	A/A
Undecanol	A		A		A		B		A	B							A						
Urea	A	A	A	A	A	A	A	B	A	NR							B						
Vinylidene chloride	NR		C		NR		NR		NR	NR							NR						
Xylene	C	NR	C	C	C	NR	NR	NR	C	NR	A	NR	C	C	B	B	NR	A/A	A/A	A/A	A/A	A/A	B/NR
Zinc chloride	A	A	A	A	A	A	A	A			A/A	A/A	A/A	A/A	B			A/A	A/A	A/A	A/A	A/A	A/A
Zinc stearate	A		A		A		A		A	A	A/A						A	A/A	A/A	A/A	A/A	A/A	A/A

11.10 IUPAC Acronyms of Polymers and Elastomers

Table 11.9. IUPAC acronyms of polymers and elastomers

Acronym	Polymer name
A/EPDM/S	acrylonitrile/ethylene-propene-diene/styrene
A/MMA	acrylonitrile/methyl methacrylate
ABS	acrylonitrile/butadiene/styrene
ASA	acrylonitrile/styrene/acrylate
CA	cellulose acetate
CAB	cellulose acetate butyrate
CAP	cellulose acetate propionate
CF	cresol-formaldehyde
CMC	carboxymethylcellulose
CN	cellulose nitrate
CP	cellulose propionate
E/EA	ethylene/ethyl acrylate
E/P	ethylene/propene
E/VAC	ethylene/vinyl acetate
EC	ethylcellulose
EP	epoxide;epoxy
EPDM	ethylene/propene/diene
FEP	perfluoro(ethylene/propene); tetrafluoroethylene/hexafluoropropene
MF	melamine-formaldehyde
MPF	melamine/phenol-formaldehyde
PA	polyamide
PAN	polyacrylonitrile
PB	poly(l-butene)
PBA	poly(butylacrylate)
PBT	poly(butyleneterephthalate)
PC	polycarbonate
PCTFE	poly(chlorotrifluoroethylene)
PDAP	poly(diallylphthalate)
PE	polyethylene
PEO	poly(ethylene oxide)
PETP	poly(ethylene terephthalate)
PF	phenol-formaldehyde
PIB	polyisobutylene
PMMA	poly(methylmethacrylate)
POM	poly(oxymethylene); polyformaidehyde
pp	polypropene
PS	polystyrene

Table 11.9. *(continued)*

Acronym	Polymer name
PTFE	poly(tetrafluoroethylene)
PUR	polyurethane
PVAC	poly(vinyl acetate)
PVAL	poly(vinyl alcohol)
PVB	poly(vinylbutyral)
PVC	poly(vinyl chloride)
PVDC	poly(vinylidene dichloride)
PVDF	poly(vinylidene difluoride)
PVF	poly(vinyl fluoride)
PVFM	poly(vinylformal)
PVK	polyvinylcarbazole
PVP	polyvinylpyrrolidinone
S/B	styrene/butadiene
S/MS	styrene/a-methylstyrene
SI	silicone
UF	urea-formaldehyde
UP	unsaturated polyester
VC/E	vinyl chloride/ethylene
VC/E/MA	vinyl chloride/ethylene/methylacrylate
VC/E/VAC	vinyl chloride/ethylene/vinylacetate

11.11 Economic Data on Polymers and Related Chemical Intermediates

11.11.1 Average Prices of Polymers

Table 11.10. Average prices of polymers (2006)

Polymer	Price (US$/kg)
Acetal (copolymer)	3.60
Acetal (homopolymer)	2.70
Acrylonitrile styrene acrylate (ASA)	3.90
Acrylonitrile-butadiene-styrene (ABS)	2.50
Acrylonitrile-butadiene-styrene (ABS) (high impact grade)	2.70
Acrylonitrile-butadiene-styrene (ABS) (medium impact grade)	1.90
Acrylonitrile-butadiene-styrene (ABS) (reinforced 30 vol.% glass)	2.80
Ethylene vinyl acetate	3.70
High density polyethylene (HDPE)	0.90

Table 11.10. *(continued)*

Polymer	Price (US$/kg)
Low density polyethylene (LDPE)	1.90
Polyamide (Nylon 11)	8.30
Polyamide (Nylon 12)	6.60
Polyamide (Nylon 46)	9.00
Polyamide (Nylon 6)	3.10
Polyamide (Nylon 66)	3.70
Polycarbonate (PC)	5.50
Polycarbonate (PC)(high impact)	1.90
Polyester	2.40
Polyethyleneterephthalate (PET)	1.25
Polymethylmethacrylate (PPMA)	4.90
Polypropylene (PP)	1.40
Polystyrene (GPPS)	1.56
Polystyrene (HIPS)	1.60
Polyvinylchloride (PVC) Flexible	3.79
Polyvinylchloride (PVC) Rigid	1.72
PPA	8.09
Silicone	9.92

11.11.2 Production Capacities, Prices and Major Producers of Polymers and Chemical Intermediates

Table 11.11. Annual production capacities and prices of polymers and related chemical intermediates

Polymer or chemical intermediate	Major producers worlwide (annual capacity /10^3 tonnes)	Price 2004 (US$/tonne)
Acrylonitrile (ACN)	British Petroleum-BP (955); Solutia (490); Sterling Chemicals (335); BASF (280); Cytec Industries (227); DSM (200); DuPont (185); Saratovorgsintez (150); Repsol (125); PetKim (92); Arpechim (75)	1400–1510
Acrylonitrile-Butadiene-Styrene (ABS)	Chi Mei (1000); Bayer (679); GE Plastics (615); LG Chem (500); BASF (460); Dow Chemical (382); Cheil Industries (330); Formosa Chemicals (240); Korea Kumo Petrochemical (200); UMG ABS (175); Thai ABS (100); Polimeri Europa (80)	1525–1830
Adipic acid	DuPont (1045); Rhodia (470); Solutia (400); BASF (260); Radici (150); Asahi Kasei (120)	1320–1460
Adiponitrile	Invista (615); Butachimie (468); Solutia (300); BASF (140); Asahi Kasei (41); Liaoyang Petrochemical (24)	
Butadiene	BP (315); Polimeri Europa (320); Dow Chemical (275); Shell (195); Oxeno (180); Basell (170); Repsol (162); Sabic (130); DSM (120); Naphthachimie (120); BASF (105); HICI (100); Huntsman Petrochemicals (100); Atofina (60)	490

Table 11.11. *(continued)*

Polymer or chemical intermediate	Major producers worlwide (annual capacity /10^3 tonnes)	Price 2004 (US$/tonne)
Caprolactam	Solutia (500); BASF (420); DSM (250); Bayer (180); Radici (130);	950–970
Cellulose		780–800
Cellulose acetate		450
Dimethyl terephthalate (DMT)	KoSa (1120); DuPont (610); Vorridian (550); Petrocel (500); Oxxynova (480); Teijin (250); Khimvolokno Mogilev (305); Bombay Dyeing (165); Elana (105); Interquisa (90)	740
Ethylene	Dow (2825); BP (2270); Polimeri Europa (2190); BASF (1520); Fina (1500); Rühr Oel (1300); Borealis (1280); Sabic (1215); Basell (1050); Atofina (1040); Shell (900); Repsol (880); Huntsman (865); Naphthachimie (725); OMV (655); Exxon (545); Noretyl (450); Copenor (380); Enichem (250)	420–440
Ethylene dichloride (EDC)	SolVin (2130); EVC (1355); Atofina (980); Hydropolymers (975); LVM (930); Shin-Etsus (840); Vestolit (590); Ineos (550); ViniChlor (490); Wacker-Chemie (410); Enichem (355); Aiscondel (272); Dow (260); BSL (255)	225–235
Ethylene-prolypene diene monomer (EPDM)	DuPont-Dow (190); Exxon Mobil Chemical (180); DSM (170); Lanxess (115); Crompton (91); Polimeri Europa (85); Société du Caoutchouc (85); Japan Synthetic Rubber (70); Mitsui (60); Sumitomo (40); Petrochima (30)	850–900
Expandable polystyrene (EPS)	Nova Chemicals (320); BASF (260); BP (175); StyroChem (105); Polimeri Europa (70); Kaucuk (70); Dwory (65); SunPor (55); Unipol (55); Dow Chemical (40)	1580–1620
Formaldehyde	Dynea (720); BASF (650); Perstorp Formox (550); Degussa (519); Borden (380); Total (370); Formol y Derivados (280); Sadepan Chimica (250); Caldic Chemie (215); Krems Chemie (170); Akzo Nobel (110)	260
High density polyethylene (hdPE)	BP (1375); Borealis (1240); Basell (1200); Dow (950); Atofina (940); Sabic (855); Polimeri Europa (390); Repsol (245)	850–880
Isophthalic acid (PIA)	BP (325); AG International (120); Eastman Chemical (68); Lonza (70); KP Chemical (60); Interquisa (50);	940–1020
Linear low density polyethylene (LldPE)	Dow Chemical (800); Borealis (600); Polimeri Europa (590); BP (530); Cipen (420); Sabic (370); BSL (210);	950–1000
Low density polyethylene (LdPE)	Basel (1030); Polimeri Europa (830); Borealis (800); Exxon Mobil (665); Atofina (590); Sabic (565); Dow Chemical (520); BP (370); Repsol (230); TDSEA (170); Specialty Polymers (130)	1145–1160
Methyl Diphenyl diisocyanate (MDI)	BASF (715); Dow Chemical (760); Bayer (710); Rubicon (390); Huntsman (300); Nippon Polyurethane Industry (170)	2320–2500
Methyl methacrylate (MMA)	Rohm and Haas (640); Lucite (525); Mitsubishi Rayon (265); Atofina (180); Cyro Industries (132); Asahi Kasei (70); Repsol (45); BASF (36)	1465–1525
Nylon	Solutia; DSM; Rhodia (280); DuPont (150); Radici (90)	9760
Phenol	Ineos Phenol (1060); Polimeri Europa (480); Ertisa (370); Borealis (130)	950–1100
Polyacetals	DuPont (150); Polyplastics (150); Ticona (150); Ultraform (70); Korea Engineering Plastics (55); Asahi Kasei (44); Mitsubishi Engineering (20); Thai Polyacetal (20); Zaklady Azotowe Tarnowie (10)	2920–3490

Table 11.11. *(continued)*

Polymer or chemical intermediate	Major producers worlwide (annual capacity /10^3 tonnes)	Price 2004 (US$/tonne)
Polyacrylamide (PAM)	SNF (166); Ciba (115); CNPC (60); Cytec (47); Sinopec (32); Nalco (30); Stockhausen (26); Dia-Nitrix (12); Harima Chemical Japan (11)	3050–4300
Polyacrylic acid	Rohm and Haas (85); BASF (70); Nalco (30); Coatex (24); Ciba (18); National Starch (18); Protex (18); Kemira (16)	1950–4920
Polyaramides	DuPont (100); Teijin (100)	
Polybutylene terephthalate (PBT)	GE Plastics (120); BASF-GE (100); DuBay Polymer (80); DuPont (70); Chang Chung Plastics (66); Shinkong Synthetic Fibers (40); DSM (30); Ticona (30); Toray (24)	n.a.
Polycarbonate (PC)	Bayer (830); GE Plastics (780); Teijin Chemical (300); Dow Chemical (205); Mitsubishi Engineering Plastics (90); Sam Yang (85); LG Dow (70); Thai Polycarbonate (60); Sumitomo (50); Asahi Kasei (50); Formosa (50); Idmetsu Petrochemical (47)	2990–3600
Polychloroprene	DuPont (100); Lanxess (65); Denki Kagaku Kogyo (48); Enichem (43); Tosoh (30); Shanxi Synthetic Rubber (25); Showa Denko (20)	4000–5000
Polyester polyols	Dow (580); Bayer (565); BASF (290); Shell (250); Repsol (200); ICI (45); DuPont (40);	1700–1830
Polyethylene terephthalate (PET)	Voridian (465); DuPont (280); Dow (270); M&G Polimeri (195); Elana (120)	1400–1464
Polymethyl methacrylate (PMMA)	Rohm and Haas (640); Lucite (525); Mitsubishi Rayon (265); Atofina (180); Cyro Industries (132); Asahi Kasei (70); Repsol (45); BASF (36)	2260–2685
Polypropylene (PP)	Basell (2700); Borealis (1500); Atofina (1260); BP (1180); Sabic (1100); Exxon (500); BSL (210)	900–1000
Polystyrene (PS)	Dow (630); BASF (605); BP (340); Polimeri Europa (335)	1040–1060 1085–1110
Polytetrafluoro-ethylene (PTFE)	DuPont (25); Daikin (11); Dyneon (9.8); Asahi Glass (7); Solvay (6.5); AGC Chemicals (4)	
Polyvinylchloride (PVC)	Solvay (815); Atofina (700); HydroPolymers (610); Vinnolit (600); Shin-Etsu (295); JSC (255); Aiscondel (200)	770–780
Propylene	BP (1925); Dow Chemical (1335); Polimeri Europa (1145); Shell (840); BASF (810); Sabic (675); Borealis (620); Atofina (600); Repsol (500); Basell (475)	750–780
Propylene glycol (PG)	Dow Chemical (570); Lyondell (400); BP (90); BASF (80); Seraya Chemicals (65); Huntsman (60); Repsol (52); SKC (50); Arch Chemicals (35); Jin Hua Chemical (20); Sasol (18)	1080–1110
Propylene oxide (PO)	Lyondell (1900); Dow (1810); Elba (500); Huntsman (240); Repsol (220); BP (205); Sumitomo (200); Shell (200); Nihon Oxirane (180); BASF (125); Asahi Glass (110)	1410–1490
Purified phthalic anhydride (PPA)	BASF (210); Lonza (110); Proviron (100); Atofina (90); Bayer (85)	850–890
Purified terephthalic acid (PTA)	BP (3390); DuPont (570); Interquisa (1275); Tereftalatos Mexicanos (1050); Voridian (940); DAK (550); Rhodiaco (285); Invista (180); Dow (180)	980–1100
Styrene	Dow Chemical (1280); BASF (1050); Atofina (720); EniChem (660); Lyondel (640); Elba (550); Repsol (480); Shell (440); BP (380)	920

Table 11.11. *(continued)*

Polymer or chemical intermediate	Major producers worlwide (annual capacity /10^3 tonnes)	Price 2004 (US$/tonne)
Styrene butadiene rubber (SBR)	Polimeri Europa (230); Petro Borzesti (150); Dow (160); Lanxess (140); Michelin (40);	1520–1830
Toluene diisocyanate (TDI)	Bayer (420); BASF (370); Dow (215); Lyondell (260); Mitsui Chemicals (243);	1950–2074
Urea	Yara (2335); SKW (1070); Zaklady Azotowe (960); Togliatti Azot (900); Chimco (800); Ammonil (720); Ege Gubre Sanayii (600); Agrolinz (570); BASF (540); Fertiberia (500); Petrochemija (495);	200–205
Vinyl chloride monomer (VCM)	EVC (1090); Solvay (925); Vinnolit (660); Atofina (615); Shin-Etsu (600); HydroPolymer (590); LVM (550); Dow (300);	740–750

References: *Chemical Week* (CheW), *European Chemical News* (ECN), *Chemical Engineering and News* (CEN), and *Chemical Engineering* (CE).

11.12 Further Reading

ASH, M.B.; ASH, I.A.(1992) *Handbook of Plastic Compounds, Elastomers, and Resins, An International Guide by Category, Tradename, Composition, and Suppliers.* VCH, Weinheim.

BOST, J. (1985) *Matières plastiques, Vol. 1* & 2. Techniques & Documentation, Paris.

ELIAS, H.-G. (1993) *An Introduction to Plastics.* VCH, Weinheim.

FIZ Chemie (1992) *Parat - Index of Polymer Trade Names, 2nd. ed.* VCH, Weinheim.

VAN KREVELEN, D.W. (1994) *Properties of Polymers.* Elsevier, Amsterdam.

Index

(z+1)-average molar mass 696
(z+1)-average relative molar mass 696
2-methylpropane 1066
2-methylpropene 1066
70/30 Pt/Ir 579

A coefficient 1050
Abbe number 36
Abbe's equation 35
abestos 801, 861
abietic acid 698
abrasion resistance 114
ABS 706
absolute density 2
absolute humidity 1054
absolute magnetic susceptibility 491
absolute refractive index 33
absolute Seebeck coefficient 544
absolute temperature coefficient of
  refractive index 36
absorbance 40
  additivity 41
absorption
  Bouger's equation 39
  Bunsen–Roscoe 39
  coefficient 39
  decadic linear coefficient 39
  Einstein coefficients 42
  Napierian linear coefficient 39
  process 41
ac magnetic permeability 506
acanthite 397, 801
acceptor 459
accessory minerals 893
acetaldehyde 1122
acetals 711
acetic
  acid 148, 1122, 1168
  anhydride 1122
acetone 700, 1122
acetonitrile 1122
acetophenone 1122
acetyl
  acetone 1122
  chloride 1122
acetylene 1065
acetylene tetrabromide 777, 1171
Acheson process 626
achondrites 917
achorite 783
acicular 758, 892
acicular iron ore 828
acid lead 198
acid-copper lead 570
acidic electrolyte 1077
acid-leaching plant (ALP) 285
acid-regeneration plant (ARP) 285
acier inoxydable 95
acmite 801, 804
acoustical properties 23
acrisols 953

acronyms of rock-forming minerals 798
acrylic acid 1122
acrylic fiber 1027
acrylics 709
acrylonitrile 706, 717, 1122
acrylonitrile-butadiene-styrene (ABS) 706, 714, 721
actinium 1204
actinium a 1204
actinium b 1204
actinium c 1204
actinium d 1204
actinium k 1204
actinium series 1202
actinium x 1204
actinolite 801
actinon 1093, 1183, 1204
actinouranium 1204
activated manganese dioxide (AMD) 156
activated titanium anode 580
activation 694
activity calculations 1207
activity of a material containing natural U and Th 1207
activity of radionuclide 1207
adamantine 760, 783
additives 692
adiabatic flame temperature 1003, 1064
admiralty brass 184
admiralty gun metal 186
adularia 845
advanced ceramics 635
  hardness scales 12
aegirine 801
aerolites 467, 914, 915
Aerosil® 595
aerosol 1180
aerospace grade 1027
aerozine 50 1013
agate 467, 782, 801
aggregates 974
aging of ferroelectrics 538
A-glass 671
air 1065, 1074
aircraft window 672
air-hardening cold-work steels 118
air-hardening tool steels 118
AISI 600 series 121
AISI designation of tool steels 116, 117
akermanite 802
AL-6X 132
AL-6XN 132
alabamine 1183
alabandite 152, 802
alabaster 829
albedo 37, 396
albite 802
Alclad® 176
alexandrite 781, 817
alfisols 946
alite 971
alkali feldspars 899
alkali metals 213, 241
  amides 1075
  azides 1075
  lithium 217
  properties 214
alkali-cellulose 701
alkaline electrolytes 1077
alkaline process 609
alkaline solutions 164, 321
alkaline-earth metals 243
  properties 245
alkanes 909
Alkrothal® 14 549
alkylcelluloses 701
allanite 802
allochromatism 433
allotriomorph 758
allotriomorphous 892
allotropism 64
allotropism of iron 64
Alloy 19® 546
Alloy 20® 546
Alloy 31 132
Alloy 904L 132
Alloy Casting Institute (ACI) 103
alloy steels 89, 90
  carburizing 90
  case-hardening 90
Alloy® 20Mo-4 132
alloys
  nickel 128
alluvial placer deposits 278
allyl
  alcohol 1122
  chloride 1122
  cyanide 1122
almandine 782, 802
almandine spinel 783
almandite 802
Alnico magnets 511
alpha silicon carbide 626
alpha titanium alloys 304

alpha-alumina 606
alpha-alumina crystallites 603
alpha-beta titanium alloys 305
alpha-boron nitride 637
alpha-cristoballite 594
alpha-ferrite 75
alpha-iron 65
alpha-nitrogen 1075
alpha-pinene 1132
alpha-quartz 594
alpha-titanium 274
alpha-tridymite 594
altered ilmenite 279
alum 164
alum process 609
Alumel® 546
alumina 164, 274, 288, 600, 663, 795, 819, 980, 1032
  brown fused 608
  calcination 168
  calcined 603, 605, 606
  fibers 1028
  fused 605, 614
  high-purity 609
    alkaline process 609
    alum process 609
    chloride process 609
    gel process 609
  hydrate 167, 603
  metallurgical-grade 168, 604
  non-metallurgical-grade 168, 604
  tabular 605, 607
  trihydrate 168, 603
  white fused 608
alumina-silica
  fibers 1028
aluminized steel 176
aluminosilicates 596
aluminum 159, 297, 1032, 1183
  alloys 10, 159, 170
    applications and uses 176
    cast alloys 171
    standard designations 171
    wrought alloys 171
  brass 184
  bronze 185, 186
  carbide 653
  cathode 563, 564
  diboride 648
  dodecaboride 648
  dross 169
  dross recyclers 178
  electrowinning 168, 573
  hydroxides 602, 603, 606, 607
  killed steels 85
  major producers 177
  nitride 658
  oxide 159, 168
  oxihydroxides 603, 605
  secondary production 169
  selected properties 160
  sesquioxide 165, 600, 606, 663
  triethyl 703
  trihydrate 167, 605
  trihydroxides 605
aluminum-phosphate minerals 433
alunite 165, 802
alushite 816
alvite 337
amalgam 260, 397, 401
amazonite 782
amazonite green 841
amber 803
amblygonite 220, 803
American Cut 789
americanites 921
amethyst 467, 759, 782
Amex process 443
amides 710
aminoplastics 713
ammonal 1015
ammonia 1065, 1075, 1168
  heptoxide 393
ammonium
  cation 962
  chloride 151, 192, 284, 413
  dimolybdate 374
  diuranate 445
  hexachloroplatinate 413, 414
  hydrogen phosphate 964
  hydroxide 445, 1168
  metavanadate 340
  nitrate 1016
  nitrate–fuel oil 1017
  paratungstate 387
  perchlorate 1016
  perrhenate 392, 393
  picrate 1017
  polyvanadate 340
  sulfate 127
  thiocyanate 329
amosite 829
Ampere's law 488
Ampere-turn (A-turn) 488

amphiboles 278, 467
amphibolites 902
amphigene 837
amygdaloidal 759
analcidite 803
analcime 803
analcite 803
anasovite type I 617
anasovite type II 617
anatase 277, 614, 616, 666, 803
andalusite 165, 597, 599, 600, 803
Andersson–Magnéli 618
Andersson–Magnéli crystal lattice 576
andesine 804
andisols 946
andosols 950
Andrade's equation 18
andradite 782, 804
ANFO 1015, 1017
angiosperms 983
angle
  of incidence 32, 34
  of refraction 32, 34
anglesite 199, 569, 804
anhedral 758, 892
anhydrite 261, 754, 756, 804
aniline 237, 1122
aniline black 342
aniline-formaldehyde 713
anisotropic materials 765
ankerite 804
annabergite 805
annealed glass 676
annelids 931
anode 561
  activated titanium 580
  dimensionally stable 580
    for oxygen 581
  electrochemical equivalents 558
  for cathodic protection 587
  hydrogen-diffusion 582
  lead and lead-alloy 569
  materials 565, 567, 568
  noble-metal-coated titanium 578
  oxide-coated titanium 580
  platinized titanium 579
  precious- and noble-metal 568
  Pt-coated 580
  ruthenized titanium 580
anodic protection 586
anolyte 561
anorthic 1211
anorthite 805
anorthoclase 853
anorthosite 279
anorthosite complexes 277
anosovite 616, 805
anosovite type II 805
anthophyllite 805
anthracite 909
anthraxylon 1004
antibonding 455
antiferromagnetic 503
  compounds 503
  elements 503
  materials 503
antiferromagnets 496
antigorite 805
antimonial lead 198, 203, 570
antimonite 858
antimony 125, 806
  bloom 864
  chloride 1122
  fluoride 1122
  glance 858
antioxidants 693
antlerite 806
Antoine's law 1110
Antonoff's rule 1116
anyolite 867
apatite 261, 754, 786, 806
aphanitic 895
aphthitalite 806
aplite 755
aplite-pegmatite veins 222
apparent
  density 2
  mass 4
  weight 4
aqua regia 401
aquamarine 248, 781, 789, 791, 809
aqueous manganous sulfate electrolytes
  electrowinning 154
aragonite 261, 806
Aralac 701
Archeans cratons 887
Archimedes theorem 3
archons 786
arenosols 950
areometers 1104
argentite 397, 801
argentum 396, 855, 1183
argillans 938

argon 123, 447, 1065, 1090, 1092
argon/oxygen decarburization (AOD) 115
argon-oxygen decarbonization vessel 103
argyria 397
argyrodite 469
aridisols 946
arizonite 279, 850
arkose 907
armalcolite 807
Armstrong's mixture 1015
arrest points 76
arsenic 125, 142, 807
arsenic chloride 1122
arsenical pyrite 807
arsenicum 807, 1183
arsenopyrite 402, 807
arsine 1065
artificial radionuclides 1202
artificial wool 701
asbestos 754
asbolane 143
ash 904
Ashby's mechanical performance indices 21
ASME Boiler and Pressure Vessel Code 16
asterism 767
asthenosphere 887
ASTM standards for testing refractories 643
Astroloy® 132
atacamite 179, 807
atmophiles 1185
atmospheric nitrogen 962
atom
  polarizability 523
atomic gyromagnetic ratio 491
attapulgite 755, 846
attenuation index 39
attenuation ratio 512
attritus 1004
Auer-gas mantle 423
augelite 807
augite 808
aurum 400, 828, 1183
austempered ductile iron 80
austenite 66, 90, 96, 103
  finish temperature 139
  stabilizers 78, 101
  start temperature 139
austenitic 97
austenitic stainless steels 101
australasites 921
autoignition temperature 1062, 1063
automated tape lay-up 1030
automorphous 892
autunite 440, 808
average degree of polymerization 695
Avogadro's constant 2
Avogadro–Ampere equation 1041
Avogadro–Ampere law 1040, 1055
awaruite 67
azacyclopentane 1133
azote 1075, 1183
azurite 179, 808

baddeleyite 328, 337, 620, 668, 808
baking soda 235
balas ruby 783, 857
ball clay 598
band theory 455
barite 264, 754, 808, 1174
barite-water 1174
barium 263, 1099
  amalgam 264
  chloride 264
  crown 674
  hexaboride 648
  oxide 264
  sulfide 264
  titanate 536
bark 983
barometric equation 1045
barylites 762, 777, 894
baryte 808
barytine 808
basalt 755, 979
bastanaesite 448
bastnaesite 425, 809
  hydrochloric acid digestion process 428
  mining and mineral dressing 427
batholiths 890
batteries 556
battery grid 203
bauxite 125, 165, 340, 600, 601, 608, 609, 682, 755, 907
  Bayer process 166, 601
  chemistry 601
  comminution 602
  diasporic 166, 601
  digestion 602

gibbsitic 166, 601
Hall–Heroult process 166
mineralogy 601
mineralogy and chemistry 166
bauxitic
digestion 167
Bayer cycle 167
Bayer process 166, 601, 607
bayerite 603, 809
bazzite 433, 434
BCS theory 482
beach sands 278
bead test 775
beam electromagnetic radiation 32
Beattie–Bridgman 1044
Becher process 283, 289
bediasites 921
Beer–Lambert's law 40
deviation 41
belite 971
bell metal ore 857
Bending's alloy 210
Benedict, Webb and Rubin 1044
Benelite process 284, 285
benitoite 781
bentonite 842
benzal chloride 1122
benzaldehyde 1122
benzene 1122
benzoyl chloride 1122
benzoyl peroxide 693
benzyl
acetate 1122
alcohol 1122
benzoate 1123
chloride 1123
berdesinskiite 809
bernstein 803
Berthelot 1044
bertrandite 248, 754, 809
beryl 248, 433, 754, 761, 781, 789, 809
properties 790
beryllia 218, 244, 663
beryllium 233, 244, 1032
boride 648
copper 184
copper cast 186
diboride 648
fluoride 248
hemiboride 648
hemicarbide 653
hexaboride 648
hydroxide 248
metal 248
monoboride 648
nitride 658
oxide 663
producers 250
beryllium-aluminum alloys 249
beryllosis 244
Bessemer screw stock 87
beta alumina 607
beta boron nitride 638
beta silicon carbide 626
beta titanium alloys 305
beta-cristoballite 594
beta-ferrite 75
beta-iron 65
beta-pinene 1132
beta-quartz 594
beta-titanium 274
B-H curve 505
B-H diagram 510
B-H hysteresis loop 504, 505
biaxial 765
bicarbonate 232
bieberite 144
binary compounds
Strukturbericht designation 1216
bindheimite 810
binding energy of the electron 553
biocompatibility 47
biogenic sedimentary rocks 909
biomaterials 47
biophiles 1185
biopoloymer 724
biotite 433, 810
Biot–Savart equation 488
birefringence 37, 765
bisbeeite 817
bischofite 810
bismuth 142, 810
bismuth fusible alloy 210
bismuth solder 210
bitter spar 838
bituminous coal 909
black ash 263
black iron 85
black jack 856
black lead 829
black opal 782
black powder 1015
black silicon carbide 628
bladed 758

blanchardite 812
blast furnace 71
blast-furnace slag 974, 976
blasting agents 1015
blende 188
blister 180
blister test 775, *see also* bead test
Bloch boundaries 501, 534
Bloch walls 504
block copolymer 697
bloedite 810
bloodstone 782
blue lead 826
blue vitriol 815
boart/bort 783, 821
boehmite 165, 601, 603, 811, 905
Bohr magneton 490
bohrium 1184
boiling point elevation 1118
Boltzmann constant 501, 1106
Boltzmann distribution 460
Bolzano process 253
Bond's index 628
bonding
  conduction band 455
  energy-band gap 455
  valence band 455
boracite 811
borates 679, 754
borax 233, 471, 754, 775, 811
borax bead 775
borax bead test *see bead test*
Borazon® 638, 658
borides 648
  properties 648
bornite 179, 811
boron 470, 586, 648, 788, 1032
  atoms 585
  carbide 637, 639, 653
    applications and uses 637
  chemical vapor deposition (CVD) 1025
  fibers 1025
  nitride (BN) 470, 637, 658, 1019
    applications and uses 638
  sesquioxide 639
  tribromide 1123
  trichloride 1025, 1065, 1123
  trifluoride 1065
borosilicate crown 674
Borstar process 704
bosh 71
botryroidal 758
Boudouard reaction 282
Bouger's law 39
boulangerite 811
bournonite 811
Boyle temperature 1046
Boyle–Mariotte law 1039
bradleyite 812
braggite 414
brannerite 440, 441, 812
brasses 182, 184
braunite 152, 812
Brauns liquor 1171
Bravais space lattices 1211
Brazilian emerald 783
breakdown voltage 522
briartite 469
brick 978, 979, 980
Bridgman–Stockbarge melt growth technique 795
Bridgman–Stockbarge process 796
Briggs logarithm 23
Brinell hardness 12, 13
brines 251
brittle 762
brittle silver ore 857
brittleness 17
brochantite 812
bromargyrite 812
bromellite 812
bromine 1123
bromine liquid 1172
bromoargyrite 397
bromobenzene 1123
bromochloromethane 1123
bromoethane 1123
bromoform 777, 1135, 1172, 1174
bromyrite 812
bronzes 182, 185
bronzite 823
brookite 277, 614, 616, 666, 813
brown corundum 608
brown fused alumina 608
brown lead 339
brown manganese 838
Brownian motion 47
brucite 251, 613, 813
building materials 967
building stones 979
  properties 980
bulk density 2
bulk modulus 8

bullion 397
Buna® 717
Bunsen absorption coefficient 1050
bunsenite 813
Bunsen–Roscoe coefficient
  of absorption 39
buoyancy forces 3, 28
Burmese rubies 794
burned alumina 606
butadiene 706, 710, 717
butadiene 1,3 1065
butadiene acrylonitrile rubber 722
butane (n-) 1009, 1065
1,3-butanediol 1123
1,4-butanediol 1123
butanoic acid 1123, 1124
1-butanol 1123
2-butanol 1123
butene-1 1065
2-butoxyethanol 1123
butyl
  benzoate 1123
  glycolate 1123
  stearate 1123
  toluene 1123
butyl rubber (IIR) 717, 721
butyric acid 1123
butyronitrile 1124
byssolite 801
bytownite 813

cabochon 767
cadmium 536
cadmium copper 184
cadmium oxide 842
caesium 241, 1183
calamine 188, 856
calaverite 813
calcareous spar 814
calcia 218, 260, 274, 610, 664, 968
  lime 610
calcinated dolomite 253
  metallothermic reductions 253
calcination 971
calcine 190
  leaching 191
calcined alumina (CA) 168, 603
calcined dolomite 611
calcined vanadium pentoxide 340
calcite 190, 261, 612, 760, 761, 814,
  905, 908
calcium 243, 260, 536
  acetylenide 262
  alloys 261
  carbide 262
  carbonate 1089
  cyanamide 262
  hexaboride 649
  hydrogen phosphate 963
  hydroxide 225, 262, 610, 972, 973
  hypochlorite 262
  oxide 260, 610, 664
  phosphate 262, 964
  producers 262
  sulfate 262
  synthetic carbonate 261
  tungstates 387
calcium-based chemicals 261
calcium-lead alloys 571
calcium-tin-lead alloys 571
callaite 862
calogerasite 355
calomel 814
cambisols 954
cambium 983
campylite 851
Canadian deuterium uranium
  (CANDU) 448
Cañon Diablo troilite (CDT) 784
caoutchouc 716
capacitance 520
  of a parallel-electrode capacitor 521
  temperature coefficient 520
capacitor 520, 539
  charging 521
  discharging 521
  electrostatic energy 522
  geometries 521
capillarity 1116
capillary 758, 1116
capillary depletion 1117
capillary rise 1116, 1117
caproic acid 1124
caratage 401
carbides 262
  properties 648
  tools 333
Carbolon® 655
carbon 117, 404, 639, 786, 797, 1032
  anodes 572
  black 72
  chemical vapor deposition
    (CVD) 1026

diamondlike (DLC) 585
dioxide 148, 167, 573, 602
dioxide (CO_2) 72, 1065, 1089
disulfide 700, 1124
easily machinable steels 87
fibers 1026
carbonization 1027
graphitization 1027
stabilization 1027
matrix 1034
monofilaments 1026
monoxide (CO) 71, 714, 1065, 1087
flammability limits 1088
steels 84, 85, 86, 156
tetrachloride 1124
tool steels 119
carbonado 783, 821
carbonates 929, 1077
carbonatites 345
carbon–carbon composites (CCCs) 1019, 1034
carbon-in-pulp process 404
carbonization 1027
carbon-manganese steels 112
carbonyl iron 64, 73
carbonyl refining process 127
carbonyl sulfide 1065, 1088
Carborundum® 626, 655
carboxyhemoglobin 1088
carboxymethyl-hydroxypropyl guar (CMHPG) 679
carburizing 90
carburizing alloyed steels 90
carburizing steels 86
carnallite 238, 240, 251, 252, 814
carnotite 265, 340, 440, 814
Carpenter® 20Cb-3 132
Carpenter® 20Mo-6 132
carrolite 143
cartridge brass 184
CAS Registry Number (CARN) 50
cascandite 433
case steels
high-hardenability 90
medium-hardenability 90
casein plastics 701
casein-formaldehyde 701, 721
casein-formaldehyde thermoplastics 701
cassiopeium 1183
cassiterite 205, 207, 346, 386, 426, 433, 814
cast aluminum alloys 171
physical properties 175
cast copper alloys 183
physical properties 186
cast irons 77, 78
classification 80
high silicon level 80
cast steels 89
categories 95
Castner cells 234
catharometer 1080
cathode 561
delithiation 559
electrochemical equivalents 558, 560
for anodic protection 586
lithiation 559
material 563, 566
aluminum cathodes 563
low-carbon steel 563
mercury 565
nickel 565
titanium 564
zirconium 565
cathodic protection 587
cathodic ray television (CRT) tubes 263
cathodoluminescence 766
minerals 766
catholyte 561
caustic potash 1169
caustic soda 235, 1169
caustic-calcined magnesia 612
CBN 658
CdTe 471
celestine 263, 815
celestite 263, 815
cell multiplicity 1229
cell volume 1230
celluloid 691, 700
cellulose 699, 984, 1026
acetate 700, 721
acetobutyrate 721
acetopropionate 721
diacetate 700
nitrate 699, 702, 721
propionate 700
triacetate 700
xanthate 700
cellulosics 699, 701
celsian 815
celtium 336, 1183

cement 630, 967
  gypsum 968
  hardening 972
  history 969
  nonhydraulic 968
  oxide components 971
cementite 75
centrifuge tube 672
ceramic
  for construction 978
  maximum operating temperature 1241
  pyrometric cone equivalent (PCE) 641
ceramic hard ferrite magnets 512
ceramic matrix composites (CMCs) 1019, 1033
  properties 1035
ceramic oxides
  fibers 1028
  pervoskite-type structure 575
  spinel-type structure 575
ceramic-grade concentrate 222
ceramics 593
  advanced 635
  calcining 593
  engineered 635
  firing 593
  properties 648
  raw materials
    properties 628
  traditional 629
ceria 664
cerianite 664
ceric 422
cerium 422, 424
cerium dioxide 664
cerium hexaboride 649
cermet 640
cerussite 199, 815
cervantite 815
cesium 239, 240, 241
  amalgam 241
  hydroxide 241
  major producers 243
  salts 242
Ceylon ruby 794
ceylonite 838, 848
C-glass 671, 673
chabasite 815
chalcanthite 815
chalcedony 782, 801
chalcocite 179, 816
chalcophile 374, 1185
chalcopyrite 125, 152, 179, 190, 408, 816, 908
chalcotrichite 821
chalk 610, 754
chamotte 599
charge transfer transitions (CTT) 759
Charles and Gay-Lussac's law 1040
Charpy test 18
chatoyancy 767
cheddites 1015
cheluviation 930
Chemical Abstract Service (CAS) 50
chemical bonding in crystalline solids 455
chemical composition of dry air 1074
chemical grade chromite 369
chemical lead 198, 202, 570
chemical manganese dioxide (CMD) 156
chemical sedimentary rocks 908
chemical vapor deposition (CVD) 797
chemically resistant glass 673
chernozems 952
chert 908, 909
chessylite 808
chiastolite 803
Chilean saltpeter 233
china 629
china clay 598
chloanthite 816
chlor-alkali process 573
chlorargyrite 816
chlorates 574
chloride process 287, 609
chloride slag 281
chloride stress-corrosion cracking 97
chlorinatable slag 281
chlorinated polyvinylchloride 705, 721
chlorine 290, 404, 573, 1065, 1090
chlorine gas 151, 227, 234, 252
chlorite (1M) 816
chloritoid (2M) 816
chloroargyrite 397
chlorobenzene 1124
1-chlorobutane 1124
2-chlorobutane 1124
chlorocyclohexane 1124
chlorodifluoromethane 707
chloroethane 1065, 1124
2-chloroethanol 1124
chlorofluorinated polyethylene 721
chlorofluorocarbons 1093
chloroform 1124

chloromethane 1065
chloromethyl methyl ether 1124
chloronaphthalene 1124
1-chloropentane 1124
chloroprene rubber (CPR) 717
chloropropane 1124
chlorosulfonated polyethylene
  (CSM) 718
chlorotrifluoromethane 1065
chlorspine 783
chondrodite 817, 915
chromate anion 367
chrome 368
chrome iron ore 817
chrome vanadium 148
Chromel® 546
chromia 370
chromian diopside 786
chromian pyrope 786
chromic acid electrolysis 371
chromic iron 817
chromite 368, 575, 754, 817
  chemical-grade 369
  foundry-grade 369
  metallurgical-grade 369
  producers 372
  refractory-grade 369
  silicothermic process 370
chromite ore
  aluminothermic process 369
  soda-ash roasting 370
chromite spinel 218
chromium (Cr) 59, 95, 96, 120, 145, 289,
  339, 367, 503, 616
  alum 371
  aluminothermic process 371
  applications and uses 372
  boride 649
  carbide 101, 102, 653
  chemicals 369
  compounds 369
  copper 184, 186
  diboride 649
  disilicide 661
  electrowinning 371
  heminitride 658
  hexavalent 367
  metal 369
  monoboride 649
  nitride 658
  oxide 664
  properties 60
  pure metal 369
  sesquioxide 370
  silicide 661
  steels 84
  trioxide 371
chromium-alum electrolysis 371
chromium-molybdenum steels 84
chromium-vanadium steels 84
chromophore 793
chrysoberyl 248, 781, 817
chrysocolla 817
chrysolite 826
chrysolite light yellowish green 845
chrysotile 817
cinnabar 191, 817
cis-platin 416
citrine 782
clarain 1005
class I dielectrics 538
class II dielectrics 539
classes of symmetry 1212
classification of cast irons 80
classification of fluids 1106
classification of fuels 1000
classification of igneous rocks 891, 899
classification of industrial dielectrics 538
classification of meteorites 914
classification of natural and synthethic
  polymers 692
classification of plastics and
  elastomers 697
classification of proppant materials 679
classification of refractories 630
clathrates 1009, 1087, 1090, 1094, 1095
Clausius 1044
Clausius–Clapeyron equation 31,
  1110, 1118
Clausius–Mosotti equation 523
clay mineral 629
clayey limestones 601
clays 467, 596, 600, 755, 909, 929
cleavage 760
Clérici's liquor 1172
cleveite 1091
clevelandite 802
Clifford's rule 786
clinker 970
  formation 971
clinohumite 818
clinorhombic 1211
clinozoïsite 818
close packed arrangements 1211

closed tube test 772
coal 909, 1004
  anthracite 1006
  ash content (AC) 1005
  bituminous 1006
  classification 1006
  fixed carbon (FC) 1005
  lignite 1006
  moisture content (MC) 1005
  petrographical classification of 1004
  properties 1007
  subbituminous 1006
  volatile matter (VM) 1005
coarse aggregates 975
cobalt (Co) 59, 117, 141
  allotropes 141
  alloys 141, 145, 146
  major producers 149
  metal 142
    electrowinning 144
  minerals 143
  properties 60
  superalloys 145
cobalt beryllium copper 184
cobalt bloom 823
cobalt hemiboride 649
cobalt monoboride 649
cobaltite 143, 575, 818
coefficient of cubic thermal
  expansion 27
coefficient of linear thermal
  expansion 26
coefficient of surface thermal
  expansion 27
coefficient of thermal expansion 26
coercitive electric field strenght 535
coercitive force 505
coercitive magnetic field strenght 505
coercivity 505
coesite 594, 786, 818
coffinite 440, 441, 818
coherent deposit process 364
coke 1004
  ash content (AC) 1005
  fixed carbon (FC) 1005
  moisture content (MC) 1005
  properties 1007
  volatile matter (VM) 1005
coke oven gas 1081
cold working 11
cold-hearth melting 297
colemanite 471, 819
colligative properties 1118
collodions 700
colloidal and dispersed systems 1180
collophane 908
colophony 697
coloradoite 819
coloration of igneous rocks 894
columbite 344, 345, 392, 433, 665, 819
columbium 344, 1183
columbotantalite 356
columbotantalite ore 346
columnar 758, 892
comburant 999, 1062
  source of ignition 1062
combustion 999
  adiabatic flame temperature 1003
  enthalpy 999, 1002
  excess of air 1002
  stoichiometric equation 1001
  stoichiometric ratios 1001
  thermodynamic properties 1004
commercially pure nickel 131
commercially pure titanium 301
commodities
  world annual production 1248
common lead 570
common nonferrous metals 159
compacted graphite cast iron 80
complete wetting 1114
composite 1019
  density 1021
  elastic moduli 1022
  loading perpendicular to fibers 1023
  material 1019
  physical properties 1021
  reinforcements 1025
  specific heat capacity 1023
  structural classification 1020
  tensile strength 1022
  thermal conductivity 1023
  thermal expansion coefficients 1024
  voids fraction 1021
compound semiconductor 457, 459
compounds
  Strukturbericht designation 1218
compressibility factor 1046
compression 7
compression modulus 8
compression test 10
compressive
  strength 10
  stress 7, 8

concentration of electric charge
  carriers 460
concentric 758
conchoidal 761
concrete 630, 967, 976, 977
  alkali-silica reaction 978
  degradation 977
  prestressed 976
  recycled 975
  steel reinforced 976
  sulfate attack 978
  typical mixtures 977
condenser 520
conduction 28
conduction band 455
conductor 456
conglomerate 907
constant stress 17
Constantan® 546, 549
constants 50
construction materials 967
  properties 980
contact angle 1113
continental crust 887
continuous fibers 1025
Continuously Closed Cup Test 1121
convection 28
conversion process 194
cooling by adiabatic demagnetization
  496
Cooper pairs 483
copolymer 697
  ethylene-chlorotrifluoroethylene 709
  ethylene-tetrafluoroethylene 709
copolymerization 694
copper 126, 179, 249, 819, 1032
  alloys 179, 181, 182
  blister 180
  carbonates 179
  cathode 574
  electrorefining 180, 564
  electrorefining byproduct 404
  electrowinning 180, 571
  hydrometallurgical process 180
  hydroxide 179
  leaching 180
  lead 198, 202
  major producers 187
  nickels 182
  pyrometallurgical process 180
  selected properties 160
  smelting 180
  sulfide 127
  UNS designations 181
copper vitriol 815
copperas 287, 840
copper-beryllium alloys 249
copper-nickel 124, 185
copper-nickel alloy 185, 304
copper-nickel ores 413
cordierite 819
core 886, 888
Corning® 0080 672
Corning® 0120 672
Corning® 0137 672
Corning® 0138 672
Corning® 0160 672
Corning® 0281 672
Corning® 0317 672
Corning® 0320 672
Corning® 0331 672
Corning® 6720 672
Corning® 7570 672
Corning® 8078 673
Corning® 9025 673
Corning® 9068 673
corona mechanism 533
corroding lead 570
corrosion resistance 108, 114
corundum 165, 329, 663, 792, 819
  properties 793
cosmogenic radionuclides 1202, 1206
cosmonuclides 1202
cotunnite 819
Coulomb's
  forces 1037
  law 519
  modulus 8
coulsonite 820
coumarone-indene plastics 702
country rock 752
coupholite 762, 777, 894
covelline 820
covellite 820
covolume 1042
CPVC 705
crack
  dimension 16
  geometry 16
creep mechanism 17
cristobalite (alpha) 820
cristoballite 594, 596
critical angle 34
critical constants 1046

critical density 477, 1047
critical magnetic field strenght 477
critical molar volume 1047
critical opalescence 1047
critical parameters 1047
critical point 65, 75, 1047
critical pressure 1047
critical temperature 477, 1047
crocidolite 853
crocoite 368, 369, 820
Cronifer® 132
cross product 1228
crotolinas 938
crotonaldehyde 1125
Crown flint 675
crude oil 909, 1008
crushing strength 10
crust 886
cryolite 165, 168, 233, 820
cryolithe 754
cryptocrystalline graphite 623
cryptomelane 152
crystal 751, 1209
  anhedral 758
  cell multiplicity 1229
  charge transfer electronic transitions 759
  color 759
  density 1228
  development 892
  dimensions 892
  euhedral 758
  external shapes 892
  field theory (CFT) 759
  glass 672
  habit 758
  lattice 1231
  morphology 1213
  proportion 892
  pulling 795
  Schoenflies–Fedorov point group 1213
  space lattice 1228
  space lattice structure 759
  structures of gas hydrates 1094
  subhedral 758
  symmetry 1213
  system 757, 1211
  theoretical density 1228
crystalline graphite 623
crystallites 607, 609
crystallochemistry 1209
crystallographic calculations 1228
crystallography 757
crystals 37, 757
  symmetry 1212
  uniaxial 765
Crystolon® 655
CSM 718
cubanite 125, 820
cubic 1211
cubic boron nitride 787
cubic expansion 1046
cubic space groups 1227
cumar gum 702
cuprite 179, 821
cupronickels 129
cuprum 179, 819, 1183
Curie point 535, 536, 538
Curie temperature 501, 508, 534, 538
Curie Weiss law 501
current density 461
CVD silicon carbide 628
cyanite 834
1-cyanobutane 1125
2-cyanoethanol 1125
cyanocobalamine 142
cyanogen 1065
cyclic stresses 18
cyclohexane 1125
cyclohexanethiol 1125
cyclohexanol 1125
cyclohexanone 1125
cyclohexene 1125
cyclonite 1017
cyclooctane 1125
cyclopentane 1125
cyclopentanol 1125
cyclopentene 1125
cyclopropane 1065
cyclotetramethylene tetranitrate 1017
cylinder glass 676
cyprine 864
Czochralski 795
Czochralski crystallization process 472
  pulling crystal growth technique 472
Czochralski method 795

d'Arcet's alloy 210
Dalton 694
Dalton's law 1041
damping capacity of solids 24
damping constant 39
damping of sound 24

Dana's classes 757
DAPEX process 443
Darcy equation 1107
Darcy–Weisbach equation 1107, 1108
dark red silver ore 850
darmstadtium 1184
datolite 821
davidite 440
dawsonite 821
dead-burned dolomite 611
dead-burned magnesia 613
Debye temperature 31
Debye's forces 1042
1-decanol 1125
decay chains 1202
decomposition 167
decyl oleate 1125
deformation phenomena 19
degree of saturation 1056
Delrin® 711
delta-ferrite 75
delta-iron 66
demagnetization 496, 505
demantoid 782, 804
dendritic 758
dense aqueous solutions of inorganic salts 1172
dense emulsions 1174
dense halogenated organic solvents 1171
dense media 777
densities of states 460
density 1, 762, *see also* mass density
  apparent 2
  bulk 2
  of mixtures 5
  tap 2
  temperature dependence 2
  theoretical 2
  x-ray 2
dental amalgam 399
Denver cell 199
depth of penetration 507
descaling 271, 272
desliming 167
dessicants 1095
  properties 1096
detritic or clastic sedimentary rocks 907
deuterium 1065, 1079, 1080
deuterium oxide 1125
deuterohydrates 1090, 1094
dew point 1057
diabase 755
diacetyl 1125
diadochy 751, 757
diagenesis 889
diallage 822
diamagnetic materials 491, 499
diamagnetism 479, 485
diamagnets 491, 499
  magnetic permeabilities 500
  magnetic susceptibilities 500
1,2-diaminoethane 1125
diammonium
  hydrogen phosphate 964
  molybdate 374
diamond 471, 585, 654, 753, 754, 783, 786, 821
  American Cut 789
  caratage 789
  clarity 789
  classification 784
  color 788
  cutting 788
  luster 760
  micro-Vickers indenter 763
  physical properties and characteristics 785
  shaping 788
  standard brilliant cut 789
  synthesis
    chemical vapor deposition (CVD) 797
    high pressure high temperature (HPHT) 797
  synthetic electrodes 585
  valuation 788
diaphaneity 760
diaspore 165, 601, 604, 821
diaspore clay 599
diatomaceous earth 595
diatomite 595, 755, 908
diazodinitrophenol 1015, 1016
diborane 1065
dibromomethane 1125, 1171
1,2-dibromopropane 1125
dibutyl
  ether 1125
  ketone 1125
dibutylamine 1125
dicalcium silicate 971
3,4-dichlorobenzotri-fluoride 1125
3,5-dichlorobenzoyl chloride 1125
dichlorodifluoromethane 1066
1,1-dichloroethane 1126

1,2-dichloroethane 1126
dichlorofluoromethane 1066
dichloromethane 321, 1126
1,2-dichloropropane 1126
dichlorosilane 1066
dichroism 37, 766
dichromate anion 367
dicyclohexylamine 1126
didynium 1183
dielectric
  absorption 524
  behavior 530
  breakdown 522, 532, 533
  breakdown voltage 524
  constant 520, 523
  electrical properties 540
  field strength 522
  heating 526
  linear 538
  losses 520, 525, 526, 532
  materials 519, 539
    electrostriction 533
  polarization 523
  properties of gases 1052
  strength 533
  thickness 528
dieterici 1044
diethanolamine 1126
diethyl
  ether 1126
  ketone 1126
  malonate 1126
  phthalate 1126
  sulfate 1126
  sulfide 1126
diethylamine 1126
diethylene glycol 1126
diethylenetriamine 1126
diffusion 462
diffusion coating 364
digestion of bauxite 602
dihexyl ether 1126
dihydrogen 1078
diiodomethane 777, 1126, 1171
diisobutyl ketone 1126
diisocyanate 715
diisopropyl ether 1126
diisopropylamine 1126
dilatometry 65, 74
dimensionaly stable anodes 580
1,2-dimethoxyethane 1126
dimethoxyethane 556
dimethoxymethane 1126, 1127
dimethyl
  adipate 1127
  carbonate 1127
  glutarate 1127
  hydrazine
    unsymmetrical 1013
  phthalate 1127
  silicon chloride 468
  sulfate 1127
  sulfide 1127
  sulfoxide 1127
  terephthalate (DMT) 712
dimethylamine 1066, 1126
2,2-dimethylbutane 1127
2,3-dimethylbutane 1127
dimethylether 1066
3,3-dimethylhexane 1127
2,2-dimethylpentane 1127
diopside 822
dioptase 822
diorite 899
1,4 dioxane 1127
diphenyl ether 1127
dipole orientation 531
dipole polarization 530
dipropyl ether 1133
dipropyl ketone 1133
dipropylene glycol monomethyl ether 1127
direct reduced iron (DRI) 72
direct smelting processes 200
discontinuous fibers 1025
dispersion 35, 766
  coefficient 36
disruptive potential 1053
dissipation factor 525
disthene 834
di-tert-butyl ketone 1127
d-limonene 1129
1-dodecanol 1127
dolime 252, 253, 611, 613
  aluminothermic reduction 253
doloma 610, 611
dolomite 251, 261, 610, 678, 755, 756, 822, 905, 908, 979
  applications and uses 611, 612
  calcined 611
  calcitic 611
  dead burned 611
  stabilized refractory 611
donor 459

dopant 458, 472
doping 457, 458
doré bullion 397
double refraction 37
Dow Chemical process 252
Downs cells 226
Downs electrolytic cells 234
dravite 783, 822
drop solder 211
drop-weight method 1117
dross 169
druse 752
drusy 758
dry air 1054
chemical composition 1074
heat capacities 1056
dry ice 1089, 1090
dry-bulb temperature 1057
drying agents 1095
properties 1096
du Nouy ring method 1118
Duane and Hunt relation 554
dubnium 1184
Duboin's liquor 1172
ductile (nodular) cast iron 79, 80
ductile-brittle transition 18
ductile-to-brittle transition temperature (DBTT) 18
Dulong's equations 1002
Dulong–Petit rule 26, 31
dunite 125, 612
duplex material 361
duplex stainless steels 102
physical properties 106
Dupré equation 1114, 1115
durain 1005
Duranickel® 132
Duriron® 80
Dwight Lloyd sintering machine 200
dyes 692
dykes 891
dynamic friction coefficient 20
dynamic viscosity 1105
dynamite 1015
dysprosia 664
dysprosium 422, 424
dysprosium oxide 664

earth
core 886, 888
core-mantle boundary (CMB) 888
crust 886, 887
interior 886
discontinuities 889
magnetic field 888
mantle 886, 887
rotation 888
transition zone 887
earthworms 943
Ebonex® 574, 576, 577
economic data for industrial minerals 1249
eddy-current losses 506
E-glass 671, 673, 1025
eglestonite 822
Einstein coefficient 42
of absorption 42, 43, 45
of emission 43
of simulated emission 44
Einstein equations 462
ekaaluminium 1183
ekaboron 433, 1183
ekacaesium 1183
ekasilicon 469
elaeolite 843
elastic 762
elastic modulus 7
elastic waves 24
elastomers 691, 692, 715
classification 698
IUPAC acronyms 745
elbaite 822
electric
dipole moment 522
discharge 533
displacement 522
field frequency 532
field strength 522
flux density 522
furnace 103
mobility 461
polarization 523
susceptibility 524
electric arc furnace (EAF) 127
manganese ore 155
electrical resistivity 508, 526, 527, 548
temperature coefficient 527, 548
electrical classification of solids 456
electrical glass 673
electrical resistance 478
electricity
price 1254
SI and cgs units used 529

electrocatalyst 562, 582, 583
electrochemical equivalence 556
electrochemical galvanic series 590
electrochemical manganese dioxide (EMD) 156
electrochemistry 561
electrode 520
  capacitance 520
  carbon-based 572
  electrochemical equivalence 556
  for corrosion protection and control 586
  material 554, 556, 561
  overpotentials 562
  suppliers and manufacturers 589
electrodialysis 561, 580
electrofused alumina-zirconia 609
electrofused magnesia 614
electrogalvanizing of steel 582
electrolyser 561
electrolysis 127
electrolysis cell 564
electrolyte 555, 561
  ionic conductivity 557
  nitric-acid-containing 573
electrolytic cell 561
electrolytic cementation 364
electrolytic iron 64, 73
electrolytic manganese metal 154
electrolytic reduction process 252
electrolytic tough pitch copper 184
electrolyzer 1083
electromagnetic induction 489
electromagnetic interferences (EMI) 512
electromagnetic radiation 33, 41, 554
electromagnetism 490, 498
  Langevin's classical theory 499
electromigration 461
electromotive force 545
electron
  binding energy 552
  color centers 759
  work function 552, 553
electron-beam melting (EB) 304
electronegativity 48
  Allred–Rochow's electronegativity 50
  Mulliken–Jaffe's electronegativity 48
  Pauling 48
electron-emitting materials 552
electronic breakdown 533
electronic polarization 530
electronic-grade silicon 468
electrons
  flux 554
  secondary emission coefficient 555
electrooxidation 561
electropolishing 271
electropositivity 48
electrorefining 127
electroslag refining (ESR) 115
electrostatic energy 522
electrostatic units (esu) 529
electrostriction 533
electrothermal-silicothermic reduction process 155
electrowinning
  alloys 203
  aluminum 168
  manganese metal 155
  of aqueous manganous electrolytes 154
  of metal 154
  of zinc 192
electrum 397, 402, 823
elemental semiconductors 457
elements
  geochemical classification 1185
Ellingham's diagram 168, 273
elongation 10
emanation 1093, 1183
embolite 823
emerald 248, 753, 781, 789, 790, 809
  shaping and treatment 791
emerald green 801
emery 754
emission 42
  einstein coefficient
emulsions and suspensions 1174
enargite 823
endogeneous rocks 890
energetic condition of Bohr 44
energy-band gap 455
engineered ceramics 635
enstatite 823
enstatite chondrites 916
enthalpy 1057
enthalpy of combustion 999
entisols 946
Eötvös equation 1113
epichloridrin rubber 721
epichlorohydrin 1127
epidote 433, 823
epoxy novolac resins 715
epoxy resin 715, 721

epsilon-iron 66
epsomite 823
equation of state of ideal gases 1041
equation of state of real gases 1042
equilibrium hydrogen 1079
erbium 422, 424
erythrite 823
erythronium 339, 1183
eskolaite 369, 664, 824
essential minerals 893
esterification 694
esters 679
etchants for iron and steels 66
etching 272
etching procedures 271
ethane 1009, 1066
ethanethiol 1127
ethanoic acid 1122
ethanol 700, 1127
ethanolamine 1127
ethenic polymers 702
2-ethoxyethanol 1127
2-ethoxyethyl acetate 1127
ethyl
  acetate 1128
  acrylate 1128
  benzoate 1128
  bromide 1123
  butanoate 1128
  butyl ether 1128
  butyrate 1128
  chloride 1124
  chloroacetate 1128
  chloroformate 1128
  cyanide 1133
  formate 1128
  mercaptan 1127
ethylamine 1128
ethylbenzene 1128
ethylcelluloses 701
ethylcyclohexane 1128
ethylcyclopentane 1128
ethylene 1066
  chloride 1128
  chlorotrifluoroethylene 722
  glycol 1126, 1128
  oxide 1128
  polymerization 703
  propylene diene rubber 722
  tetrafluoroethylene 722
ethylene propylene rubber (EPR) 718
ethylene-chlorotrifluoroethylene copolymer (ECTFE) 709
ethylenediamine 232, 237, 1128
ethylene-propylene rubber 722
ethylene-tetrafluoroethylene copolymer (ETFE) 709
ethylpropylether 1128
ettringite 972
eucolite 824
eucryptite 220
eudialyte 328, 824
euhedral 758, 892
Eurodiff 445
europium 422, 424, 664
europium oxide 664
eutectoid steel 76
euxenite 345, 355, 433, 824
excitation 46
excitation timelife 46
excluded volume 1042
ex-PAN 1026
ex-Pitch 1026
explosion pressure 1063
explosive limit 1062
explosives 999, 1015
  properties 1016
explosivity limits 1062
exponential equation 1106
exrinsic semiconductors 458
extinction coefficient 45
extruded polystyrene 1093
extrusion of polymer fibers 1024
extrusives rocks 891

falcondoite 824
false galena 856
fanning friction factor 1107
Faraday constant 558
Faraday's law 489
fassaite 808
fatigue 18
faujasite 824
fayalite 825
fayalite-forsterite 888
F-center 759
feldspar 222, 228, 261, 279, 467, 598, 601, 754, 777
  index 897
  plagioclases 914
feldspathoids 777, 899

felsic magmas 891
FEP 708
ferberite 386, 825
fergusonite 825
Fermi gas 501, 523
Fermi level 456, 460
ferralsols 951
ferric iron 154
ferrimagnetic 504
ferrimagnetic materials 504
ferrite 65, 96, 575
  hot (high)-acid leach (HAL) 193
ferrite stabilizers 78
ferritic stainless steels 97
  physical properties 100
ferroaxinite 825
ferrochrome 369
  carbothermic process 369
  high-carbon-grade 369
  low-carbon 370
  producers 372
  properties 370
ferrochromium 103
ferroelectric 539
  aging 538
  domains 534, 535
  hysteresis loop 534
  materials 534
  properties 536
ferromagnesian
  minerals 891
  silicates 596, 914
ferromagnetic 494, 497, 501, 504, 510
  compounds 502
  elements 502
  ferrites 502
  garnets 502
  materials 491, 501, 510
  nonretentive 507
ferromagnetism 64
ferromagnets 491, 496
  remanence 505
  retentivity 505
ferromanganese 151, 155, 611
  alloy 153
ferromolybdenum 103, 375
ferronickel alloy 127
ferroniobium 345
ferropseudobrookite 825
ferrosilicon 580, 596, 608, 1174
ferrosilicon-water 1174
ferrosilite 845
ferrotitanium 296
  commercial grades 296
  producers 296
ferrotungsten 387
ferrous chloride 284
ferrous metals 59
ferrous oxide (FeO) 72, 284, 866
ferrous sulfate heptahydrate 840
ferrovanadium 341
ferrum 59, 1183, *see iron*
fertilizers 961
  chemical 961
  mineral 961
  mixed 961
  nitrogen 962
  phosphorus 963
  potassium 964
  straight 961
fiber
  carbonization 1027
  graphitization 1027
  stabilization 1027
fiber reinforced polymers 1019
fiberization 1026
fibrolite 855
fibrous chrysotile 805
Fick's law 462
field-induced isothermal magnetic entropy change 496
field-induced magnetostriction 494
filiform 758
filler 692, 693, 1019
filter materials 629
fine aggregate 976
fire assays 768
fire resisting glass 674
fireclays 597
  applications and uses 597
fired bricks 629
fired ceramics 978
Fischer–Tropsch process 1082
Fischer–Tropsch reaction 1088
flake graphite 623
  applications and uses 625
flame coloration tests 769
flame fusion 795
flame test 768
flammability limits 1062
flammability of gases and vapors 1062
flammability of liquids 1121
flammability range 1062
flash point 1121

flash powder 1015
flat glass 676
Flint 467
Flint clay 599
float glass 673, 676
Float Zone (FZ) method 472
floating dredge 206
floating zone 796
fluid
  classification 1106
  friction pressure losses 1106
  laminar flow in circular pipes 1107
  Magneto-Archimedes effect 1175
  mass density 1103
  pressure drop 1106
  shear rate 1105
  shear stress 1105
  turbulent flow in rough pipes 1107
  viscosities 1104
fluidity 1105
fluor spar 825
fluorescence 45, 46, 766
  delayed 47
  minerals 766
fluoride anions 583
fluorinated ethylene propylene (FEP) 708, 722
fluorinated polyolefines 707
fluorination 444
fluorine 1066, 1090
fluorine gas 708
fluorite 261, 759, 825
fluoro crown 674
fluorobenzene 1128
fluorocarbons 707, 708
fluoroelastomers 719
fluorspar 71, 261, 754
fluvisols 949
flux 629, 797
flux growth technique 797
fluxons 481
fly ash 976
foliated 758
Fool's Gold 850
foote minerals 220
footwall 752
forced convection 28
formaldehyde 711
formamide 1128
formic acid 1128, 1168
forsterite 826
Foucault-current losses 506
foundry grade chromite 369
Fourier's first law 29
Fourier's second law 29
frac fluids 679
fracture 761
  property 17
  toughness 16, 17
fracturing techniques 677
framesite 783
francium 243
Franck–Condon of the transition 46
Franck–Condon transitions 41
franklinite 188, 190, 826
free convection 28
free settling 1109
free surface energy 1111
free-settling ratios 1110
freezing point depression 1119
Freons® 1093
friction 19
frictional force 19
froth flotation 199, 206, 263
fuchsite 842
fuel 999, 1062
  Dulong's equations 1002
  gaseous
    Wobbe Index (WI) 1003
  gross heating value (GHV) 1002
  high heating value (HHV) 1002
  liquid 1008
  low heating value (LHV) 1002
  net heating value (NHV) 1002
  source of ignition 1062
  stoichiometric coefficients 1000
fuel cells 556
fuels
  classification 1000
  gaseous 1009
  hypergolic 1012
  petroleum 1012
fulgurites 920
fuller's earth 846
fullerenes 482
fully halogenated hydrocarbons 1093
fully stabilized zirconia 621
fulvalenes 482
fumed silica 595
furan 1128
furan plastics 715
furfural 1128
furfuraldehyde 715
furfuryl alcohol 715

fusain 1004, 1005
fused alumina 614
fused silica 594, 596
fused vanadium pentoxide 340
fused zirconia 622
fusibility test 770
fusible alloys 209
  low-melting-point 210

GaAs 471
gabbro 899
gabbrodolerite 408
gadolinia 664
gadolinite 392, 433
gadolinium 422, 424
gadolinium oxide 664
gahnite 826
Galathite® 701
galaxite 826
galena 152, 188, 190, 199, 826
galena-water 1174
gallium atoms 459
gallium-arsenide 472
galmei 856
gamma-austenite 75
gamma-iron 65
gangue 752
gangue minerals 70, 126
  acid leaching 443
GaP 471
garnet 433
garnets 754, 760
garnierite 126, 824
gas 1037, 1054
  absolute 1043
  barometric equation 1045
  compressibility factor (Z) 1046
  conditions 1040
  density 1044
  dry air 1054
  explosivity limits 1062
  flammability range 1062
  humidity 1054
  hydrates 1087, 1090, 1094
    crystal structures 1094
  hygrometry 1054
  isobaric 1040
  isotropic volumic expansion 1046
  moist air 1054
  molar heat capacity 1049
  molecular mass 1045
  molecules
    mean free path 1048
    mean velocity 1048
    microscopic properties 1048
  Paschen curve 1053
  permeability coefficients 1052
  permeability of polymers 1051
  pressure 1037, 1043
  producers 1100
  psychrometry 1054
  scale height 1045
  water vapor 1054
gas-atomization process 301
gas-atomized iron powders 123
GaSb 471
gas-cooled fast breeder reactors (GCFRS) 448
gaseous fuel 1009
  combustion related properties 1010
gaseous fuel-oxidant mixture
  adiabatic flame temperature 1064
gases
  A coefficient 1050
  autoignition temperature 1063
  closed tube test 773
  critical temperature 1048
  dielectric properties 1052, 1053
  disruptive potential 1053
  dynamic viscosity 1049
  ignition energy 1063
  industrial 1074
  L coefficient 1050
  maximum explosion pressure 1063
  maximum rate of pressure rise 1063
  properties 1065
  solubility 1050, 1051
  specific gravity 1045
  threshold limit averages 1064
  toxicity 1064
gas–liquid-solid interface 1115
gauge length 10
gaylussite 826
gehlenite 827
geikielite 827
gel process 609
gelisols 947
gems
  floating zone (FZ) melt growth technique 796
  hydrothermal growth technique 796
  skull melting melt growth technique 796

gemstones 753, 756, 781
Bridgman–Stockbarge melt growth technique 795
Czochralski (CZ) melt growth technique 795
flux growth technique 797
properties 800
sol–gel growth techniques 797
synthetic
from melts 795
from solutions 796
verneuil melt growth technique 795
general characteristics of the three natural and the artificial radioactive decay series 1203
genthite 824
geobarometers 911
geochemical classification of the elements 1185
geological time scale 1256
georgite 921
geosphere 886
geothermal gradients 911
geothermometers 911
germanite 469
germanium 457, 469, 470, 472
applications and uses 470
dioxide 469
monocrystal 458
gersdorffite 125, 827
getters 1099
properties 1099
giant magnetocaloric effect (GMCE) 497
Gibbs free enthalpy 1111
Gibbs molar enthalpy 274
gibbsite 165, 168, 601, 603, 604, 827
dehydration 606
gilding metal 185
Ginzburg–Landau theory 478
giobertite 838
glaserite 806
glass 593, 671
fibers 1025
physical properties 672
raw materials
properties 628
tanks 671
transition 671, 697
transition temperature 671, 697
glass–ceramic-matrix composites (GMCs) 1019
glass-grade material 222
glass-to-metal seal 211
glassware 672
glassy 760
Glauber salt 224, 235, 827
glauberite 827
glaucodot 827
glauconite 828
glaucophane 828, 853
glazes 630
gleysols 949
glucinium 244, 1183
glucose 1120
glutaraldehyde 1128
glycerol 1128
gneiss 595, 912
goethite 828, 905, 908, 1081
goethite process 194
gold 152, 400, 414, 828
alloys 404
properties 405
applications and uses 406
as a byproduct 404
bullion 404
caratage 401
carbon-in-pulp process (CIP) 404
electrodeposits 404
extraction
cementation method 403
cyaniding process 403
merryl-crowe process 403
placer or gravity separation method 403
leaf 400
mineral 402
mining 402
panning 403
plating 582
producers 406
refining induction 404
sluice box 403
gold-cadmium alloy 139
goshenite 781, 789, 792, 809
goslarite 828
graft polymer 697
granite 595, 755, 899, 979, 980
granodiorite 755, 979
graphite 387, 471, 551, 572, 573, 574, 623, 627, 654, 707, 708, 754, 829, 909
applications and uses 625
graphitization 1027
gravel 756, 975
gray antimony 858

gray cast iron 79, 80
  physical properties 81
gray nickel pyrite 827
green gold 405, 406
green lead ore 851
green silicon carbide 628
green vitriol 840
greenalite 829
greenockite 829
greseins 426
grey tin 204
greyzems 952
Grimm–Sommerfeld rule 459
grog 597
gross caloric value 1063
gross heating value 1002
grossular 782, 829
grossularite 829
groutite 828
grunerite 829
guadalcazarite 841
guano 962
guar 679
gum rosin 698
gumbelite 832
gummite 441, 829
gun metal 186
guncotton 1017
Gutenberg discontinuity 888
Gutta Percha 716
gymnosperms 983
gypsum 261, 262, 754, 756, 829, 908, 963, 972, 973, 1081
  cement 968

Haber–Bosch process 1075, 1086
habitus 758
hackly 761
hafnium 326, 329, 336, 445, 664
  carbide 337, 654
  diboride 649
  dioxide 664
  disilicide 661
  Kroll process 337
  monoboride 649
  nitride 659
  oxychlorides 329
  producers 337
  tetrachloride 337
hafnon 337, 830
Hagen–Poiseuille equation 1107
Hagen–Poiseuille law 1106
hahnium 1184
halite 233, 791, 830
Hall coefficient 463
Hall effect 462
Hall field 462, 494
Hall–Heroult process 164, 166, 168, 169, 563, 573, 601
halloysite 830
halocarbons 1093
halogenated hydrocarbons 1093
halogens 354
halons 1093
hamartite 809
hand lay-up of prepreg 1030
hanging wall 752
hanksite 830
hard clay 599
hard magnetic materials 510
hardhead 207
hardmetal 639
  properties 640
hardness 11, 762
hardwood 983, 985
  properties 991
Harper's alloy 210
hassium 1184
Hastelloy® 132, 133
hausmannite 830
haüyne 830
hauynite 830
Haynes® 133, 146
Haynes®1233 146
Haynes®214 133
Haynes®230 133
Haynes®242 133
Haynes®25 146
Haynes®556 133
HDPE 703
heartwood 983, 985
heat
  capacity 25
  flux 28
  transfer fluids
    properties 1178
  transfer processes 28
heating alloys 548
heating by adiabatic magnetization 496
heating values 1063
heat-treated slag (HTS) 285
heavy liquids 777, 1171
heavy media 776, 1171

heavy metals
  inorganic salts
    saturated aqueous solutions 1172
heavy spar 264, 754, 808
heavy water 1080, 1121, 1125
  physical properties 1167
hedenbergite 831
heliodor 248, 781, 789, 792, 809
helions 1091
helium 447, 483, 1066, 1090, 1091, 1094
hematite 68, 70, 277, 296, 617, 831, 936
hematite process 194
hemicellulose 984
hemimorphite 188
hemoglobin 1078
hemoilmenite 277, 278, 279
Henry's law 1050
heptafluorotantalate 345
1-heptanol 1129
hercynian granite 222
hercynite 831
Hermann–Mauguin 1213
Hermann–Mauguin notation 757
Hess's law 1001
hessite 831
hessonite 782, 829
1-heptane 1129
heterogeneite 144
heterogeneity index 696
heteropolymer 724
heulandite 831
Hevea brasiliensis 716
hexachloroiridic acids 579
hexachloroplatinic acid 579
hexafluoropropylene 719
hexagonal 1211
hexagonal boron nitride (HBN) 319, 638
hexagonal space groups 1226
hexahydroxybenzene 237
hexamethylene diamine (HMD) 710
hexamethylolmelamine 713
hexanitrostilbene 1017
1-hexanol 1129
hexavalent chromium 368
1-hexene 1129
hexogene 1017
hiddenite 783, 857
high copper alloys 182, 184
high density polyethylene 703
high explosives 1015
high heating value 1002, 1063
high modulus grade 1027
high nickel alloys 131
high pressure high temperature (HPHT) 797
high temperature resistors 551
high tensile brass 186
high thermochemical decomposition of water (HTDW) 1084
high-alumina refractories 600
high-carbon grade 369
high-carbon steels 85, 87
high-duty fireclay 597
high-field superconductors 480
high-hardenability case steels 90
highly oriented polyethylene 1027
high-purity alumina 609
high-silicon 78
high-silicon cast irons 80
high-speed-tool steel 90
high-strength glass 673, 1025
high-strength low-alloy steels (HSLA) 112
  mechanical properties 114
high-temperature electrolysis (HTE) 1084
high-test peroxide 1013
Highweld process 341
historical names of the elements 1181, 1183
histosols 947, 949
HMX 1017
HNS 1017
hole color centers 759
holmium 422, 424
hololeucocrates 894
holomelanocrates 894
homocyclonite 1017
homopolymer 724
hongquiite 617, 831
Hooke's law 7, 9, 28, 1022
horizons 927, 931
horn quicksilver 814
horn silver 397
hornfels 912
hortonolite 825
hot briquetted iron (HBI) 72
hot dip galvanizing 195
hot isostatic pressing (HIP) 145
hot isostatically pressed silicon nitride 636
hot-acid leach (HAL) 193
hot-pressed silicon nitride 636
hot-work tool steels 118

HSLA steels
selected grades 113
huebnerite 386, 832
human bandwidth 23
Hume–Rothery rules 458
humid heat 1056
humidity 1054
humidity ratio 1054, 1056
humification 929
humite 832
Humphrey's spirals 280, 427
humus 929
Hunter process 291, 292
hyacinth 783, 867
hyaline 892
hyaline igneous rocks 895
hydrargillite 603
hydrargyrum 840, 1183
hydrated lime 610
hydrates of gases 1094
hydraulic bronze 186
hydraulic diameter 1108
hydraulic lime 969
hydrazine 1012, 1129
hydride/dehydride process 299
hydrides 242
hydrobromic acid 1168
hydrobutyl terephtlate 712
hydrocarbons 169, 573, 1085, 1112
halogenated 1093
partial oxidation 1083
hydrocassiterite 205
hydrochloric acid (HCl) 148, 150, 196, 768, 1168, 1241
hydrochloricauric acid ($HAuCl_4$) 404
hydrochlorofluorocarbons 1093
hydrocyclones 280
hydrofluoric acid 1168, 1243
hydrofluoric-nitric acids 637
hydrofluoric-sulfuric acids 637
hydrofluorination 444
hydrofluorocarbons 1093
hydrogen 1066, 1078, 1099
azide 1075
chloride (HCl) 1241
cyanide (HCN) 403, 714, 1129
flammability limits 1085
fluoride (HF) 1129, 1243
gas 123
halides 1081
hexachloroplatinate 415
Messerschmidtt process 1081
peroxide 1013
pressure swing absorption (PSA) 1083
sulfide 396, 409
hydrogenium 1078
hydrogenocarbonate 235
hydroiodic acid 1169
Hydrolite® 1081
hydrometallurgical process 126
hydrometallurgy 561
hydromica 832
hydromuscovite 832
hydronium 1081
hydrophiles 1185
hydrostatic balance 4
hydrostatic stress 8
hydrothermal growth technique 796
hydrotimeter scales 1104
hydrotimeters 1104
hydroxy-terminator polybutadiene (HTPB) 1014
hygrometry 1054
Hypalon® 718
hypergolic 1012
hyperosmotic 1121
hypersiliceous magmae 891
hypertectoid steel 76
hypidiomorphous 758, 892
hyposiliceous 891
hyposmotic 1121
hypotectoid steel 76
hysteresis 9, 24
hysteresis loop 506, 535
hysteresis losses 508

IACS 179
ice 912
physical properties 913
polymorphs 914
ideal gas 1037
equation of state 1041
idiomorphous 892
idocrase 864
igneous rocks 426, 889, 890
ignition energy 1063
ignition synthesis 141
IIR 717
illinium 1183
illite 596, 629, 832, 905
ilmenite 276, 277, 278, 279, 287, 296, 328, 448, 617, 832
beneficiation techniques 280

EARS process 285
ERMS roasting process 285
grain 283
Murso process 285
smelting 282
impactites 594, 920
Imperial smelting process 192
impressed current anode materials 588
InAs 471
inceptisols 947
Incoloy® 134
Incoloy®800 134
Incoloy®825 134
Incoloy®902 134
Incoloy®903 134
Incoloy®907 134
Incoloy®909 134
Incoloy®925 134
Inconel® 134, 135
Inconel®600 134
Inconel®601 134
Inconel®617 135
Inconel®625 135
Inconel®686 135
Inconel®718 135
index of refraction 32, 33, 39
indianite 805
indicatrix 36, 37, 765
indicolite 783
indium fusible alloy 210
induction heating 507
industrial anode materials 565
industrial cathode materials 563
industrial ceramics 635
industrial minerals 753, 754
industrial rocks 753, 754
inert gases 1090
infrasounds 23
ingot iron 64, 73, 84
initial magnetic permeability 506
injection molding 1031
inner core 888
InP 471
InSb 471
insertion 559
insulation resistance 526, 528
insulator 456, 539
electrical properties 540
thermal instability 533
insulator-to-metal transition 533
intercalation 559, 582
intercalation compounds 559
intercombination 46
intermediate modulus grade 1027
internal conversion 46
internal discharge 533
internal frictions 25
international annealed copper standard (IACS) 179
interplanar spacing 1231
intrinsic semiconductors 457
intrusives rocks 890
Invar® 136
Invar®42 136
inverse magnetostriction 494
iodargyrite 832
iodine-sulfur cycle 1084
iodoargyrite 397
iodobenzene 1129
iodomethane 1129, 1171
iodyrite 832
ionic polarization 530, 531
ionic polymerization 694
ionic solutions 556
ionicity
degree 48
ionium 1203
ionizing energy 552
ionophores 556
ions 555
iridescence 767
iridium 407, 414, 416, 568, 833
dioxide 415, 583
iridosmine 413, 833
iron (Fe) 59, 165, 616, 832, 894
allotropes 65
allotropism 64
alloys 64
alpha-iron 65
beta-iron 65
carbide 74, 121
carbonyl process 71
cementite 74
critical point 65
delta-iron 66
direct reduction 72
ductility 79
epsilon-iron 66
gamma-iron 65
hydroxide 127
hydroxides 68
malleable 79
metallographic etchants 67
metallurgy 73

meteoric 67
meteorites 67, 918
mining 70
native 67
ore 68
oxides 908
pelletizing 70
properties 60
pure 64
siderites 67
sintering 70
smelting reduction 72
sponge-reduced 123
terrestrial 67
transition temperature 65
iron diboride 649
iron monoboride 649
iron powder 122
gas-atomized 123
water-atomized 122
iron-based superalloys 121
iron-carbon 73
iron-carbon phase diagram 74, 77
arrest points 76, 77
iron-carbon system 74
iron-cementite 73
iron-chromium-carbon 97
ironmaking
blast-furnace process 71
iron-nickel alloy 66
ironstone 908, 909
irregular 761
Isasmelt process 201
isinglass 842
isobaric coefficient of cubic expansion 1046
isobutanol 1129
isobutanolamine 1129
isobutyl
acetate 1129
heptyl ketone 1129
isobutyrate 1129
isobutyraldehyde 1129
isobutyric acid 1129
isochore compressibility 23
isocumene 1133
isometric 1211
isopentyl alcohol 1131
isopropanol 1129
isopropanol amine 1129
isopropyl alcohol 1133
isopropyl chloride 1129
isopropylamine 1129
isopropylbenzene 1129
isosmotic 1121
isostrain 1022
isotactic polymer 697
isotherm 1046
isotherm of a real gas 1046
isothermal entropy density change 496
isothermal magnetic entropy change 496
isothermal specific entropy change 496
isotonic 1121
isotope-effect exponent 482
isotopes 1202
isotropic 765
isotropic material 36
IUPAC acronyms of polymers and elastomers 745
ivoirites 921

Jablonski diagram 46
Jablonski photophysical diagram 45
jacobsite 152, 833
jadeite 782, 833
jardin 790
jargon 783
jarosite 833
jarosite process 194
jasper 467
jennite 973
jervisite 434
joliotium 1184
josephinite 67
Josephson-effect 485
Joule effect 253
Joule's heating 506, 507, 562
Joule's magnetostriction 494
JS-700 136
juonniite 434
Jurin's Law 1116, 1117

kainite 251, 833
kalium 1183
kallium 237
Kanthal® 550, 551
Kanthal® 52 549
Kanthal® 70 549
kaolin 221, 598, 600
kaolin clay 682
kaolinite 165, 596, 629, 834, 905
karelianite 834

karrooite 834
karrooite-pseudobrookite series 282
kastanozems 952
Keesom's forces 1042
Kel-F® 709
kennedyite 807
kernite 471, 834
kerolite 859
kerosene 329, 346, 356, 444, 450, 1012
Kevlar® 710, 1027
kidney ore 831
kieselguhr 595, 755
kiesserite 251
kimberlites 786, 787
kinematic viscosity 1105
Kivcet process 200
Klein's liquor 1172
knebelite 825
Knoop hardness 12, 764
kolbeckite 434
korloy 197
krennerite 834
kristiansenite 434
Kroll process 276, 288, 290, 291, 292, 299, 330, 578
krypton 1066, 1090, 1092
kunzite 220, 783, 857
kupfernickel 124
kurchatovium 1184
kyanite 165, 597, 599, 600, 754, 834
kyzylkumite 835

L coefficient 1050
labradorescence 767
labradorite 782, 835
laccoliths 891
lactic acid 1129
lacustrine magnesite 612
Lamé coefficients 23
lamellar 892
laminated glass 676
lamproite 786, 787
Landé's factors 491
Lanes process 1082
langbeinite 251, 835
Lanital 701
lanthania 665
lanthanic contraction 422
lanthanides 326, 422
  discovery milestones 425
  physical and chemical properties 424
lanthanum 422, 424
  dicarbide 655
  dioxide 665
  flint 675
  hexaboride 650
  oxide 423
lanthanum-barium copper oxide 481, 484
lapidary 781
lapilli 904
lapis lazuli 782, 836
Laplace's law 499
Laporte rule 46
larnite 835, 971
lascas 467, 594, 796
lasurite 836
latent enthalpy 30, 31
laterites 144, 166, 906
laumontite 835
laurite 409
Laves phases 145
lawrencium 1184
lawsonite 836
lazulite 836
lazurite 836
LDPE 703
Le Chatelier's principle 1050
lead 196
  acid-copper 570
  alloys 196, 198, 201
  anodes 569, 571
  antimonial 198, 570
  azide 1015, 1016
  bullion 200, 201
  chemical 198, 570
  conventional blast furnace process 200
  copper 184
  corroding 570
  dioxide 573, 574
  glance 826
  Imperial smelting process 200
  Isasmelt process 201
  Kivcet process 200
  ore 856
  Outokumpu flash smelting process 201
  physical properties 202
  plumbate 196
  QSL process (Queneau–Schuhmann–Lurgi) 200
  roasting 199
  selected properties 160

sintering 199
slag 201
spar 804
styphnate 1015, 1016
tellurium 198
tin 203
tin bath 211
vitriol 804
lead-calcium-tin 570
lead-silver 570
lead-tellurium copper 202
leakage current 528
ledeburite 75
Lehmann discontinuity 888
Lely process 627
Lennard–Jones equation 1042
Lenz's law 490, 499
lepidocrocite 836
lepidolite 220, 221, 222, 223, 240, 248, 836
lepidomelane 810
less common minerals 893
lessivage 930
leucite 240, 837
leucocrates 894
leucoxene 277, 279, 280, 287, 328, 850
Lexan® 711
Lichtenberg's alloy 210
light water 1121
lignine 984
lignite 909
lime 610, 664, 968
applications and uses 610
hydrated 968
hydraulic 969
slaked 968
limestone 200, 261, 610, 678, 756, 908, 909, 970, 979
dolomitic 611
limewater 262
limonite 68, 125, 760, 828, 837, 908, 936
linalool 1129
linear combination of atomic orbitals 455
linear dielectrics 538
linnaeite 143, 837
linneite 837
linotype 203
Lipowitz's alloy 210
liquid
calculation of major losses 1108
capillarity 1116
capillary rise 1116, 1117
chemical reagents
selected properties 1168
contact angle 1113
drop-weight method 1117
du Nouy ring method 1118
dynamic viscosity 1105
flammability 1121
flash point 1121
free settling 1109
fuel 1008
properties 1008
hot metal 72
hydrogen 1012
hydrometer scales 1104
intrinsic fluid property 1104
kinematic viscosity 1105
mass density 1103
maximum bubble pressure 1117
metals
physical properties 1175
oxygen 1012
pressure 1037
propellants 1011, 1013
sedimentation 1109
sessile drop 1118
specific gravity 3, 1103
surface tension 1110, 1112
temperature 1112
vapor pressure 1110
viscosities 1104
wetting 1113
Wilhelmy plate 1118
work of adhesion 1114
work of cohesion 1114
litharge 837
lithcoa 219
lithiated intercalation compounds 229
lithiation 559
lithiation reaction 559
lithification 889, 905
lithine 229
lithium 217
applications and uses 229
battery-grade ingot 227
brine 223
carbonate 220, 223, 230
from brines 224
from LiOH 225
major producers 226
catalyst-grade traps 227
cations
intercalation 559

chloride 220, 225, 227, 229
chloride electrolysis 225
deintercalation 559
fluoride 229
hydride 219
hydroxide 217, 218, 219
hypochlorite 229
ingot producers 231
isotopes 218
isotopic fractionation process 219
metal producers 230
mineral 230
molten-salt electrowinning 226
nitride 217
stearate 229, 693
sulfate 224
technical-grade traps 227
thermal properties 217
traps 227
lithium-carbonate equivalent 222
lithium-metal traps 225
litholites 67, 914
lithology 885
lithophiles 1185
lithopone 264, 286
lithosiderites 67, 914, 919
lithosols 949
lithosphere 885, 888, 905
lithotypes 1005
livingstonite 837
lixiviation 930
lodestone 838
log decrement 25
logarithm decrement 25
London forces 1042
long-wave infrared (LWIR) 244, 249
loparite 426, 837
mining and mineral dressing 427
Lorentz equation 35
Lorentz force 462
loss coefficient 24
loss tangent 525
low brass 185
low carbon steel cathodes 563
low density polyethylene 702
low explosives 1015
low heating value 1002, 1063
low melting point 209
low temperature of molten inorganic salts 1174
low-alloy steels 89
low-alloy tool steels 118
low-carbon ferrochrome 370
low-carbon steels 85, 563
low-duty fireclay 597
lower explosive limit 1062
lower flammability limit (LFL) 1062
lower mantle 888
LST 1173
lubricants 693
lubricating action
of liquids 20
of molecules 20
lubricating properties 19
luminescence 45, 766
lunar caustic 400
luster 760
metallic 760
nonmetallic 760
lutetium 422, 424
luvisols 953
lyosol 1180

machinable glass 673
machining tools 115
MacLaurin's power series 1043
Macor® 673
macromolecules 691, 693, 694
mafic igneous rocks 339
mafic magmas 891
magamatic hard rock deposits 277
magbasite 434
maghemite 837
magma 887, 890
anatexy process 910
felsic 891
hypersiliceous 891
mafic 891
magmatic rocks 889, 890
magnesia 218, 612, 665, 847
dead burned 613
electrofused 614
sintered 613
synthetic 613
magnesia-chrome bricks 369
magnesiochromite 838
magnesioferrite 838, 848
magnesite 250, 251, 612, 754, 838
applications 612
metallothermic reductions 253
magnesium 243, 250, 290, 293, 894, 1032
alloys 250, 255
physical properties 256

standard ASTM designations 255
amalgam 251
applications and uses 255
boride 470
chloride 251, 252
drosses 255
electrolytic reduction 252
fluoride 248
hydroxide 613
IG Farben process 251
nonelectrolytic processes 252
oxide 252, 612, 613, 665, 679
oxychloride 613
oxysulfate 613
producers 254, 259
refining 253
scrap 255
tungstates 387
magnet steels 511
magnetic
dipole 490
domains 501
energy density 493
energy loss 506
entropy change 496, 497
field 487, 488, 489
coercitive force 505
flux 489, 512
flux density 488
force 492, 493
hard materials
properties 513
induction 488, 512
induction at saturation 508
iron ore 838
materials
applications 516
classification 498
physical quantities 487
metals
properties 508
moment 490
permeability 489, 506, 508
permeability of vacuum 488
physical quantities 487
pyrite 851
refrigeration 496
resonance imaging 484
shield
attenuation ratio 512
efficiency 512
susceptibility 491, 492
magnetism
Maxwell's theory 499
magnetite 68, 151, 280, 329, 575, 786, 838, 1174
magnetite-water 1174
magnetizability
atomic or molecular 491
magnetization 491
intensity 491
spontaneous 501
magnetocaloric effect (MCE) 495
magnetomotive force 488
magnetoresistance 494
magnetostriction 494
fractional change in length 495
major losses 1106
majority carriers 459
malachite 179, 838
malacon 783, 867
malaia 782
malleable 762
malleable cast iron 79, 80
Malotte's metal 210
mammillary 758
manganese 117, 149, 289, 503, 571
(alpha-Mn) 150
(beta-Mn) 150
(delta-Mn) 150
(gamma-Mn) 150
allotropes
physical properties 150
cations 572, 575
dioxide 150, 151, 443, 572, 575, 754
industrial uses 157
major producers 157
metal 153
metallurgical uses 156
metallurgy 155
mining 153
nodules 153
nonmetallurgical uses 156
ores 152, 153
arc smelting 154
electrothermal-silicothermic reduction 154
properties 60
stainless steels 101
manganese (Mn) 59
manganese-based alloys 149
Manganin® 548, 549
manganite 152, 156, 838
manganophyllite 810

manganosite 839
manganotantalite 355
manganous salts 150
mannacanite 276, 832
mantle 886, 887
maraging steels 120
  physical properties 120
marble 610, 756, 912
marcasite 68, 839
margarite 839
marginal reserves 752
marialite 839
MAR-M509 146
marsh gas 1086
martensite 103, 121
  finish temperature 139
  start temperature 139
  thermoelastic transformation 139
martensite-to-austenite transformation 139
martensitic stainless steels 97
martite 831
mass average molar mass 696
mass average relative molar mass 696
mass density 1, 3, 1103, *see also* density
mass fraction 1056
mass magnetic susceptibility, 492
massicot 839
massive 892
master alloy 297
masurium 1183
material
  activity 1207
  anisotropy 16
  breakage ability 17
  corrosion rate 1242, 1243, 1244
  corrosion rates 1241
  hardness 11
  isotropic 36
  mass density 1
  mechanical properties 22
  physical properties 1
  professional societies 1257
  thermal properties 32
  toughness 15
matrix 1019
matte 180
Matthiessen's equation 527
maximum allowable stress 15
maximum bubble pressure 1117
maximum explosion pressure 1062, 1063
maximum kinetic energy 553
maximum magnetic permeability 506
maximum rate of pressure rise 1062, 1063
Maxwell equation 493
Maxwell relation 496
Maxwell's laws 483
Maxwell–Boltzmann distribution 44
Mayer's equation for ideal gases 1049
m-chlorobenzotrifluoride 1124
McKelvey diagram 753
m-cresol 1124
m-dichlorobenzene 1125
mean free path 1048
mean square velocity of gas molecules 1048
mean velocity of gas molecules 1048
measurements of surface tension 1117
medium density polyethylene 702
medium permitivity 519
medium-carbon steels 85, 86
medium-duty fireclay 597
medium-hardenability case steels 90
megacrystals 892, 895
meionite 840
Meissner–Ochsenfeld effect 483
meitnerium 1184
melaconite 859
melamine-formaldehyde 713, 722
melanite 804
melanochalcite 859
melanocrates 894
melanterite 287, 840
melilite 840
melinite 1017
Mendeleev's periodic chart 1182
mendelevium 1184
meniscus 4
mercuric
  chloride 191
mercury 191, 242, 840, 1174
  cathode 260, 565
  fulminate 1015, 1016
  iodide 191
  removal 191
  superconductivity 483
mercury-bromoform 1174
Merryl-Crowe process 403
merwinite 840
Mesh-on-Lead® 571
mesitylene 1129
mesocrates 894
mesosiderites 919

mesosphere 887
mesothorium 1204
Messerschmidtt process 1082
metacinnabar 841
metakaolin 596, 597
metal hydride reduction 301
metal matrix composites (MMCS) 1020, 1031
  properties 1033
metal maximum operating temperature 1237
metallic 760
metallic character 457
metalliding process 364
metalloids 457
metallurgical-grade alumina 604
metallurgical-grade chromite 369
metallurgical-grade silicon 468
metals
  hardness scales 12
  platinum-group *see* PGM
  rare-earth 422
  refractory 266
metamorphic
  facies 912
  grade 911
  rocks 889, 910
metamorphism 910
metavanadate anion 338
meteoric iron 67
meteorites 125, 786, 889, 914
  glassy 920
  modern classification 915
methane 586, 1009, 1066, 1082, 1086
methanesulfonic acid 1129
methanethiol 1129
methanoic acid 1128
methanol 709, 1089, 1130
2-methoxyethanol 1130
methyl
  acetate 1130
  acetoacetate 1130
  acrylate 1130
  alcohol 1130
  amyl ketone 1130
  benzoate 1130
  ethyl
    ketone (MEK) 1130
    ketoxime 1130
  formate 1130
  isoamyl ketone 1130
  isobutanoate 1130
  isobutenyl ketone 1130
  isobutyl ketone 1130
  isocyanate 1130
  isopropyl ketone 1130
  laurate 1130
  myristate 1130
  n-propyl ketone 1130
  phenyl
    amine 1130
    ether 1130
    ketone 1130
  pivalate 1130
  propionate 1130
  salicylate 1131
  tert-butyl
    ether 1131
    ketone 1131
methyl hydrazine 1012
methyl iodide 1129
methyl isobutyl ketone (MIBK) 329, 346, 356
methyl mercaptan 1129
2-methyl pentane 1130
4-methyl pyridine 1131
2-methyl-1,3-butadiene 1131
2-methyl-1-butanol 1131
3-methyl-1-butanol 1131
2-methyl-1-butene 1131
3-methyl-2-butanol 1131
2-methyl-2-butene 1131
4-methyl-2-pentanol 1131
methylal 1126
2-methylaminoethanol 1131
2-methylbutane 1131
methylcelluloses 701
methylcyclohexane 1131
methylcyclopentane 1131
methylene bromide 1125, 1171
methylene chloride 1126
methylene iodide 777, 1126, 1171
2-methylheptane 1131
2-methylhexane 1131
4-methylmorpholine 1131
1-methylnaphtalene 1131
2-methylnaphtalene 1131
3-methylpentane 1131
methyltrichlorosilane (CH_3SiCl_3) 1028
micaceous 758
micas 467, 598, 601, 629, 754, 761, 893
microcline 238, 841
microcosmic salt 775
microcracks 621

microfibrils 984
microlites 355, 895
micronutrients 961, 965
microscopic magnetic dipole moment 490
microscopic properties of gas molecules 1048
microsheet glass 673
microsilica 596
mild steel 84, 85
milk of lime 610, 613, 968
milkstone 701
mill scale 123
Miller process 404
millerite 125, 841
mineraloids 751
minerals 751
  accessory 893
  admixtures 976
  bead test with borax 775
  bead test with microcosmic salt 776
  Bowen's crystallization series 893
  charge transfer electronic transitions 759
  chatoyancy 767
  chemical reactivity 767
  cleavage 760
  closed tube test 772
  composition 893
  crystallization sequence 894
  Dana's class 757
  Dana's classification 779
  density 762, 894
  ferro-magnesian 894
  ferromagnetic 766
  fracture 761
  hardness 762, 764
  index of refraction 765
  industrial 753
    economic data 1249
  jarosite-type 194
  Kobell's fusibility scale 770
  metamorphic rocks 911
  Miller indices 760
  miscellaneous properties 767
  modal composition 893
  open tube test 774
  parting 761
  phosphorus-rich 964
  play of colors 767
  potassium-rich 965
  properties 800, 801
  pyrognostic tests 768
  radioactivity 767
  rock forming
    ima acronyms 798
  sink-float techniques 776
  streak 761
  Strunz classification 778
  Strunz's class 757
  synonyms 868
  tenacity 761
  tests with cobalt nitrate and sulfur iodide 771
  transmission of light 760
minimum ignition energy 1062, 1063
minium 199, 841
minor losses 1106
minority carriers 459
minsands 280
mischmetal 430, 1183
mispickel 402, 807
mixed metal oxides (MMO) 580
mixing ratio 1055
mixture
  density 5
MnLow 549
mock 856
modal composition 893
moder 930
modified Lely process 627
modulus
  of elasticity 15
    volumetric 8
  of resilience 15, 24
  of rigidity 8
  of toughness 15
Moho 887
Mohorovicic discontinuity 887
Mohs hardness 762
Mohs scale 764
Mohs scale of hardness
  mineral 762
moissanite 626, 655
moist air 1054
  refractivity 1058
moisture content 985, 1055
MOL anode 572
molar heat capacity 25
molar magnetic susceptibility 492
molar mass 694
  (z+1)-average 696
  mass-average 696
  number-average 695

z-average 696
molar refraction 35
molar refractivity 35
mold steels 119
molecular molar mass 694
molecular sieves 1095, 1099
molecular spectroscopy
rotation 42
rotation-vibration 42
molecule
polarizability 523
mollisols 947
molten aluminum 170
molten iron 66, 72
molten potassium hydrogenosulfate 419
molten salt 797, 1174
container material 1238
physical properties 1177
molten sodium hydrogenocarbonate 419
molten sodium tetraborate 419
molten titanium 274
molten-salt electrolysis 234, 248
molybdenite 373, 374, 392, 393, 841
molybdenum 117, 297, 373, 392, 409, 1032
alloys 373
carbide-strengthened 375
properties 376
applications and uses 380, 381
bending 377
boride 650
brazing 378
carbide 655
cleaning 380
corrosion resistance 373
deep drawing 377
descaling 381
diboride 650
disilicide 551, 661, 708
drilling 379
electrical discharge machining 380
etching 380, 381
face milling 379
forming 377
grinding 379
hemiboride 650
hemicarbide 655
heminitride 659
joining 377
Lurgi design 374
machining 378
metal 375
metal powder 375
metalworking 377
Nichols-Herreshoff 374
nitride 659
pickling 380, 381
producers 385
punching 377
roaster-flue dusts 393
sawing 380
shearing 377
spinning 377
stamping 377
steels 84
threading 379
trioxide 374, 375
turning 378
welding 377
molybdenum-alloy high-speed tool steel 119
molybdic acid 374
molybdic ochre 841
monazite 278, 280, 328, 425, 448, 841
alkali digestion 427
caustic soda digestion process 449
hydrometallurgical concentration processes 427, 449
mining and mineral dressing 427
ore concentration 449
ore-beneficiation concentration 427
sulfuric acid digestion process 428, 449
Mond process 1088
Monel® 136
Monel® 450 136
Monel® K500 136
monergol 1014
monochromatic radiation
decadic molar extinction coefficient 40
Napierian molar extinction coefficient 40
monoclinic 1211
monoclinic space groups 1221
monoethylene glycol (MEG) 712
monofilaments
extrusion of polymer fibers 1024
pyrolytic conversion of precursor fibers 1024
monographies on major industrial gases 1074
monoisotopic 1202
monolithic refractories 597
monomers 691

monomethyl hydrazine 1012
mononuclidic elements 1202
monopropellant 1014
monosilane 467
monotropic conversion 666
monotype 203
monteponite 842
monticellite 786, 842
montmorillonite 596, 598, 629, 842
montroydite 842
Moody chart 1107
moonstone 760, 782
mor 930
morganite 781, 789, 792, 809
morpholine 1131
mortar 630, 976
Moseley's rule 337
mossite 355
mottled cast iron 80
MP35N 146
m-toluidine 1134
mudstone 907
mull 930
mullanite 811
mullite 597, 600, 842
  electrofused 600
  sintered 600
mullite-forming minerals 599
multiplicity of the cell 1229
Munsell notation 937
Muntz metal 185
muriatic acid, 1168
muscovite 165, 433, 842
Muthmann's liquor 1171
m-xylene 1136

n,n,n',n'-tetramethylenediamine 1134
n,n-dimethylaniline 1127
n,n-dimethylformamide 1127
nahcolite 755, 843
NaK 232
names of transfermium elements 1184
n-amyl acetate 1122
naphthalene 232
Napierian logarithm 7, 23, 24, 25, 39, 461
nascent chlorine gas 150
native gold 402
native iron 67
natrium 232, 1183
natroborocalcite 863
natrocalcite 826
natrolite 843
natronite 233
natural convection 28
natural decay series of uranium-235 1204
natural decay series of uranium-238 1203
natural gas 909, 1009
natural ilmenite 279
natural magnesia 612
natural manganese dioxide (NMD) 156
natural rubber (NR) 716, 722
natural silica 594
natural specific activity 1208
natural strain 7
natural-gas hydrates 1009
naturally occurring radioactive material 1206
naval brass 185
n-butanol 1123
n-butyl acetate 1123
n-butylamine 1123
n-butylaniline 1123
n-butylbenzene 1123
n-butylcyclohexane 1123
n-butyllithium 227
n-butyraldehyde 1123
n-butyric acid 1124
n-decane 1125
n-dodecane 1127
near alpha titanium alloys 305
necking 9
needle iron stone 828
Néel temperature 503
neocolmanite 819
neodymium 422, 424
neodymium iron boron magnets 511
neohexane 1127
neon 1066, 1090, 1091
Neoprene® 717
nepheline 165, 843
nepheline syenite 228
nephelite 843
nephrite 782, 801
neptunium
  series 1202
neptunium-237 1202
Nernstian theoretical 562
net caloric value 1063
net heating value 1002
net polarization 535
Nevindene 702
nevyanskite 833
New Jersey zinc process 192

Newton's alloy 210
Newton's law 1109, 1113
Newtonian fluid 1105, 1106
Nextel® 1028
n-heptadecane 1128
n-heptane 1128
n-hexadecane 1129
n-hexane 1129
niccolite 125, 843
Nichrome 60-15 550
Nichrome 70-30 550
Nichrome 80-20 550
Nichrome® 549
nickel (Ni) 59, 96, 103, 120, 124, 793
  alloys 124, 127, 128, 145
    class 129
    physical properties 131
  bloom 805
  carbonyl 1088
  cast irons 78
  cathodes 565
  chloride solution 126
  electrodeposits 127
  ferromagnetism 124
  from lateritic ores 127
  from sulfide ores 126
  glance 827
  major producers 141
  matte 126
  metallurgy 126
  oxide 126, 127
  processing 144
  properties 60
  silver 125, 185
  steels 84
  sulfide 126
  sulfide ores 414
  superalloys 128, 130
Nickel 200 131
Nickel 201 131
Nickel 205 131
Nickel 211 131
Nickel 233 131
Nickel 270 131
Nickel 290 131
nickel-bearing
  laterite deposits 125
  sulfide orebodies 125
nickel-beryllium alloys 249
nickel-chromium steels 84
nickel-chromium-molybdenum steels 84
nickeline 843
nickel-molybdenum steels 84
nickel-titanium 139
nickel-titanium naval ordnance
  laboratory
  shape memory metal alloy 139
nickel-titanium naval ordnance
  laboratory (NiTiNOL) 139
nicols 814
Nicrosil® 546
nielsbohrium 1184
Nimonic® 136, 137
Nimonic® 105 137
Nimonic® 115 137
Nimonic® 263 137
Nimonic® 81 137
Nimonic® 90 137
Nimonic® 901 137
niobia 665
niobiate 345
niobio-tantalates 355
niobite 344, 345, 819
niobite-tantalite 345
niobium 327, 343, 355, 574, 578, 616
  alloys 343
  boride 650
  carbide 356, 655
  carbothermic reduction 347
  cleaning 349
  corrosion resistance 344
  diboride 650
  disilicide 661
  drilling 347
  etching 349
  hemicarbide 655
  heptafluorotantalate 346
  hydrogenofluoride 346, 356
  hydroxide 346
  joining 349
  machining 347
  machining and forming facilities 352
  metallothermic reduction 347
  metalworking 347
  nitride 659
  pentaoxide 665
  pentoxide 344, 346
  pickling 349
  producers 345, 352
  properties 348
  screw cutting 349
  spinning 349
  turning 347
  welding 349

niobium-tantalum concentrates
  processing 346
nioccalite 355
Nisil® 546
NIST polynomial equations for
  thermocouple 547
nital 67
niter 238, 843
NiTiNOL 139
  austenitic 139
  self-propagating high-temperature
    synthesis (SPHS) 141
  shape memory effect 140
  superelasticity 140
niton 1093, 1183
nitosols 953
Nitrasil® 659
nitrate 576, 754
  anion 962
nitratine 844
nitratite 233, 844
nitric acid 148, 150, 164, 204, 1131, 1169,
  1242
  inhibited red-fuming 1013
nitric oxide 1066, 1075
nitride 242
  properties 648
nitrile rubber (NR) 717
nitrobenzene 1131
nitrocellulose 1014, 1015, 1017
nitroethane 1131
nitrogen 102, 123, 788, 962, 1067, 1075,
  1099
  dioxide 1067, 1075
  pentoxide 1075
  tetroxide 1013
  trifluoride 1067
nitroglycerine 198, 1017
nitroguanidine 1017
nitromethane 1017, 1131
nitronatrite 844
1-nitropropane 1131
2-nitropropane 1131
nitrotriazolone 1017
nitrous oxide 1067, 1075
n-methylformamide 1131
n-methylpyrrolidone 1131
n-nonane 1131
nobelium 1184
noble gases 1090
  properties 1090
noble metal coated titanium (NMCT) 578
noble metals anodes 568
n-octane 1132
Nomex® 710, 1027
nonanol 1132
nonbonding orbital 455
nonferrous metals 159
nonmetallic 753, 760
non-metallurgical-grade alumina 604
nonretentive 507
nonsparking 164
nonwetting 1114
norbergite 844
Norbide® 653
Nordhausen's acid 260, 1170
nordstrandite 603
NORM 1206
normal and standard conditions 1040
normal composition 893
normal hydrogen 1079
Norsk-hydro process 252
northupite 844
nosean 844
noselite 844
Novolac® 714
Novolac® resin 714
n-pentadecane 1132
n-pentane 1132
n-propyl
  acetate 1133
  formate 1133
n-propylbenzene 1133
NR 716, 717
n-tetradecane 1134
NTO 1017
n-tributyl phosphate (TBP) 450
n-tridecane 1135
n-type semiconductors 458
nuclear decay series 1202
nuclear fuel cycle 446
nuclear magnetic resonance (NMR) 484
nuclear magnetism 499
nuclear magneton 490
nuclear series
  decay chains 1202
nuclear spin angular momemtum 490
number average molar mass 695
n-undecane 1135
nu-number 36
nutrients 961
n-valeric acid 1135
n-vinyl-2-pyrrolidone 1136
Nylon® 710

obsidian 904
oceanic crust 887
o-chlorobenzaldehyde 1124
o-chlorobenzylchloride 1124
o-chlorotoluene 1124
o-cresol 1124
octafluoro propane 1067
1,3-octanediol 1132
1-octanol 1132
2-octanol 1132
1-octene 1132
octogene 1017
o-dichlorobenzene 1125
o-diethylbenzene 1126
Ohm's law 461, 478, 526, 528
ohmic drop 562, 573
oil 909
oil-hardening tool steels 119
oil-well production 677
  hydraulic fracturing 677
  pressure acidizing 677
oleic acid 1132
oleum fumans 1170
oleyl alcohol 1132
oligoclase 845
oligoelements 961, 965
olivine 278, 755, 786, 826, 845, 891, 914
olkhonskite 844
onofrite 841
onyx 782
oolitic 758
opal 467
opalescence 767
opaque 760
open tube test 774
ophtalmic glass 673
optical
  density 40
  extinction 40
  properties 32, 765
  pumping 42
  susceptibilities 524
orangite 860
ore 751
  deposit 752
  metallography 766
  microscopy 766
  minerals 752
orebody 752
organic heavy media
  density 1171
  mineralogy 1171
  refractive index 1171
organogermanium 470
orpiment 845
orthobrannerite 812
orthoclase 165, 221, 238, 596, 629, 845
orthoferrosilite 845
ortho-hydrogen 1079
orthorhombic 1211
orthorhombic space groups 1222
orthose 845
osmiridium 409, 414
osmium 407, 414, 416, 583
osmolarity 1120
osmosis 1120
osmotic pressure 1120
o-toluidine 1134
ottrelite 816
outer core 888
Outokumpu flash smelting process 201
Outokumpu zinc 194
oxalates 576
oxidation resistance 122, 148
oxide-coated titanium anode 580
oxides 242, 663, 1077
  properties 648
oxidizer 285, 999
  hypergolic 1013
oxisols 947
oxygen 1067, 1076, 1090, 1099
  atomic 1077
  magneto-archimedes effect 1076
  steelmaking 1078
oxyhemoglobin 1078, 1088
o-xylene 1136
oxyliquits 1015
Oxylite® 1077
oyster shells 610
ozone 1067, 1076

PA 710
pai-t'ung 124
paleotemperatures 1076
palladium 313, 314, 407, 409, 413, 414, 416, 584, 845, 1080
pallasites 919
palongs 206
palygorskite 846
panchromium 339, 1183
panclastites 1015
p-anisaldehyde 1122

parachor 1113
paraelectrics 534
parahydrogen 1079
paramagnetic 497
paramagnetic liquid oxygen 1175
paramagnetic materials 491, 500
paramagnets 491, 500
partial oxidation 1083
partial pressure 1041
partial wetting 1114
partially stabilized zirconia 620
particles 1025
parting 760
Paschen curve 1053
Paschen's law 1053
patronite 340, 846
Pauling electronegativity 48, 49, 1076
Pauling's diadochy rules 433
PbTe 471
p-chlorotoluene 1124
p-cymene 1125
P-E diagram 534
pearceite 846
pearly 760
Pearson's notation 757
peat 909
pebbles 907
pectolite 846
pedogenesis 927, 929
pedology 927, 928
pegmatite 219, 221, 248, 402, 791, 792, 796, 895
pegmatitic 895
Peng–Robinson 1044
Pensky–Martens Closed Cup Test 1121
pentaerythritol tetranitrate 1017
pentane 1009
1-pentanol 1132
3-pentanone 1132
1-pentene (a-amylene) 1132
pentlandite 125, 408, 846
peptide formation 694
perchlorates 574
perchloric acid 1169
Percus–Yevick 1044
perfluorinated alkoxy (PFA) 708, 722
perfluoroalkoxy 708
performance index 21
perhydrol 1013
periclase 665, 847
peridot 826, 845
peridotite 125, 784, 902
peridots 467
peristerite 782
perlite 75, 76, 755
permanent magnets 510
permanganate 156, 572
permeability coefficients of most common polymers 734
permeability of vacuum 488
permitivity
  of a medium 519
  of a vacuum 519
  relative 520
perovskite 277, 575, 847, 888
peroxodisulfuric acid 580
petalite 219, 220, 223, 847
PETN 1017
petrography 885
petrolatum 1008
  specific gravity 1008
petroleum 909
petroleum products 407
petrology 885
petzite 847
pezzottaite 847
PFA 708
PGMs *see* platinum-group metals
  alloys 416
  applications and uses 420
  arsenides 408
  corrosion resistance 417
  producers 421
  sulfides 408
  tellurides 408
phaeozems 952
phaneritic 895
phanerocrystals 892
pharmaceutical glass 673
phenakite 248
phenocrystals 892, 895
phenol-formaldehyde 714, 722
phenol-formaldehyde resins 691
phenolics 714
phenylethene 706
phenylethyl alcohol 1132
phlogopite 786, 847
phonons 482
phosgene 1067, 1088
phosphate 679
phosphate crown 675
phosphate rocks 756
phosphine 1067
phosphomimetite 851

phosphor bronze 185
phosphorescence 45, 47, 766
  minerals 766
phosphoric acid 1132, 1169
phosphorite 756, 908
phosphorus 610, 963
  bromide 1132
  chloride 1132
  pentafluoride 1067
photocathode materials 553, 554
photoconductivity 458
photoelectric effect 553, 554
photoelectric quantum yield 553
photoelectrons 553
photoemission 554
photoluminescence 45
photolysis 693
photovoltaic 458
phyllosilicates 596
physical characteristics of earth's interior 889
physical properties of polymers 720
PI 710
pickling 271
2-picoline 1132
3-picoline 1132
4-picoline 1132
picotite 838
picral 67
picric acid 1017
picroilmenite 277, 279
Pidgeon's magnetherm process 252
Pidgeon's process 253
piedmontite 848
piemontite 848
piezoelectric materials 534
piezoelectricity 534, 766
  minerals 766
pig iron 73
pigeon blood 794
piperidine 1132
PIR 716
pirssonite 848
pisolitic 758
pistanite 840
pitch 1026
pitchblende 265, 440, 848
  grinding 442
Plaggen cultivation 928
plagioclase feldspars 895
plagioclases 165, 596, 629
plain carbon steels 85
  typical chemical composition 87
Planck constant 491
Planck radiation formula 44
Planck's constant 460
Planck–Einstein equation 41
plane angle between lattice planes 1230
planosols 951
plasma melting 304
plaster of Paris 968
plastic deformation 9
plasticizers 692
plastics 1015
platinized titanium 579
platinized titanium anodes 579
platinum 407, 413, 414, 415, 428, 546, 551, 552, 568, 578, 848
  alloys
    physical properties 416
    tensile strength and elongation 417
  cleaning labware 419
  metal and alloy suppliers 421
  ores 409
platinum-10 rhodium 546
platinum-13 rhodium 546
platinum-30 rhodium 546
platinum-5 molybdenum 546
platinum-6 rhodium 546
platinum-cobalt 511
platinum-cobalt magnets 511
platinum-group metals (PGMs) 407
platinum-iron magnets 511
platonician regular polyhedrons 1210
plattnerite 569, 573, 848
pleochroism 37, 765
pleonaste 848
Plexiglas® 709
plumbago 625, 829
plumbous chloride 819
plumbum 196, 1183
plumose 758
plutonic rocks 890
  classification 901
plutonium 436, 437, 438, 448, 452
  allotropes 453
  dioxide 454
  isotope 454
  isotopes 452
  radionuclides 453
  tetrafluoride 454
plutons 910
PMCs 1029
  processing 1030

PMMA 709
PMP 704
podzols 951
Podzoluvisols 952
point groups 757, 1212
Poisson's ratio 8, 10, 23, 64
polarizability 523
polarization 523, 530
  dipole 531
  effect of frequency 531
  electronic 530
  ionic 531
  mechanisms 532
  space charge 531
  spontaneous 534
polethylene fibers 1027
polianite 850
pollucite 240, 242, 849
polonium 1203
polyacetals (PAc) 711
polyacrylic butadiene rubber 722
polyacrylonitrile (PAN) 1026
polyamide (PA) 710
  nylon 723
  nylon 11 722
polyamide-imide 722
polyaramid fibers 1027
polyaramide (PAR) 710, 723
polyarylate resins 723
polybasite 849
polybenzene-imidazole 723
polybutadiene 717, 1014
  rubber 716, 723
  terephtalate 723
polybutadiene acrylic acid acrylonitrile (PBAN) 1014
polybutylene (PB) 704, 723
polybutylene terephtalate (PBT) 712
polycarbonates (PC) 711, 723
polychloroprene rubber 723
polychlorotrifluoroethylene (PCTFE) 708
polycholoroprene 717
polycondensation 694
polycrystalline silicon 467, 468, 472
polydiallyphthalate (PDP) 713
polyester sulfone (PSU) 711
polyether
  ether ketone 723
  imide 723
  sulfone 723
polyethylene
  fibers 1027
  high density 703
  highly oriented 1027
  low density 703
  naphtalate 723
  oxide 724
  terephtalate 724
polyethylene (PE) 702, 704, 723, 1093
polyethylene terephtalate (PET) 470, 712
polyhalite 849
polyhedrons 1210
polyhydroxybutyrate 724
polyimides (PI) 710, 724
polyisocyanurate 1093
polyisoprene 724
polyisoprene rubber 716
polylactic acid 724
polymer matrix composites (PMCs) 1019, 1029
  properties 1031
polymerization 691, 693
  average degree 695
  by addition 693
  by free-radicals 693
polymers 691, 695
  additives 692
  atactic 697
  chemical resistance 735
  classification 692
  fillers 692
  fluorinated 707
  gas permeability 734, 1051
  isotactic 697
  IUPAC acronyms 745
  physical properties 720, 727
  syndiotactic 697
  tacticity 697
polymetallic nodules 152
polymethyl methacrylate (PMMA) 709, 724
polymethylpentene (PMP) 704, 724
polymignite 867
polymorphism 64
polyolefins 702
polyoxymethylene 724
polyphenylene
  atactic 724
  oxide 724
  sulfide 724
polyphenylene oxide (PPO) 712
polyphenylene sulfide (PPS) 712
polyphenylsulfone 711
polypropylene (PP) 703, 704, 725

polysilane 1028
polysiloxane 719, 726
polystyrene (PS) 706, 725
polysulfide rubber 718, 725
polysulfides 233
polysulfone (PSU) 711, 725
polytetrafluoroethylene (PTEE) 707, 708, 725
polythene 702
polytrifluorochloroethylene 725
polyurethane 725, 1014, 1093
polyurethanes (PUR) 715
polyvinyl
  alcohol 725
  dichloride 705
polyvinyl acetate (PVA) 705, 725
polyvinyl butyral (PVB) 676
polyvinyl chloride (PVC) 704, 705, 725
polyvinylidene chloride (PVDV) 705, 725
polyvinylidene fluoride (PVDF) 706, 725
Populus balsamifera 986
porcelain 629
porcelain bricks 633
porcelain enamels 630
porpezite 414
porphyritic 895
porphyritic rhyolite 895
porphyritic texture 895
porphyrocrystals 892
porphyroid 895
porphyry copper 392
Portland cement 968, 969, 976
  chemical composition 970
  chemistry 971
  nomenclature 973
  processing 970
  raw materials 969
Portland clinker 611
portlandite 973, 978
potash 755, 908
potash mica 842
potash soda lead glass 672
potassium 237, 345, 964
  amalgam 237
  applications and uses 239
  chloride 170, 220, 238
  dichromate 368, 574
  fluoride 337, 356
  heptafluorotantalate 356
  hydroxide 238, 1169
  oxalate 237
  perchlorate 297
  permanganate 151, 156
  salt 238
  sulfate 238
Pourbaix diagram 587
powder metallurgy 145
  apparent density 122
  bulk density 122
  of titanium 299
  pore-free density 122
  theoretical density 122
powellite 374, 849
pozzolan 968, 976
PP 703
PPS 712
praseodymium 422, 424
precious and noble metals anodes 568
precious gemstone 781
precipitated silica 595
precipitation of secondary phases 11
prepregging 1030
pressure 1037
  non-SI units 1038
  normal and standard temperature 1041
  of the standard atmosphere 1039
pressure acidizing 678
pressure drop 1106
prestressed concrete 976
pretulite 434
prices of pure elements 1245
primary explosives 1015
primers 1015
primordial radionuclides 1201, 1205
principal refractive indices 37, 765
principle of corresponding states 1048
producer gas 282
production of proppants 687
profile 927
promethium 422, 424
proof strength 10
propadiene 1,2 1067
propagation 694
propanal 1132
propane 1067
1,2-′propanediol 1132
1,3-propanediol 1132
1,3-propanethiol 1133
propanoic acid 1133
1-propanol 1133
2-propanol 1133
propanone 1133
propanoxypropane 1133

propargyl alcohol 1133
propellant
  liquid 1011
  solid 1014
propellants 999, 1011
  cryogenic 1012
  hypergolic 1012
  petroleum-based 1012
propene 1067
propergols 1011
properties of cobalt alloys 145
properties of composites 1021
properties of gases 1037
properties of ice 913
properties of ice polymorphs 914
properties of industrial graphite grades 624
properties of liquids 1103
properties of molybdenum alloys 375
properties of proppants 679
properties of selected commercial explosives 1016, 1017
properties of selected ferroelectric materials 536
properties of selected gold alloys 405
properties of selected silver alloys 399
properties of semiconductors 464
properties of the elements 1185
properties of thorium, uranium and plutonium 436
properties of tungsten alloys 387
properties of water 1121
properties of woods 985
propionaldehyde 1132
propionic acid 1133
propionitrile 1133
proppants 677, 678
  atomization 683
  classification 679
  commercial 683
    properties 684, 686
  fire polishing 683
  flame spraying 683
  materials 678
  producers 687
  production 687
  properties 680
  synthetic 682
    pelletizing 682
    sintering 682
  testing laboratories 689
propping agent 678
proprionic acid 1133
propyl alcohol 1133
propyl chloride 1124
propyl mercaptan 1133
propylamine 1133
propylene 703, 710, 1067
propylene carbonate 1133
1,2-propylene glycol 1133
1,2-propylene oxide 1133
propylene oxide 1133
propylene-vynilidene hexafluoride 725
protium 1079
protoactinium 1204
protolith 910, 911
protore 752
proustite 849
PS 706
pseudobrookite 849
pseudocumene 1135
pseudoelasticity 140
pseudorutile 279, 850
psilomelane 152, 850
PSR 718
PSU 711
psychrometric charts 1058
psychrometric equations 1058, 1061
psychrometric properties 1054
PTFE 707
Pt-wire 768, 776
p-type semiconductors 459
pulling crystal growth technique 472
pultrusion 1030
pumice 755, 904
PUR 715
pure copper 184
pure elements
  price 1245
  Strukturbericht designation 1215
pure iron 64
  grades 73
pure substances
  nist molar thermodynamic properties 1195
PUREX process 446
purified terephthalic acid (PTA) 712
PVA 705
PVC 704
PVDC 705
PVF 706
p-xylene 1136
pycnite 861
pycnometer

four-mass method 5
three-mass method 4
pyrargyrite 850
Pyrex® 671
Pyrex®0211 673
Pyrex®7059 673
Pyrex®7070 673
Pyrex®7740 673
Pyrex®7789 673
Pyrex®7799 673
Pyrex®7800 673
Pyrex®7913 674
Pyrex®plus 674
pyridine 1133
pyrite 68, 188, 190, 402, 760, 850, 908
pyrochlore 345, 355, 440, 850
pyroclastic 904
pyroclastic igneous rocks 904
pyroclastic sedimentary rocks 907
pyroelectricity 766
minerals 766
pyrognostic tests 768
pyrohydrolysis 445
pyrolitic boron nitride 638
pyrolusite 151, 152, 443, 850
pyrolytic conversion of precursor fibers 1024
pyrolyxin 699
pyrometallurgical process 126
pyrometallurgy 261
pyrometric cone equivalent 597, 641
pyromorphite 851
pyrope 782, 851
pyrophanite 152, 851
pyrophillite 755
pyrophoricity 273
refractory metals 273
pyrophyllite 851
pyrosphere 888
pyrotechnic mixtures 1015
pyroxene megacrystals 279
pyroxenes 278, 467, 891, 914
pyroxenites 902
pyrrhotite 125, 402, 408, 851
pyrrolidine 1133
2-pyrrolidinone 1133

qandilite 851
Q-factor 25, 532
QSL process 200
quadratic 1211
quartz 264, 467, 594, 595, 598, 665, 755, 760, 771, 777, 796, 852, 899, 905
quartzite 403, 595, 756, 912
quaternary compounds
Strukturbericht designation 1218
Queneau–Schuhmann–Lurgi process 200
Quercus virginiana 986
quicklime 260, 261, 610, 613, 968
quicksilver 840
Quinn's equation 17

radial blende 866
radiated 758
radiation 28
electromagnetic 38
spectrum 38
radioactinium 1204
radioactive 1206
radioactive decay series 1203
radioisotopes 265
radiolarite 908
radiolysis 693
radionuclides 1202
cosmogenic 1206
decaying 1207
non-series primordial 1205
primordial 1201
radiothorium 1204
radium 264, 329, 440, 1203
radon 329, 442, 1067, 1090, 1092, 1093
rammelsbergite 125, 852
ramsdellite 852
rankers 950
Raoult's cryoscopic constant 1120
Raoult's ebullioscopic constant 1119
Raoult's law 1118
Raoult's law and freezing point depression 1119
Raoult's law of tonometry 1118
rare earths
applications and uses 430
physical and chemical properties 424
producers or processor 431
purification or refining 428
rare gases 1090
rare-earth metals 422
Ames laboratory process 428
applications and uses 429
liquid–liquid extraction process 429
metallothermic reduction 428
rasorite 834

rayon 700, 1026
RDX 1017
reaction bonded silicon nitride (RBSN) 635
reactive metals 266
  properties 267
real density 2
real gases 1037
  covolume 1042
  critical molar volume 1047
  critical opalescence 1047
  critical point 1047
  critical pressure 1047
  critical temperature 1047
  equation of state 1044
  excluded volume 1042
  isotherm 1046
  isothermal virial coefficients 1043
  Van der Waals equation of state 1042
realgar 760, 852
reciprocal lattice 1232
red beryl 789
red brass 185
red gold 405, 406
red lead 368
red lead oxide 841
red mud 167, 602
red rubicelle 857
red zinc oxide 867
Redlich–Kwong 1044
Redlich–Kwong–Soave 1044
Redlich–Kwong–Soave–Gibbons–Laughton 1044
reduced iron 64, 73
reductant 999
reduction on charcoal 771
reduction test on charcoal 771
refined silver 399
reflection coefficient of the surface 552
reflective index 37
refractive index (RI) 33, 765
  temperature coefficient 36
refractive index of moist air 1058
refractivity 35
refractory 593, 630
  classification 630
  fireclays 597
  grade chromite 369
  manufacturers 634
  properties 631
  raw materials
    properties 628
refractory metals 266
  corrosion resistance 271
  descaling procedures 272
  etching 272
  properties 267
  pyrophoricity 273
regosols 950
regular-grade silicon 468
Reichert cones 280
reinforced concrete 976
reinforcement material 1019
reinforcing bars 976
relative density 762, 1103
relative dielectric permittivity 520
relative humidity 1056
relative index of refraction 33
relative magnetic permeability of a material 489
relative molar mass 694
relative molecular molar mass 694
relative permittivity 520
relative refractive index 33
relative Seebeck coefficient 544
relative temperature coefficient of refractive index 36
remanent magnetic induction 505
remanent polarization 535
rendzinas 950
Rene® 41 137
Rene® 95 137
reniform 758
Repetti discontinuity 887
reserve base 753
reserves 752
residual clays 907
residual sedimentary rocks 906
residues 691
resilience 15
resin
  formulation 1030
  transfer molding 1030
resin-based composites 1019
resin-coated sand
  producers 688
resinous 760
resistance alloys 548
resistance temperature detectors (RTD) 552
resistance thermal devices 552
resistor 548, 549, 550
resistor alloy 10 549
resistor alloy 15 549

resistor alloy 30 549
resistor alloy 5 549
resonance factor 25
reticulated 758
Retjers' liquor 1173
Reynolds number 1107
rhenium 391, 392
  alloys 378
  applications and uses 393
  catalysts 393
  cold isostatic pressing (CIP) 391
  heptoxide 374, 393
  powder injection molding (PIM) 391
  sulfide oxidizes 393
rheostats 549
rhizalites 921
rhodite 413
rhodium 398, 407, 413, 416, 578
rhodizite 240, 242
rhodochrosite 152, 852
rhodolite 782, 851
rhodonite 152, 853
rhombohedral 1211
Richard's rule 31
Richardson constant 552
Richardson–Dushman equation 552
Ridgeway 763
Ridgeway scale 764
riebeckite 853
right-hand rule 488
rimmed steels 85
ringwoodite 853, 888
Robax® 674
rock crystal 467, 782
rock forming minerals 751
rock salt 233, 756, 759, 830, 908, 909
rock texture 895
rocks 885
  extrusive 891
  fluid flow characteristics 921
  foliated 911
  igneous 889, 890
    acidity 897
    alkalinity 897
    aphanitic 895
    chemical composition 898
    chemistry 896
    classification 899, 900
    coloration 894
    crystallinity 896
    glassy 895
    hyaline 895
    mineralogy 892
    pegmatitic texture 895
    petrographic classification 891
    phaneritic 895
    porphyritic 895
    porphyroid texture 895
    QAPF-diagrams 899
    saturation 897
    Streckeisen's diagrams 899
  intrusive 890
  magmatic 889
  mechanical behavior 921
  metamorphic 910
    contact 911
    regional 911
    thermal 911
  non-foliated 911
  phaneritic texture 891
  plutonic 890, 901
  properties 922
  pyroclastic 904
    classification 904
  sedimentary 889, 904
    biogenic 909
    carbonaceous 909
    chemical 908
    deposition 905
    detritic 907
    diagenesis 905
    lithification 905
    pyroclastic 907
    residual 906
    sedimentation 905
    transportation 905
    weathering/erosion 905
  terrigenous
    clastic 907
  texture 895
  ultramafic 902
  volcanic 891, 903
Rockwell hardness 12, 13
roentgenium 1184
rolled zinc 197
romanechite 152, 850
roscoelite 340
rose quartz 782
Rose's alloy 210
rosenbuschite 853
rosin 697
Rosival 763
Rosival scale 763, 764
rostfrei Stahl 95

rotary-kiln furnaces 127
rotating electrode process 299
roughness 20
round silica sand 595
 producers 688
rubber 692, 715
rubellite 783
rubicelle 783
rubidium 239
 hydroxide 239
 major producers 241
ruby 753, 781, 794, 819
 shaping and treatment 794
ruby silver 849
ruby silver ore 850
ruby spinal 857
Russian processes 252
rustless 95
rustproof iron 95
ruthenium 314, 407, 409, 416, 580, 581
 dioxide 409, 582
rutherfordium 1184
rutile 275, 277, 279, 280, 283, 292, 328, 427, 448, 614, 615, 667, 853
 pigments 288

saccharose 1118
sacrificial anode 588
sacrificial anodes materials 587
safety glass 676
Saffil® 1028
safflorite 143
salpeter 1075
salpeter nitre 843
salt cake 170
salt of phosphorus 775
salt spirit 1168
saltpeter 238, 962
samaria 665
samarium 422, 424
samarium cobalt magnets 511
samarium oxide 665
samarskite 355
sand 756
sand dune placer deposits 277
sandstone 595, 907, 909, 979
sandy clay loam 940
Sanicro® 28 137
sanidine 853
sapphire 663, 753, 781, 793
 glass 674
 "kashmir" sapphire 794
 thermal treatment 794
sapphirine 854
sapwood 983, 985
Saran® 705
satin spar 829
saturation magnetic induction 504
saturation polarization point 535
saukovite 841
SBR 717
scalar product 1228
scale height 1045
scandiobabingtonite 434
scandium 422, 424, 433
 alloys 435
 applications and uses 434
 chemicals 435
 metal 435
 sesquioxide 433, 434
 trifluoride 434
scavengers 1099
 properties 1099
scheelite 386, 387, 854
schiller 767
schists 912
Schoenflies–Fedorov 1213
Schorl 854
Schott® 674, 675
schreyerite 854
Schröder's liquor 1172
scleroscope hardness number 13
scoria 904
scorodite 854
scorzalite 836
scrutinyite 573, 854
seaborgium 1184
seawater magnesia clinker 613, 614
sec-butylamine 1123
secondary electrons 554
secondary emission coefficient 555
secondary explosives 1015
sectile 762
Securit® 676
sediment 906
sedimentary rocks 889, 904
sedimentation 1109
Seebeck
 coefficient 544
 effect 543
 electromotive force 543
seed lac 699
Seger's pyrometric cone 630

selected properties of molecular sieve 1098
selenite 829
selenium 233
Sell–Meier formula 35
semiconductors 455, 456
  applications 467
  classes 457
  compound 457, 459
  concentration of acceptors 460
  concentration of donors 460
  concentration of electric charge carriers 460
  densities of states 460
  doping 457
  electric mobility 461
  electromigration 461
  Grimm–Sommerfeld rule 459
  intrinsic 457
  materials 455
  metal-oxide 474
  n-type 458, 459, 460, 474
  p-n junction 475
  properties of 464
  p-type 459, 460, 474
  type-n 457
  type-p 457
  wafer processing 471
semigraphite 572
semimetals 457
semiprecious gemstone 781
senarmontite 855
separator 561
sepiolite 755
serpentine 611, 786
serpentinite 612
sessile drop 1118
S-glass 673, 1025
Shabaeva's liquor 1173
shale 907, 970
shape memory alloys (SMAs) 139
  nickel-titanium solid 140
shape memory effect 139, 140
shaped refractories 597
shear
  modulus 8
  rate 1105
  strain 8
  stress 7, 1105
sheet lead 198
shellac 699
Sherritt ammonia pressure leaching 127
Sherritt-Gordon ammonia leaching process 144
shielding efficiency 512
shock-resisting tool steels 119
short term exposure limit 1064
shortite 855
shunts 549
SiAl 887
sialon (SiAlON) 636
  applications 636
Siberian red lead 368
siberite 783
siderite 68, 612, 855, 914, 918
siderolites 67, 914
siderophiles 1185
siderose 855
siderurgy 73
siegenite 143
silane 1067
silica 70, 165, 594, 665, 908, 929, 976, 979, 980
  bricks 629, 633
  fumed 595
  fused 594, 596
  gel 595
  natural 594
  precipitated 595
  sand 468
  specialty 594
  vitreous 596
silicates 70, 1077
silicides
  properties 648
silicium dioxide 665
silicomanganese 155
  calcium-carbide furnace 156
silicon 117, 463, 596, 719
  aluminum oxynitride (SiAlON) 636
  applications and uses 468
  brass 186
  bronze 185
  carbide 468, 625, 655, 1028, 1032
    Acheson process 626
    grades 628
    Lely process 627
    polymorphism 626
    polytypism 626
  dioxide 463
  hexaboride 650
  hydrogenated amorphous 468
  hyperpure 468
  killed steels 85

monocrystal 458
nitride 635, 659, 787
hot isostatically pressed 636
hot-pressed 636
sintered 636
rubber 719, 726
single-crystal ingots 472
tetraboride 651
tetrachloride 329, 467, 595, 1067, 1133
metallothermic reduction 468
tetrachlorosilane 595
tetrafluoride 467, 1067
silicon carbide 551
fibers 1028
fibres 1028
silicon-manganese steels 84
silicothermic reduction 253
silky 760
sill 891
sillimanite 165, 329, 448, 597, 599, 600, 755, 855
silt 907
siltstone 907
silver 396, 855
alloys
applications and uses 398
properties 399
chloride 400
fulminate 400
nitrate 400
silver alloys 396, 398
silver bearing copper 184
silver electroplating 125
silver glance 801
silver magnesium alloy 399
silver-palladium 399
SiMa 887
Simplex process 370
singlet states 46
sinhalite 855
sink-float separations 776
sintered alumina 607
sintered magnesia 613
sintered silicon nitride 636
Sirosmelt lance 201
siserskite 833
sizing agent 1027
skarns 912
skin depth 528
skin effect 528
skobolite 355
skutterudite 143, 856
slagging 281
slaked lime 610
slate 756, 912
sliding friction coefficient 20
smalt 142
smaragdite 801
smelter gas 282
smithsonite 188, 856
smokeless powder 1015
smoky quartz 759, 782
S–N plots 18
Snellius–Descartes law 32, 33
soapstone 859
soda ash 233, 235
soda ash roasting 370
soda lime glass 673
soda niter 233, 844, 962, 1075
soda-lime-silica 415
sodalite 856
sodamide 232
sodium 232
aluminate 167, 602, 679
aluminate liquor 602
amalgams 233
applications and uses 235
bicarbonate 843
carbonate 225, 232, 234
chlorate 341, 443
chloride 170, 291, 607
chromate 371
dichromate 371
electrolysis 226
hexachloroplatinate 579
hydrogenocarbonate 232
hydroxide 167, 234, 413, 414, 427, 679, 1169
hydroxide film 232
hypochlorite 580, 699
major producers 236
molten-salt electrowinning 234
nitrate 297
polysulfide 718
sulfate dekahydrate 224
tetrahydroxyaluminate 602
triphosphate 427
tungstate 1173
xanthate 223
zeolite 1095
sodium-cesium alloy 242
sodium-d-line 36
soft ferromagnetic materials 506, 507
soft quick solder 211

Index

soft superconductors. 478
softwoods 983
  properties 991
soil 927
  acidity 945
  alteration 929
  ASTM civil engineering classification 956, 957
  ASTM standards 960
  attributes 937
  cementation 944
  cheluviation 930
  classification 928
  clay minerals 929
  coloration 936
  consistency 944
  effervescence 945
  erosion 929
  FAO classification 948
  formation 943
  French classification 954
  horizons 927, 931
    boundaries 936
    international nomenclature 932
    subdivision 932
  humification 929
  identification 957
  ISO standards 958
  lessivage 930
  lixiviation 930
  micronutrients 965
  mineralization 929
  Munsell color chart 937
  organic matter 943
  organic matter (SOM) 936
  physical properties 961
  Plaggen cultivation 928
  plant roots 945
  profile 927, 931, 941
  properties 936
  redoximorphic features (RMF) 937
  structure 941, 942
  taxonomy 945
  terminology for rock fragments 939
  texture 938, 939, 940
  USDA classification 945
  weathering 929
solar evaporation process 224
solenoid 487, 488, 489
sol–gel growth techniques 797
sol–gel silica 595
solid fuels 1004
  properties 1007
solid ion conductors 556
solid material
  anisotropic 37
  biaxial 37
  compression 7
  elasticity 8
  linear strain 7
  mechanical behavior 6
  resilience 15
  siffness 8
  slidding 19
  tension 7
  uniaxial 37
solid oxide fuel cells (SOFCs) 621
solid oxide membrane (SOM) 1084
solid propellant 1014
solid solutions 11
solids
  dispersion 11
  electrical classification 456
  heat of fusion 1057
  mass density 4
  sessile drop 1114
  specific damping capacity 24
  specific gravity 3
  strengthening mechanisms 11
  x-ray density 2
solonchaks 949
solonetz 952
solubility of gases in liquids 1050
solutes
  nonvolatile
    colligative properties 1118
solutions
  boiling point elevation 1119
  hyperosmotic 1121
  hyposmotic 1121
  isosmotic 1121
  isotonic 1121
solvents 692
Sorel slag® 285
Soremetal® 73
Souchine–Rohrbach's liquor 1173
sound 23
  attenuation 24
  damping 24
  intensity 23
  longitudinal velocity 23
  point source 24
  powers 23
  pressures 23

source of ignition 1062
space charge polarization 530, 531
space groups 757, 1221
  cubic 1227
  hexagonal 1226
  monoclinic 1221
  orthorhombic 1222
  tetragonal 1223
  triclinic 1221
  trigonal 1225
space lattice
  Bravais 1212
  parameter 1209
  parameters 757, 1209
    rhombohedral 1211
  plan angle 1230
  structure type 757
  unit cell volume 1230
  volume 1230
specialty silicas 594
specific
  activity 1207
  enthalpy 1057
  gravity 3, 762, 1103
  heat capacity 26, 1049
  humidity 1056
  latent enthalpy 30
  magnetization 492
  molar extinction coefficient 40
  refractivity 35
  weight 3
spectral emissivity 30
spectrolite 835
specularite 831
spent fireclay 597
spent lime 262
spent magnesia 613
sperrylite 413
sperylite 856
spessartine 782, 856
spessartite 152, 856
sphalerite 152, 188, 190, 199, 469, 856
sphene 277, 861
spin
  multiplicity 45
spinel 165, 575, 786, 857
spinnerette 1024
splintery 761
spodosols 948
spodumene 219, 220, 221, 222, 224, 228, 248, 857
sponge iron 123
sponge-reduced iron 123
spongolite 908
spontaneous magnetostriction 494
spontaneous polarization 534
spreading coefficient 1114, 1115
Sprengel explosives 1015
spurrite 857
stabilised refractory dolomite 611
stabilization 1027
stabilized zirconia (CSZ) 620, 621
stabilizers 693
stainless steel 95, 98
  application guidelines 109
  austenitic 101
    manganese-bearing 101
    nitrogen-strengthened 101
    physical properties 104
  cast heat-resistant 103
  classification 96
  corrosion resistance 96, 108
  fabrication 108
  ferritic 97
  martensitic 97
  mechanical strength 108
  melting process 103
  P-H
    physical properties 107
  precipitation-hardening 103
  scrap 103
  simplified selection 108
stalactites 261
stalactitic 758
stalagmites 261
standard calcined aluminas 607
standard mean ocean water (SMOW) 784
standard pee dee belemnite (SPDB) 784
stannite 205, 857
stannum 204, 1183
star sapphire 793
static electricity 766
static friction coefficient 19, 20
staurolite 329, 755, 857
staurotide 857
steam reforming 1082
steam-iron process 1082
steatite 859
steel 59, 84
  aluminum-killed 85
  carbon designation 84
  carburizing 86
  case-hardening 86
  eutectoid 76

high-carbon 85, 87
hypertectoid 76
hypotectoid 76
low-alloy designation 84
low-carbon 85
medium-carbon 85, 86
metallographic etchants 67
mill scale 85
rimmed 85
scrap 103
silicon-killed 85
stainless 95
ultra-high-strength 115
steel reinforced concrete 976
steelmaking 71
Stefan–Boltzmann equation 30
stellated 758
stellite 145
alloys 148
corrosion resistance 148
grades 148
Stellite® 147
Stellite®1 147
Stellite®100 147
Stellite®12 147
Stellite®20 147
Stellite®21 147
Stellite®3 147
Stellite®306 147
Stellite®6 147
Stellite®7 147
Stellite®8 147
stephanite 857
stereotype 203
sterling silver 398, 399
stibine 402
stibiopalladinite 414
stibium 806, 858, 1183
stibnite 771, 858
stick lac 699
stimulated emission 42
Einstein coefficient 44
stishovite 594, 858
stoichiometric ilmenite 277
stoichiometric rutile 275
Stokes' law 1109
stoneware 629
stony iron meteorites 919
stony meteorites 915
storage capabilities for hydrogen 1086
strain 7
strain hardening 11
exponent 11
strain rate 17
Stratcor process 341
streak 761
streak plate 761
strengite 864
strength hardening coefficient 11
strength-to-weight ratio 112
stress 7
stress cycles 18
stress-intensity factor 16
stress–strain curve 8, 15
striction 9
stromeyerite 858
strontianite 263, 858
strontium 262, 536
carbonate 263
oxide 263
sulfide 263
titanate 263
structure of polymers 697
structure of the Earth's interior 886
Strukturbericht 757, 1215
Strunz's classes 757
struverite 355
styrene 706
styrene (vinylbenzene) 1133
styrene butadiene rubber (SBR) 717
styrene-butadiene styrene rubber 726
subautomorphous 758, 892
subeconomic resources 753
subhedral 758, 892
sublimates
closed tube test 773
subsoil
horizons 937
structures 943
substrate glass 673
succinite 803
sucrose 1118, 1120
sulfatable titania slag 281
sulfate anions 978
sulfate slag 281
sulfide ores 126, 392
sulfolane 1133
sulfur 125, 610, 755, 858, 1183
dichloride 1133
dioxide 150, 374, 396
dioxide gas 1085
monochloride 1134
sulfuric acid 148, 150, 248, 1134, 1169, 1244

electrolyte 154
fuming 1170
roast process 223
sulfuryl chloride 1134
sulphate process 286
sulphur
dioxide 1067
hexafluoride 1067
trioxide 1067
sunstone 782, 845
superalloys 128, 145, 1032
iron-based 121
superconductors 477
BCS theory 482
high critical temperature 481
high-magnetic-field applications 485
low-magnetic-field applications 485
organic 482
Type I 478
Type I
properties 479
Type II 480
properties 480
vortex state 481
supercooled liquid 671
super-duty fireclay 597
superelasticity 140
superheavy water 1080, 1121
physical properties 1167
superphosphate 963
surface alloying 364
surface electrical resistivity 528
surface resistivity
skin depth 528
skin effect 528
surface tension 1110
surfactants 1112
suspensions 1174
Sutherland's equation 1049
syenite 899
sylvanite 859
sylvinite 238, 859, 964
sylvite 859
symmetry elements 1210
syndiotactic polymer 697
synthetic diamond electrodes 585
synthetic gas 1082
synthetic gemstones 795
synthetic isoprene rubber 726
synthetic magnesia 613
synthetic mullite 600
synthetic rutile 283, 286
Becher process 283
Benelite process 284
enhancement process (SREP) 286
producers 284

tabular 892
tabular alumina 607
Tachardia lacca 699
taconite 908
tactic polymer 697
tacticity 697
Tag Closed Cup Test 1121
talc 755, 859
tantala 666
tantalite 221, 666, 859
tantalum 326, 327, 344, 346, 353, 428, 468, 574, 578
alloys 353
physical properties 358
annealing 359
anodic electroetching 360
applications and uses 365
boride 651
carbides 356
cathodic sputtering deposition 363
chemical coating 363
chemical vapor deposition 363
cladding 361, 362
cleaning 360
coating techniques 361
coherent deposit process 364
corrosion resistance 353
deep drawing 357
degreasing 360
descaling 360
diboride 651
disilicide 661
electrochemical coating 363
electrodepositing 364
electroplating 364
etching 360
explosive bonding 362
fluoride 346
forming 357
grinding 359
grit blasting 360
hemicarbide 656
heminitride 659
hot rolling 362
hydrogenofluorides 356
joining 359

loose lining 361
machining 359
machining and forming facilities 367
metal 356
metalliding 364
metallurgy 355
metalworking 357
nitride (e) 659
pentaoxide 666
pentoxide 353, 355, 584
physical coating 363
physical vapor deposition 363
pickling 360
powder
hydride-dehydride process 357
producers 366
punching 357
roll bonding 362
silicide 661
spinning 359
stamping 357
thermal spraying 362
turning and milling 359
vacuum deposition 363
welding 359
Tantung G 147
tanzanite 783, 867
tap density 2
tapiolite 345, 355, 859
TATB 1018
technetium 392
technologically-enhanced naturally occurring radioactive material 1206
Technora® 710, 1027
Teflon® 707
tektites 467, 914, 920, 921
geographical locations 921
telluric iron 67
telluric silver 831
tellurium 233, 859
tellurium atoms 459
tellurium copper 184
tellurium lead 198
temperature 1057
dry bulb 1057
wet-bulb 1057
temperature coefficient of capacitance 520
temperature coefficient of thermal conductivity 528
temperature dependence of surface 1112
temperature dependence of the dynamic viscosity 1106
temperature of colour 641
tempered glass 676
tenacity 761
tenorite 859
TENORM 1206
tension 7
tephroite 152, 859
terbium 422, 424
terlinguaite 860
termination 694
ternary compounds
Strukturbericht designation 1217
terpene 698
terpolymer 706
terra rossa 907
terrestrial iron 67
terrigenous rocks 907
tert-amyl methyl ether 1122
tert-butanol 1123
tert-butyl
acetate 1123
chloride 1123
mercaptan 1123
tert-butylamine 1123
tertiary explosives 1015
testing refractories
ASTM standards 643
ISO standards 645
tetrabromo-1,1,2,2-ethane 777
tetrabromoethane 1171
1,1,1,2-tetrabromoethane (acetylene tetrabromide) 1134
1,1,2,2-tetrabromoethane (acetylene tetrabromide) 1134
tetracalcium aluminoferrite 972
tetracalcium aluminum monosulfate hydrate 978
1,1,2,2-tetrachloroethane 1134
1,1,2,2-tetrachloro-ethylene 1134
tetrachlorosilane 595, 1133
tetradymite 860
tetraethylene
glycol 1134
pentamine 1134
tetrafluoro methane 1067
tetrafluoroethylene 707
tetragonal 1211
tetragonal space groups 1223
tetragonal zirconia polycrystal 620
tetragonal β-spodumene crystals 224

tetrahedrite 860
tetrahydrofuran 1134
tetrahydrofurfuryl alcohol 1134
tetrahydroxoaluminate anion 164
tetralin 1134
tetramethylsilane 1134
tetraoxide 414
tetrazene 1015, 1016
tetryl 1017
texture 895
Thai ruby 794
theoretical density 2
thermal
  conductivity 28, 29
  diffusion 28
  diffusivity 29
  discharge 533
  energy 25, 28
  expansion 26, 527
  fatigue resistance 145
  properties 25, 28
  radiation 30
  shock resistance 27
thermal conductivity device 1080
thermochemical reduction process 253
thermochemistry 1000
thermocouples 544
  basic circuit 543
  materials 543
  NIST polynominal equations 547
  properties 545, 546
thermodynamic cell voltage, 562
thermoelectric power 544
  conductor 544
thermoelectronic 552
thermoionic emission 552
thermoionic emitters 552
thermoluminescence 766
  minerals 766
thermoplastics 692, 697, 1029
  classification 698
thermosets 692, 713, 1027, 1029
  classification 698
thermosetting plastics 692
thermosetting polymers 713
Thiokol® 718
thionyl chloride 1134
thiophene (thiofuran) 1134
thoreaulite 355
thoria 274, 666
thorianite 448, 860
thorite 448, 860
thorium 278, 329, 422, 426, 427, 436, 437, 438, 447, 1204
  applications and uses 451
  carbide 450, 656
  chloride 450
  dicarbide 656
  dioxide 450, 666
  disilicide 662
  fluoride 450
  hexaboride 651
  hydroxide 428, 450
  metal 450, 451
  mining and mineral dressing 449
  nitrate 450
  nitride 659
  nitride 660
  oxalate 428
  oxalate dihydrate 450
  purification 450
  pyrophosphate 449
  refining 450
  series 1202
  specific activity 1208
  tetraboride 651
  tetrachloride 450, 451
  tetrafluoride 450
  tetraiodide 451
thorium-232 1202
  natural decay series 1204
thoron 1093, 1183, 1204
thortveitite 434
thorutite 860
Thoulet's and Sondstadt's liquor 1173
threshold limit averages 1064
threshold limit value 1064
threshold sound power level 24
thulite 867
thulium 422, 424
tialite 795, 860
Ticle 288
tieilite 861
tiemannite 861
tigers eye quartz 760
timber 983, 997
time attenuation coefficient 25
time weighted average 1064
tin 204, 208, 297
  alloys 204, 208
  beneficiation 206
  bronze 186
  chloride 1134
  electric arc furnace (EAF) 207

electrorefining 208
gravel pump mining 206
nuclide 204
ore 814
Pest 204
pyrites 857
refining 207
roasting 207
selected properties 160
smelting 207
suction dredging 206
tetrahydride 204
underground mining 206
use in sold 208
tincal 471, 811
tinplate 208
titan
dioxide 608
titania 165, 286, 601, 614, 639, 667
slag 281, 289
worldwide 281
titanite 277, 861
titanium 251, 274, 326, 327, 342, 409, 571, 574, 577, 1032, 1099
(alpha-Ti) 274
(beta-Ti) 274
alloy powders
hydride/dehydride process (HDH) 299
alloys 274, 302, 584
alpha 305
alpha-beta 305
applications 308
ASTM designation 306
beta 305
chemical equivalents 305
copper-based 313
corrosion resistance 313
mechanical properties 310
melting techniques 297
near alpha 305
strength-to-weight ratios 304
thermal and electrical properties 312
annealing 320
anodizing 321
applications and uses 322
bending 319
blasting 320
boride 651
carbide 275, 609, 656
carbochlorination process 287
castings 320
cathodes 564
chemical etching 321
chloride 291, 1134
chloride process 287
colloidal oxyhydrate 287
commercially pure grades 301
conferences 325
corrosion resistance 275
degreasing 321
descaling 321
diboride 470, 637, 638, 639, 651
applications and uses 639
dihydride 301
dioxide 274, 278, 281, 283, 286, 287, 614, 666, 667
sulfate process 286
disilicide 662
etching 320
grade 306
grades 564
grinding 320
hemioxide 618
hongquiite 617
immunity 275
joining 320
Kroll process 330
machining 320
metal ingot 297
cold-hearth melting 298
producers 298
metal powder 298
metallurgical classification 304
metalworking 319
monoxide 617
nitride 660
oxides
properties 619
pickling 321
powder 298
gas-atomization process 301
hydride/dehydride process (HDH) 299
industrial processes 300
Kroll process 299
metal hydride reduction (MHR) 301
rotating electrode process (REP) 299
producers 302
punching 320
sesquioxide 282, 617, 667
shearing 320

slag 281
sponge 288, 290, 291
commercial specifications 293
Kroll process 288
producers 294, 295, 324
sponge producers 295
superplastic forming 319
tetrachloride 276, 283, 287, 289, 301, 703, 1067
carbochlorination 288
Hunter process 291
Kroll process 290
tetraiodide
Van Arkel–deBoer process 293
trisilicide 662
uses and applications 323
world producers 324
titanium-palladium alloy 314
titanium-ruthenium alloys 314
titanomagnetite 277
titanowodginite 434
titanyl sulfate 287
TNT 1018
tobermorite 973
tolite 1018
toluene 1134
tonicity 1121
tool and machining steel 115
tool steels 115
AISI designation 116
carbon 117
chromium 120
cobalt 117
manganese 117
molybdenum 117
nickel 120
physical properties 118
silicon 117
tungsten 117
vanadium 117
topaz 861
topazolite 782, 804
Tophel® 546
torbernite 440, 861
tosudite 816
total reflection 34
Total-Alkali-Silica diagram 902
toughened glass 676
toughness 15
tourmaline 329
toxicity of gases 1064
traditional ceramics 629
trans-1,4-polyisoprene rubber 716
transfermium elements 1184
transition alumina 606
transition temperatures 65
transition zone 887
translucent 760
transparent 760
travertin 908
travertine 814
tremolite 861
trevorite 861
tribological 19
triboluminescence 766
minerals 766
tribromoacetaldehyde 1172
2,2,2-tribromoacetal-dehyde (bromal) 1134
tribromomethane 777, 1135, 1172
tributyl phosphate 1135
tributylamine 1135
tricalcium aluminate 972
tricalcium silicate 971
1,1,1-trichloroethane 1135
1,2,4-trichlorobenzene 1135
1,1,2-trichloroethane 1135
trichloroethylene 321, 349, 1135
trichlorofluoromethane 1067, 1135
trichloromethane 1124
1,2,3-trichloropropane 1135
trichlorosilane 467
hydrogen reduction 468
1,1,2-trichlorotrifluoro-ethane 1135
trichroism 37, 766
triclinic 1211
triclinic space groups 1221
tridymite 862
triethanolamine 1135
triethyl
phosphate 1135
phosphite 1135
triethylamine 1135
triethylene glycol 1135
triethylenetetramine 1135
2,2,2-trifluoroethanol 1135
trigonal 1211
trigonal space groups 1225
triisopropyl borate 1135
trimethyl orthoformate 1135
2,2,4-trimethyl pentane 1135
2,4,4-trimethyl-1-pentene 1135
2,4,4-trimethyl-2-pentene 1135
1,2,4-trimethylbenzene 1135

1,3,5-trimethylbenzene 1135
trinitrobenzene 1017
trinitrophenol 1017
trinitrotoluene 1018
triphylite 862
triphyllite 220
triplet states 46
tripropylene glycol 1135
trititanium pentoxide 668
tritium 225, 1080
tritium gas 218
troilite 862, 914, 920
trommels 207
trona 755, 862
troostite 865
troty 1018
Trouton's first empirical rule 31
Trouton's second empirical rule 31
Trouton's third rule 31
true density 2
true strain 7
tsavorite 782
tuff 904
tungsten 117, 385, 1026, 1032
  alloys 385
    properties 388
  boride 651, 1026
  carbide 638, 639, 657
    applications and uses 640
    powder 640
  carbon black 387
  chalcogenide 387
  dinitride 660
  disilicide 662
  hemiboride 651
  hemicarbide 657
  heminitride 660
  hexachloride 387
  inert gas (TIG) 349
  monocarbide 387
  nitride 660
  oxide in 386
  powder 387
  producers 389
  silicide 662
tungsten-alloy high-speed tool steel 119
tungsten-chromium steel 84
tungsten-Re 547
turpentine 698
turquoise 760, 783, 862
tuyeres 71
Twaron® 710, 1027
Type 3A 1095
Type 4A 1095
Type 5A 1095
type-n semiconductors 457
type-p semicontuctors 457

Udimet®500 137
Udimet®700 138
UGS process 285
ulexite 471, 863
ullmanite 125, 863
ultamarine 836
ultimate tensile strength (UTS) 9
ultisols 948
ultra-high molecular weight polyethylene (UHMWPE) 703
ultra-high molecular weight polyethylene 703
ultra-high strength steel 115
ultrahigh-molecular weight polyethylene (UHMW) 1027
ultra-high-strength structural steels 115
ultramafic 339
ultramafic rocks
  classification 902
ultrasounds 23
ulvite 863
ulvospinel 863
unalloyed copper 184
uniaxial 765
uniaxial tensile test 8, 9
unified numbering system (UNS) 1181
unplastified polyvinyl chloride 726
unsaturated polyester 726
unstabilized zirconia 620
upgraded titanium slag (UGS) 285
upper explosive limit 1062
upper flammability limit (UFL) 1062
upper mantle 887
uranides 436
uraninite 265, 440, 442, 668, 863, 1091
uranium 265, 278, 329, 340, 341, 436, 437, 438, 442, 454, 1202, 1203, 1204, 1207
  anion exchange 443
  carbide 657
  cations 440
  concentration by leaching 442
  crushing 442
  depleted 439
  diboride 652
  dicarbide 657

dioxide 444, 668
preparation 445
disilicide 662
dodecaboride 652
fissionable isotope 444
hexafluoride 445, 1067
leaching 442
metal
preparation 445
minerals 243
mining 441
nitride 660
oxide 442
purification 443
radionuclides 439
recovering from leach liquors 443
refining 443
series 1202
silicide 662
solvent extraction 443
tetraboride 652
tetrafluoride 445
trioxide 444
uranium-235 446
uranium-235 1202
natural decay series 1204
uranium-238 1202
natural decay series 1203
uranophane 440, 863
uranothorite 440, 863
uranotile 863
uranyl cations 439
uranyl nitrate
crystals 444
urea-formaldehyde 713, 726
URENCO 445
uvarovite 782, 863
UX1 1203
UX2 1203

vacuum
permitivity 519
vacuum bagging and autoclave curing 1031
vacuum-arc remelting (VAR) 294, 297, 304
vacuum-arc-remelt (VAR) process 115
vacuumdistillation process (VDP) 290
valence band 455
valentinite 864
Van Arkel–deBoer process 293, 445
zirconium 331
Van der Waals 1044
Van der Waals constants 1043
Van der Waals equation of state 1042, 1043, 1047
Van't Hoff equation 1050
Van't Hoff law 1120
vanadinite 339, 340, 864
vanadium 117, 289, 297, 338, 616
alloys 338
aluminothermic reduction 342
calciothermic reduction 341
carbide 657
carbothermic reduction 342
diboride 652
disilicide 662
foil 342
hemicarbide 657
Highweld process 341
metal 338, 341
natural 338
nitride 660
pentoxide 148, 338, 339, 340, 341, 342
producers 343
silicide 663
steel 342
stratcor process 341
trichloride 339
vanadium-50 338
Xstrata process 340
vanadium (IV) chloride 1136
vanadyl ion 338
vanadyl trichloride 1136
vapor 1054
autoignition temperature 1063
explosivity limits 1062
flammability range 1062
ignition energy 1063
maximum explosion pressure 1063
maximum rate of pressure rise 1063
pressure 1110
pressure of water 1054
variscite 864
vector position 1228
vector product 1228
vein
deposits 752
graphite 625
walls 752
velocity of sound 23

verdilite 783
vermicullite 755
Verneuil melt growth technique 795
Verneuil method 795
Verneuil technique 608
Verneuil's flame fusion method 614
vertisols 948, 949
vestium 409
vesuvianite 864
vibration
  maximum amplitude 25
Vickers hardness 12, 13, 17
villiaumite 864
vinyl
  acetate 1136
  chloride monomer (VCM) 705
  ethyl ether 1136
  trichloride 1135
4-vinylcyclohexene 1136
vinylidene chloride 1136
vinylidene fluoride 719
2-vinylpyridine 1136
violarite 864
virginium 1183
virial 1044
virial coefficients 1043
virial equation of state 1043
viridine 855
Viton® 719
Viton® fluoroelastomers 719
vitrain 1005
vitreous 760
vitreous silica 596
vitriol oleum 1169
vivianite 865
volcanic rocks 904
  classification 903
volcanoes 891
volume expansion on melting 27
volume magnetostriction 494
volume resistivity 527
von Hauer's alloy 210
von Kobell's fusibility scale 770
vug 752, 757
vulcanization 716
vulcanization process 716
Vycor® 671

wad 152, 850
wadsleyite 865, 888
Waelz process 195
wafer 471, 586
  assembly 475
  cleaning 474
  dielectric deposition 474
  doping 474
  electrical test 474
  etching 473, 474
  inspection 474
  lapping 473
  masking 474
  metallization 474
  passivation 474
  polishing 473
  production 473
  slicing 473
  thermal oxidation or deposition 473
Walden's equation 1113
Walden's rule 1113
walls 752
Waspaloy® 138
waste fuels
  properties 1007
water 1136, 1174
  electrolysis 1083
  gas 1082
  latent heat of vaporization 1058, 1063
  lime 969
  opal 782
  physical properties 1167
  properties
    temperature dependence 1167
  splitting 1083
  vapor pressure 1055
water vapor 1054
  degree of saturation 1056
  heat capacities 1056
  mass fraction 1056
  relative humidity 1056
  saturation 1055
  specific humidity 1056
water-atomized iron powders 122
wave propagation 23
wavellite 865
waxy 760
wear resistance 122
weathered ilmenite 279
Weiss domains 501, 504, 508, 534, 538
Westphal balance 4
wet filament winding 1030
wet-bulb depression 1057

wet-bulb temperature 1057
wetting 1113
wheel ore endellionite 811
whiskers 1025
white cast iron 79, 80
white fused alumina 608
white gold 405, 406
white graphite 638
white lead ore 815
white nickel 816
white opal 782
white tin 204
whitewares 630
Widia® 657
wiikite 434
Wilhelmy plate 1118
willemite 865
window material
  electromagnetic transparency range 1236
  optical properties 1233
witherite 264, 865
Wobbe index 1003
wodginite 355
wohl 1044
wolfram 385, 1183
wolframite 385, 386, 387, 433, 434, 865
wollastonite 866, 973
wood 983
  applications 997
  chemical resistance 997
  decay resistance 990
  density 986
  drying 987
  durability 990
  electrical properties 989
  flammability 989
  fracture toughness 988
  Hankinson's equation 987
  heating value 989, 990
  mechanical properties 987
  moisture content 985
  physical properties 985
  shrinkage 987
  specific gravity 986
  specific heat capacity 989
  strength 987
  structure 984, 985
  thermal properties 988
  tin 814
  Young's modulus 988
Wood's alloy 210
Wood's light 766
work of adhesion 1114
work of cohesion 1114
world annual production of commodities 1248
wrought aluminum alloys 171
  physical properties 173
wrought copper alloys 183
  physical properties 184
wrought iron 73
wrought steels 95
wulfenite 374, 866
wurtzite 866
wustite 866

xenomorph 758
xenomorphous 892
xenon 1067, 1090, 1092
xenotime 280, 329, 425, 866
  mining and mineral dressing 427
xerosols 953
X-ray 244, 249
X-ray density 2
Xstrata process 340
2,4-xylenol 1136
xylosol 1180

yellow beryl 792
yellow brass 185
yellow gold 405
yellow lead ore 866
yermosols 953
yield strength (YS) 9
Young's equation 1113, 1114, 1115
Young's modulus 8, 17, 29, 64, 128
Young–Laplace equation 1116, 1117
ytterbite 423
ytterbium 422, 423, 424
yttria 218, 274, 668
yttric rare-earths 422
yttrium 422, 424
yttrium aluminum garnet 218
yttrium oxide 668

zaffre 142
zamak 197
z-average molar mass 696

z-average relative molar mass 696
zeolites 261, 467, 1081, 1095
  calcium form 1095
  potassium form 1095
  sodium form 1095
zero magnetic field 505
zero polarization 535
ziconia stabilized 551
Ziegler–Natta catalyst 704
zinc 187
  alloys 187, 196, 197
  applications and uses 195
  blende 469, 856
  Bolchem process 191
  Boliden–Norzink process 191
  chloride 192
  deposition 564
  electrolytic process 191
  electroplating 561
  electrowinning 192, 563, 572, 577, 582
  ferrite 190
  ferrite residue 192
  galvanizing 188
  hot-dip galvanizing 195
  hydrometallurgical process 191
  mercury iodide process 191
  metal ingots 192
  ore 189
  Outokumpu process 191
  oxide 189
  powder 338
  properties 197
  pyrometallurgical process 192
  roasting process 190
  selected properties 160
  spar 856
  thiocyanate-sulfide process 191
zincite 867
zinkblende 189
zinnwaldite 240, 867
Zircadyne® 331
zircon 278, 280, 328, 329, 337, 448, 618, 867
  carbochlorination reaction 329
  chlorination 329
  sands 329
zirconia 274, 618, 619, 620, 796
  fully stabilized 621
  fused 622
  partially stabilized 620
  preparation by alkaline leaching 622
  producers 622
  stabilized 622
  unstabilized 620, 621
zirconium 251, 297, 326, 336, 445, 571, 679
  alloys 326, 331
  applications and uses 335
  carbide 619, 657
  cathodes 565
  cleaning 334
  copper 184
  corrosion resistance 326, 333
  descaling 334
  diboride 652
  dioxide 618, 668, 669
  disilicide 663
  dodecaboride 653
  electropolishing 327, 334
  etching 334
  hydroxide 622
  ingot 330
  Kroll process 330
  machining 333
  nitride 661
  nuclear grades 331
  oxide films 327
  oxychlorides 329
  physical properties 332
  pickling 327, 334
  producers 336
  sponge 330
  tetrachloride 328, 329, 595, 621
  tetraiodide 328
  Van Arkel–deBoer process 331
  welding 334
zirconolite 867
zirconyl
  sulfate 330
zirconyl chloride dihydrate 621
zirkelite 867
zoïsite 867